A TEXTBOOK OF
ENGINEERING MATHEMATICS

A TEXTBOOK OF

ENGINEERING MATHEMATICS

For

B.TECH. 1st YEAR, *SEMESTER I & II*

KERALA UNIVERSITY, KERALA

(*Strictly according to the latest revised syllabus*)

By

N.P. BALI
Former Principal
S.B. College, Gurgaon
Haryana

Dr. T.K. MATHEW VARKEY
Assistant Professor
Department of Mathematics
T.K.M. College of Engineering, Kollam
Kerala

UNIVERSITY SCIENCE PRESS

(An Imprint of Laxmi Publications (P) Ltd.)

BANGALORE • CHENNAI • COCHIN • GUWAHATI • HYDERABAD
JALANDHAR • KOLKATA • LUCKNOW • MUMBAI • RANCHI
NEW DELHI • BOSTON, USA

Published by:

UNIVERSITY SCIENCE PRESS

(An Imprint of Laxmi Publications (P) Ltd.)

113, Golden House, Daryaganj,
New Delhi-110002
Phone: 011-43 53 25 00
Fax: 011-43 53 25 28
www.laxmipublications.com
info@laxmipublications.com

First Edition: 2014

OFFICES

✆ **Bangalore**	080-26 75 69 30	✆ **Jalandhar**	0181-222 12 72
✆ **Chennai**	044-24 34 47 26	✆ **Kolkata**	033-22 27 43 84
✆ **Cochin**	0484-237 70 04, 405 13 03	✆ **Lucknow**	0522-220 99 16
✆ **Guwahati**	0361-254 36 69, 251 38 81	✆ **Mumbai**	022-24 91 54 15, 24 92 78 69
✆ **Hyderabad**	040-24 65 23 33	✆ **Ranchi**	0651-220 44 64

UEM-9706-195-ATB ENGG MATH-I&II(KU)-VAR **C—**

Typeset at: Excellent Graphics, Delhi. *Printed at*:

CONTENTS

PREFACE

This book is a part of the original book **"A Textbook of Engineering Mathematics"** by N.P. Bali running into eighth edition and very well received by the teachers and students of all Indian Universities. The rapid sale of the eighth edition bears testimony to the over-whelming response. We thank them all for the appreciation.

The present form of the book contains only six chapters and covers the entire revised New syllabus for the students of semester I & II of Kerala University. Latest problems taken from Kerala University Question papers are added with solutions in this edition. More worked out problems were given as examples.

There is no dearth of books Engineering Mathematics but the students find it difficult to solve most of the problems in the exercise in the absence of an adequate number of solved examples. An outstanding and distinguishing feature of the book is the large number of typical solved examples followed by well-graded problems. Many examples and problems have been selected from recent papers of various engineering examinations.

—AUTHORS

SYLLABUS

KERALA UNIVERSITY, KERALA

13.101 ENGINEERING MATHEMATICS I

(L-T-P; 2-1-0 Credicts: 6)

(2013 scheme)

MODULE I

Differential Calculus: Hyperbolic functions and their derivatives, Leibnitz's theorem (without proof), Curvature-Radius of Curvature-Circle of Curvature-Evolute Indeterminate forms-Evaluation of limits using L' Hospital rule.

Partial Differentiation: Partial derivatives-Euler's theorem on homogeneous functions, Total derivatives, chain rule, Jacobians, Maxima and minima of functions of two variables-Lagrange's method.

MODULE II

Multiple Integrals: Evaluation of double and triple integrals (Cartesian only), Change of order of integration, change of variables (Cartesian to polar), Applications of multiple integrals, area enclosed by plane curves (Cartesian only) Triple integrals–Evaluation of triple integrals in Cartesian coordinates–Volume as triple integrals.

MODULE III

Laplace Transforms: Transform of elementary functions, shifting property, inverse transforms, transform of derivatives and integrals, transform of functions multiplied by t and divided by t, convolution theorem(without proof), transform of unit step function, unit impulse function, periodic function, second shifting theorem, solution of ordinary differential equations with constant coefficients using Laplace transforms.

Differential Equations: Linear differential equations with constant coefficients, method of variation of parameters, Cauchy and Legendre equations, Simultaneous linear equations with constant coefficients.

MODULE IV

Matrices: Rank of a matrix, elementary transformations, equivalent matrices. Echelon form, consistency, solution of system of linear equations, non-homogeneous and homogeneous equations, LU decomposition method, linear dependence and independence of vectors, eigen values and eigen vectors, properties of eigen values and eigen vectors, Caley-Hamiltion theorem (without proof) Diagonalization, Quadratic forms, reduction to canonical forms.

Note: The question paper shall consist of two parts. Part A (20 marks) 5 compulsory questions of 4 marks each. Part B (4 Modules) students must answer 1 out of 2 from each module. Each question carry 20 marks.

1

Differential Calculus

1.1. HYPERBOLIC FUNCTIONS

The exponential expressions $\frac{1}{2}(e^x - e^{-x})$ and $\frac{1}{2}(e^x + e^{-x})$ are hyperbolic sine and hyperbolic cosine respectively of the "argument" x and y are denoted by sinh x and cosh x. The hyperbolic sine and cosine are combined in various ways to define the hyperbolic tangent, hyperbolic cotangent, hyperbolic secant and hyperbolic cosecant as follows:

$$\sinh x = \frac{e^x - e^{-x}}{2}; \qquad \cosh x = \frac{e^x + e^{-x}}{2}$$

$$\tanh x = \frac{\sinh x}{\cosh x} = \frac{e^x - e^{-x}}{e^x + e^{-x}}; \qquad \coth x = \frac{\cosh x}{\sinh x};$$

$$\operatorname{sech} x = \frac{1}{\cosh x}; \qquad \operatorname{cosech} x = \frac{1}{\sinh x}.$$

Derivatives of Hyperbolic Functions

$$\frac{d}{dx}(\sinh x) = \frac{d}{dx}\left(\frac{e^x - e^{-x}}{2}\right) = \frac{e^x + e^{-x}}{2} = \cosh x.$$

Similarly, it may be proved that,

$$\frac{d}{dx}(\cosh x) = \sinh x$$

$$\frac{d}{dx}(\tanh x) = \operatorname{sech}^2 x$$

$$\frac{d}{dx}(\coth x) = -\operatorname{cosech}^2 x$$

$$\frac{d}{dx}(\operatorname{sech} x) = -\operatorname{sech} x \tanh x$$

$$\frac{d}{dx}(\operatorname{cosech} x) = -\operatorname{cosech} x \cot x.$$

1.2. DERIVATIVES OF INVERSE HYPERBOLIC FUNCTIONS

The equation $x = \sin y$ may be written as $y = \sin^{-1} x$, the equation $x = \sinh y$ may be written as $y = \sinh^{-1} x$, the function $\sinh^{-1}(x)$ being called inverse hyperbolic sine of x. Clearly there are five other inverse hyperbolic functions, the proper symbols for which should be evident.

As the hyperbolic functions are defined in terms of the exponential function, the inverse hyperbolic functions are expressible in terms of the inverse of the exponential function.

$$\frac{d}{dx}(\sinh^{-1} x) = \frac{1}{\sqrt{x^2+1}}$$

Let $\sin^{-1} x = y$, then $x = \sinh y$

Then $$\frac{dx}{dy} = \cosh y = \sqrt{1+\sinh^2 y} = \sqrt{1+x^2}$$

$\therefore$ $$\frac{dy}{dx} = \frac{1}{\sqrt{1+x^2}}.$$

i.e., $$\frac{d}{dx}(\sin^{-1} x) = \frac{1}{\sqrt{x^2+1}}.$$

Similarly, we can prove the others.

$$\frac{d}{dx}(\cosh^{-1} x) = \frac{1}{\sqrt{x^2-1}}; \quad \frac{d}{dx}(\tanh^{-1} x) = \frac{1}{1-x^2}$$

$$\frac{d}{dx}(\coth^{-1} x) = \frac{-1}{x^2-1}; \quad \text{where } |x| > 1.$$

$$\frac{d}{dx}(\text{sech}^{-1} x) = \frac{-1}{x\sqrt{1-x^2}}; \quad \frac{d}{dx}(\text{cosech}^{-1} x) = \frac{-1}{x\sqrt{1+x^2}}.$$

1.3. LEIBNITZ'S THEOREM (n[th] Derivative of the Product of Two Functions)

If u and v be any two functions of x, then $D^n(uv) = u_n v + {}^nC_1 u_{n-1} v_1 + {}^nC_2 u_{n-2} v_2 + \cdots + {}^nC_r u_{n-r} v_r + \cdots + uv_n$, where $D \equiv \frac{d}{dx}$ and the suffixes denote the order of the derivatives of u and v.

Its proof is beyond the scope of this book.

ILLUSTRATIVE EXAMPLES

Example 1. *Find the nth derivative of $x^2 e^{3x}$.*

Sol. Here $u = e^{3x}$ and $v = x^2$

then $u_n = 3^n e^{3x}$ and $v_1 = 2x,\ v_2 = 2,\ v_3 = v_4 = \cdots = 0$ $\quad | \ D^n(e^{ax}) = a^n e^{ax}$

$\therefore$ By Leibnitz's theorem,

$$\begin{aligned} D^n(uv) &= u_n v + {}^nC_1 u_{n-1} v_1 + {}^nC_2 u_{n-2} v_2 + \cdots + {}^nC_n uv_n \\ &= 3^n e^{3x} \cdot x^2 + {}^nC_1 3^{n-1} e^{3x} \cdot 2x + {}^nC_2 3^{n-2} e^{3x} \cdot 2 + 0 + \cdots \\ &= 3^n e^{3x} x^2 + 2nx\, 3^{n-1} e^{3x} + n(n-1) 3^{n-2} e^{3x} \\ &= 3^{n-2} e^{3x} [9x^2 + 6nx + n(n-1)]. \end{aligned}$$

Example 2. *Find the 10th derivative of $x^3 \sin x$.*

Sol. Here $u = \sin x$ and $v = x^3$

then $$u_n = \sin\left(x + \frac{n\pi}{2}\right) \quad \text{and} \quad v_1 = 3x^2,\ v_2 = 6x,\ v_3 = 6,\ v_4 = v_5 = \cdots = 0$$

$\therefore$ By Leibnitz's theorem,

$$\left| \begin{array}{l} D^n [\sin(ax+b)] \\ = a^n \sin\left(ax + b + \frac{n\pi}{2}\right) \end{array} \right.$$

$$D^n (uv) = u_n v + {}^nC_1 u_{n-1} v_1 + {}^nC_2 u_{n-2} v_2 + \cdots + {}^nC_n uv_n$$

$$= \sin\left(x + \frac{n\pi}{2}\right) \cdot x^3 + {}^nC_1 \sin\left(x + (n-1)\frac{\pi}{2}\right) \cdot 3x^2 + {}^nC_2 \sin\left(x + (n-2)\frac{\pi}{2}\right) \cdot 6x$$
$$+ {}^nC_3 \sin\left(x + (n-3)\frac{\pi}{2}\right) \cdot 6 + 0 + \cdots$$

Put $n = 10$, then

$$D^{10} (\sin x \cdot x^3) = \sin\left(x + \frac{10\pi}{2}\right) \cdot x^3 + {}^{10}C_1 \sin\left(x + \frac{9\pi}{2}\right) \cdot 3x^2$$
$$+ {}^{10}C_2 \sin\left(x + \frac{8\pi}{2}\right) \cdot 6x + {}^{10}C_3 \sin\left(x + \frac{7\pi}{2}\right) \cdot 6$$
$$= -\sin x \cdot x^3 + 10 \cdot \cos x \cdot 3x^2 + 270\, x \sin x - 720 \cos x.$$

Example 3. *Find the nth derivative of $x^2 \log 3x$.* (K.U. May 2009, K.U. May 2012)

Sol. Here $u = \log 3x$ and $v = x^2$

then $$u_n = \frac{(-1)^{n-1} (n-1)!\, 3^n}{(3x)^n}, \quad v_1 = 2x,\ v_2 = 2,\ v_3 = v_4 = \cdots = 0$$

$$\left| \begin{array}{l} D^n [\log(ax+b)] \\ = \frac{(-1)^{n-1} (n-1)!\, a^n}{(ax+b)^n} \end{array} \right.$$

$\therefore$ By Leibnitz's theorem,

$$D^n(uv) = u_n v + {}^nC_1 u_{n-1} v_1 + {}^nC_2 u_{n-2} v_2 + \cdots + {}^nC_n uv_n.$$
$$D^n (x^2 \log 3x) = D^n (\log 3x \cdot x^2)$$

$$= \frac{(-1)^{n-1} (n-1)!\, 3^n}{(3x)^n} x^2 + {}^nC_1 \frac{(-1)^{n-2} (n-2)!\, 3^{n-1}}{(3x)^{n-1}} \cdot 2x + {}^nC_2 \frac{(-1)^{n-3} (n-3)!\, 3^{n-2}}{(3x)^{n-2}} \cdot 2$$

$$= \frac{(-1)^{n-3} \cdot (n-3)!}{x^n} \left[(n-1)(n-2) \cdot x^2 - 2n(n-2)\, x^2 + \frac{n(n-1)}{1} \cdot x^2\right]$$

$$= \frac{(-1)^{n-3} \cdot (n-3)!}{x^{n-2}} [n^2 - 3n + 2 - 2n^2 + 4n + n^2 - n]$$

$$= \frac{(-1)^{n-3} \cdot (n-3)!}{x^{n-2}} \cdot 2 = \frac{2(-1)^{n-3} (n-3)!}{x^{n-2}}.$$

Example 4. *If $y = \cos(m \sin^{-1} x)$, prove that*

(i) $(1 - x^2)\, y_2 - xy_1 + m^2 y = 0$ *(ii) $(1 - x^2)\, y_{n+2} - (2n+1)\, x\, y_{n+1} + (m^2 - n^2)\, y_{n} = 0$.*

Sol. Here $y = \cos(m \sin^{-1} x)$

Differentiating, $y_1 = -\sin(m \sin^{-1} x) \cdot m \cdot \frac{1}{\sqrt{1-x^2}}$

$\therefore \quad \sqrt{1-x^2}\, y_1 = -m \sin(m \sin^{-1} x)$

Again differentiating, $\sqrt{1-x^2}\, y_2 + y_1 \cdot \frac{1}{2\sqrt{1-x^2}}(-2x) = m\left[-\sin(m \sin^{-1} x)\frac{m}{\sqrt{1-x^2}}\right]$

$$(1-x^2)\,y_2 - xy_1 = -m^2 y$$

$$(1-x^2)\,y_2 - xy_1 + m^2 y = 0$$

Differentiating n times using Leibnitz's theorem,

$$y_{n+2}(1-x^2) + {}^nC_1\, y_{n+1}(-2x) + {}^nC_2\, y_n(-2) - [y_{n+1}\, x + {}^nC_1\, y_{n\cdot 1}] + m^2 y_n = 0$$

$$(1-x^2)\,y_{n+2} - (2n+1)\,x\,y_{n+1} - (m^2 - n^2)\,y_n = 0.$$

Example 5. *If $x = \sin t$, $y = \sin pt$, prove that*

(i) $(1-x^2)\,y_2 - xy_1 + p^2 y = 0$ *(ii) $(1-x^2)\,y_{n+2} - (2n+1)\,x\,y_{n+1} - (n^2 - p^2) = 0$.*

Sol. Here $x = \sin t$, $y = \sin pt$

$$\frac{dx}{dt} = \cos t = \sqrt{1-\sin^2 t} = \sqrt{1-x^2}$$

$$\frac{dy}{dt} = p\cos pt = p\sqrt{1-\sin^2 pt} = p\sqrt{1-y^2}$$

$$\therefore \quad y = \frac{dy/dt}{dx/dt} = \frac{p\sqrt{1-y^2}}{\sqrt{1-x^2}}$$

$$\Rightarrow \quad y_1\sqrt{1-x^2} = p\sqrt{1-y^2}$$

Squaring both sides, $y_1^2(1-x^2) = p^2(1-y^2)$

Differentiating, $(1-x^2)\,2y_1 y_2 + y_1^2 - 2x = p^2 \cdot -2yy_1$

$$(1-x^2)\,y_2 - xy_1 + p^2 y = 0$$

Differentiating n times using Leibnitz theorem,

$$(1-x^2)\,y_{n+2} + {}^nC_1\, y_{n+1}(-2x) + {}^nC_2\, y_n(-2) - [y_{n+1}\cdot x + {}^nC_1\, y_{n\cdot 1}] + p^2 y_n = 0$$

$$(1-x^2)\,y_{n+2} - (2n+1)\,xy_{n+1} - (n^2-p^2)y_n = 0.$$

Example 6. *If $2x = y^{1/m} + y^{-1/m}$, prove that*

$(x^2-1)\,y_{n+2} + (2n+1)\,xy_{n+1} + (n^2 - m^2)\,y_n = 0.$ (K.U. May 2008)

Sol. Here $2x = y^{1/m} + \frac{1}{y^{1/m}}$

i.e., $y^{2/m} - 2xy^{1/m} + 1 = 0$

$(y^{1/m})^2 - 2x\,(y^{1/m}) + 1 = 0$, which is a quadratic in $y^{1/m}$.

i.e., $y^{1/m} = x \pm \sqrt{x^2-1}$

$$\therefore \quad y = \left(x + \sqrt{x^2-1}\right)^m$$

Differentiating, $y_1 = m\left(x + \sqrt{x^2-1}\right)^{m-1}\left(1 + \frac{1}{2\sqrt{x^2-1}}\cdot 2x\right)$

$$= \frac{m\left(x+\sqrt{x^2-1}\right)^m}{\sqrt{x^2-1}} = \frac{my}{\sqrt{x^2-1}}$$

$$\sqrt{x^2-1}\, y_1 = my \qquad \text{...(1)}$$

Again differentiating,

$$\sqrt{x^2-1}\, y_2 + y_1 \cdot \frac{1 \cdot 2x}{2\sqrt{x^2-1}} = my_1$$

i.e., $(x^2-1)\, y_2 + xy_1 = my_1 \sqrt{x^2-1} = m \cdot my,$ | Using (1)

$$(x^2-1)\, y_2 + xy_1 - m^2 y = 0$$

Differentiating n times, using Leibnitz's theorem,

$$y_{n+2}(x^2-1) + {}^nC_1\, y_{n+1}(2x) + {}^nC_2\, y_n(2) + [y_{n+1}\, x + {}^nC_1 y_n \cdot 1] - m^2 y_n = 0$$

$$y_{n+2}(x^2-1) + (2n+1)\, xy_{n+1} + y_n(n^2 - n + n - m^2) = 0$$

$$y_{n+2}(x^2-1) + (2n+1)\, xy_{n+1} + (n^2 - m^2)\, y_n = 0.$$

Example 7. *If* $y = \dfrac{\sinh^{-1}(x)}{\sqrt{1+x^2}}$, *prove that* $(1+x^2)\, y_{n+2} + (2x+3)\, xy_{n+1} + (n+1)^2 y_n = 0.$

(K.U. May 2011)

Sol. Here $\quad y = \dfrac{\sinh^{-1}(x)}{\sqrt{1+x^2}}$

$$\Rightarrow \quad y\sqrt{1+x^2} = \sin^{-1} x$$

Differentiating, $\quad y_1\sqrt{1+x^2} + y \cdot \dfrac{1}{2\sqrt{1+x^2}} \cdot 2x = \dfrac{1}{\sqrt{1+x^2}}$

i.e.,
$$(1+x^2)\, y_1 + xy = 1$$

Differentiating, $(1+x^2)\, y_2 + y_1 \cdot 2x + xy_1 + y \cdot 1 = 0$

$$(1+x^2)\, y_2 + 3xy_1 + y = 0$$

Differentiating n times, using Leibnitz's theorem,

$$y_{n+2}(1+x^2) + {}^nC_1\, y_{n+1}(2x) + {}^nC_2\, y_n(2) + 3\,[y_{n+1} \cdot x + {}^nC_1\, y_n \cdot 1] + y_n = 0$$

i.e.,
$$(1+x^2)\, y_{n+2} + (2n+3)\, x\, y_{n+1} + (n+1)^2 y_n = 0.$$

EXERCISE 1.1

1. Find the nth derivative of the following:
   (*i*) $e^x (2x+3)^3$ (*K.U. April 2013*) (*ii*) $x^3 \log(10x+5)$
   (*iii*) $\dfrac{\log x}{x}$ (*iv*) $(2x+3)^3 \sin(4x+5)$.
2. Find the 10th derivative of $x^3 e^{4x}$.
3. If $u = \dfrac{x^5}{Ax^2 + 2Bx + C}$, show that
   $(Ax^2 + 2Bx + C)\, u_n + 2n\,(Ax + B)\, u_{n-1} + n(n-1)\, Au_{n-2} = 0, \quad$ if $n > 5$.

4. If $y = (1-x)^{-a} \cdot e^{-ax}$, prove that $(1-x)y_{n+1} - (n+ax)\,y_n - nay_{n-1} = 0$.

5. If $y = a\left(x+\sqrt{a^2+x^2}\right)^n + b\left(x+\sqrt{a^2+x^2}\right)^{-n}$, show that

$$(x^2+a^2)\,y_{m+2} + (2m+1)\,xy_{m+1} + (m^2-n^2)\,y_m = 0.$$

6. If $y = x^{n-1}\log x$, prove that $xy_1 = (n-1)\,y + x^{n-1}$. By differentiating this result $(n-1)$ times, show that $y_n = \dfrac{(n-1)\,!}{x}$.

7. If $y = \dfrac{\log x}{x}$, prove that, $x^2\,y_{m+2} + (2m+3)\,xy_{m+1} + (m+1)^2\,y_m = 0$.

8. If $y = e^{a\sin^{-1}x}$, prove that $(1-x^2)\,y_{n+2} - (2n+1)\,xy_{n+1} - (n^2+a^2)\,y_n = 0$.

9. If $y = \dfrac{\sin^{-1}x}{\sqrt{1-x^2}}$, show that $(1-x^2)\,y_{n+1} - (2n+1)\,x\,y_n - n^2\,y_{n-1} = 0$.

10. If $y = (x^2-1)^n$, prove that $(x^2-1)\,y_{n+2} + 2xy_{n+1} - n(n+1)\,y_n = 0$.

Answers

1. (*i*) $e^x\,\{(2x+3)^3 + 6n\,(2x+3)^2 + 12n\,(n-1)\,(2x+3) + 8\,n\,(n-1)\,(n-2)\}$.

(*ii*) $$\frac{(-1)^{n-1}\,(n-1)\,!\cdot 10^n}{(10x+5)^n}\cdot x^3 + n\cdot\frac{(-1)^{n-2}\,(n-2)\,!\cdot 10^{n-1}}{(10x+5)^{n-1}}\cdot 3x^2$$
$$+\frac{n\,(n-1)\,(-1)^{n-3}\,(n-3)\,!\cdot 10^{n-2}}{1.\,2\,(10x+5)^{n-2}}\cdot 6x + \frac{n\,(n-1)\,(n-2)}{1\cdot 2\cdot 3}\cdot\frac{(-1)^{n-4}\,(n-4)\,!\cdot 10^{n-3}}{(10x+5)^{n-3}}\cdot 6.$$

(*iii*) $$\frac{(-1)^n\,n\,!}{x^{n+1}}\left[\log x - 1 - \frac{1}{2} - \frac{1}{3} - \dots - \frac{1}{n}\right].$$

(*iv*) $$4^{n-3}\left\{64\,(2x+3)^3\sin\left(4x+5+\frac{n\pi}{2}\right) + 96n\,(2x+3)^2\cdot\sin\left(4x+5+(n-1)\frac{\pi}{2}\right)\right.$$
$$\left. + 48n\,(n-1)\,(2x+3)\cdot\sin\left(4x+5+(n-2)\frac{\pi}{2}\right) + 8n(n-1)\,(n-2)\sin\left(4x+5+(n-3)\frac{\pi}{2}\right)\right\}$$

2. $e^{4x}\,4^{n-3}\,[64x^3 + 48nx^2 + 12n\,(n-1)\,x + n\,(n-1)\,(n-2)]$.

1.4. CURVATURE

Let P and Q be two neighbouring points on the curve. Let arc AP = S, where A is a fixed point on the curve. Let ψ be the angle made by the tangent at P with the x-axis.

Then $\dfrac{d\psi}{dS}$ is called the curvature of the curve at point P.

Thus, the curvature is defined as the rate of turning of the tangent w.r.t. the arc length.

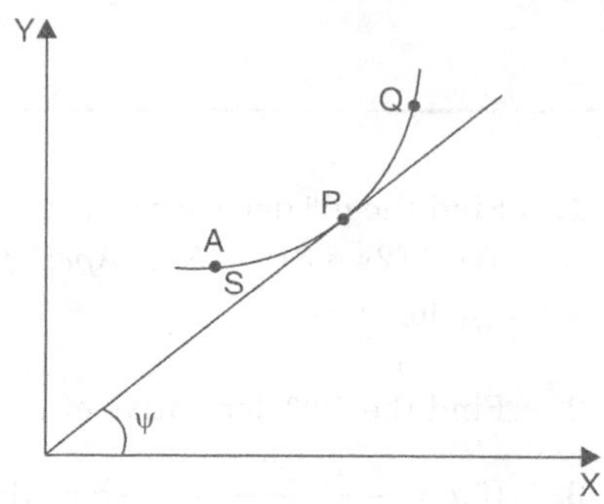

Fig. 1.1

Radius of Curvature–Cartesian Form

Radius of curvature of the curve at any point is defined as the reciprocal of the curvature at that point and is denoted by ρ.

$$\rho = \frac{ds}{d\psi}$$

We know that $\qquad \dfrac{dy}{dx} = \tan\psi$

Differentiating w.r.t. x,

$$\frac{d^2y}{dx^2} = \sec^2\psi\,\frac{d\psi}{dx}.$$

$$= \sec^2\psi\,\frac{d\psi}{ds}\cdot\frac{ds}{dx}$$

$$= (1+\tan^2\psi)\,\frac{d\psi}{ds}\cdot\frac{ds}{dx}$$

$$= \left[1+\left(\frac{dy}{dx}\right)^2\right]\cdot\frac{1}{\rho}\cdot\left[1+\left(\frac{dy}{dx}\right)^2\right]^{1/2}$$

Hence, $$\rho = \frac{\left[1+\left(\dfrac{dy}{dx}\right)^2\right]^{3/2}}{\dfrac{d^2y}{dx^2}}.$$

Note. The radius of curvature ρ is positive or negative according as $\dfrac{d^2y}{dx^2}$ is positive or negative. This indicates that the curve is either concave or convex.

Since the value of ρ is independent of the choice of axes, interchanging x and y, we get

$$\rho = \frac{\left[1+\left(\dfrac{dx}{dy}\right)^2\right]^{3/2}}{\dfrac{d^2x}{dy^2}}.$$

This formula is useful when $\dfrac{dx}{dy} = 0$, *i.e.*, the tangent is perpendicular to x-axis.

ILLUSTRATIVE EXAMPLES

Example 1. *Find the radius of curvature at the point (1, 1) on the curve $x^3 + y^3 = 2xy$.*

Sol. Here $\quad x^3 + y^3 = 2xy \qquad$ ...(1)

Differentiating Eq. (1) w.r.t. x,

$$3x^2 + 3y^2\,\frac{dy}{dx} = 2\left(x\,\frac{dy}{dx} + y\cdot 1\right)$$

$$\therefore \qquad \frac{dy}{dx} = \frac{2y-3x^2}{3y^2-2x} \qquad ...(2)$$

$$\left[\frac{dy}{dx}\right]_{(1,\,1)} = -1.$$

Again differentiating (2), we get

$$\frac{d^2y}{dx^2} = \frac{(3y^2 - 2x)\left[2\dfrac{dy}{dx} - 6x\right] - (2y - 3x^2)\left[6y\dfrac{dy}{dx} - 2\right]}{(3y^2 - 2x)^2}$$

$$\therefore \quad \left[\frac{d^2y}{dx^2}\right]_{(1,\,1)} = -16$$

$$\therefore \quad \rho_{(1,\,1)} = \frac{\left[1+\left(\dfrac{dy}{dx}\right)^2\right]^{3/2}}{\dfrac{d^2y}{dx^2}} = \frac{(1+1)^{3/2}}{-16} = -\frac{1}{4\sqrt{2}}.$$

Example 2. *Show that the radius of curvature at the point 'θ' of the curve,* *$x = 3a \cos \theta - a \cos 3\theta$; $y = 3a \sin \theta - a \sin 3\theta$ is $3a \sin \theta$.*

Sol. Here $x = 3a \cos \theta - a \cos 3\theta$;

$y = 3a \sin \theta - a \sin 3\theta$.

$$\therefore \quad \frac{dx}{d\theta} = -3a \sin \theta + 3a \sin 3\theta = 3a\,(\sin 3\theta - \sin \theta)$$

$$= 6a \cos 2\theta \sin \theta.$$

$$\frac{dy}{d\theta} = 3a \cos \theta - 3a \cos 3\theta = 3a(\cos \theta - \cos 3\theta)$$

$$= 6a \sin 2\theta \sin \theta.$$

$$\therefore \quad \frac{dy}{dx} = \frac{6a \sin 2\theta \sin \theta}{6a \cos 2\theta \sin \theta} = \tan 2\theta$$

$$\frac{d^2y}{dx^2} = \frac{d}{dx}\left(\frac{dy}{dx}\right) = \frac{d}{dx}(\tan 2\theta)$$

$$= \frac{d}{d\theta}(\tan 2\theta) \cdot \frac{d\theta}{dx}$$

$$= 2\sec^2 2\theta \cdot \frac{1}{6a \cos 2\theta \sin \theta} = \frac{\sec^3 2\theta}{3a \sin \theta}$$

$$\therefore \quad \rho = \frac{\left[1+\left(\dfrac{dy}{dx}\right)^2\right]^{3/2}}{\dfrac{d^2y}{dx^2}} = \frac{(1+\tan^2 2\theta)^{3/2}}{\dfrac{\sec^3 2\theta}{3a \sin \theta}} = 3a \sin \theta.$$

Example 3. *Show that the radius of curvature at any point of the cycloid*

$x = a(\theta - \sin \theta)$ and $y = a\,(1 - \cos \theta)$ is $\left(-4a \sin\left(\dfrac{\theta}{2}\right)\right)$.

Sol. Here $x = a(\theta - \sin \theta)$; $y = a\,(1 - \cos \theta)$

$\therefore \quad \dfrac{dx}{d\theta} = a(1 - \cos\theta), \dfrac{dy}{d\theta} = a \sin\theta$

$$\therefore \quad \frac{dy}{dx} = \frac{a \sin\theta}{a(1-\cos\theta)} = \frac{2\sin\frac{\theta}{2}\cos\frac{\theta}{2}}{2\sin^2\left(\frac{\theta}{2}\right)} = \cot\frac{\theta}{2}$$

$$\frac{d^2y}{dx^2} = -\text{cosec}^2\frac{\theta}{2}\cdot\frac{1}{2}\cdot\frac{1}{a(1-\cos\theta)}$$

$$= \frac{-\text{cosec}^2\frac{\theta}{2}}{2a\cdot 2\sin^2\frac{\theta}{2}} = \frac{-1}{4a}\text{cosec}^4\frac{\theta}{2}$$

$$\therefore \quad \rho = \frac{\left[1+\left(\frac{dy}{dx}\right)^2\right]^{3/2}}{\frac{d^2y}{dx^2}} = \frac{\left[1+\cot^2\frac{\theta}{2}\right]^{3/2}}{-\frac{1}{4a}\text{cosec}^4\frac{\theta}{2}}$$

$$= \frac{-\left(\text{cosec}^2\frac{\theta}{2}\right)^{3/2}\cdot 4a}{\text{cosec}^4\left(\frac{\theta}{2}\right)} = \left(-4a\sin\frac{\theta}{2}\right).$$

Example 4. *Find the radius of curvature at $(-2a, -2a)$ on the curve $x^2y = a(x^2 + y^2)$.*

Sol. Here $x^2y = a(x^2 + y^2)$

Differentiating, $x^2\dfrac{dy}{dx} + y \cdot 2x = a\left[2x + 2y\dfrac{dy}{dx}\right]$

$$\therefore \quad \frac{dy}{dx} = \frac{2ax - 2xy}{x^2 - 2ay}$$

$$\therefore \quad \left[\frac{dy}{dx}\right]_{(-2a,\,-2a)} = \infty$$

So we consider $\dfrac{dx}{dy} = \dfrac{x^2 - 2ay}{2ax - 2xy}$

$$\therefore \quad \left[\frac{dx}{dy}\right]_{(-2a,\,-2a)} = 0$$

$$\frac{d^2x}{dx^2} = \frac{(2ax - 2xy)\left(2x\frac{dx}{dy} - 2a\right) - (x^2 - 2ay)\left(2a\frac{dx}{dy} - 2y\frac{dx}{dy} - 2y\right)}{(2ax - 2xy)^2}$$

$$\therefore \quad \left[\frac{d^2x}{dx^2}\right]_{(-2a,\,-2a)} = \frac{-8a^3}{16a^4} = \frac{-1}{2a}.$$

$$\therefore \qquad \rho_{(-2a,\,-2a)} = \frac{\left[1+\left(\frac{dx}{dy}\right)^2\right]^{3/2}}{\frac{d^2x}{dy^2}} = \frac{(1+0)^{3/2}}{\frac{-1}{2a}}$$

$$= -2a.$$

Example 5. *Find the radius of curvature of the parabola $y^2 = 4ax$ at any point (x, y).*

Sol. Here $\qquad y^2 = 4ax$

$$2y\frac{dy}{dx} = 4a$$

$$\therefore \qquad \frac{dy}{dx} = \frac{2a}{y}$$

$$\frac{d^2y}{dx^2} = \frac{-2a}{y^2}\cdot\frac{dy}{dx} = \frac{-2a}{y^2}\cdot\frac{2a}{y} = \frac{-4a}{y^3}.$$

$$\therefore \qquad \rho = \frac{\left[1+\left(\frac{dy}{dx}\right)^2\right]^{3/2}}{\frac{d^2y}{dx^2}} = \frac{\left[1+\frac{4a^2}{y^2}\right]^{3/2}}{\frac{-4a^2}{y^3}}$$

$$= -\frac{(y^2+4a^2)^{3/2}}{4a^2} = -\frac{(4ax+4a^2)^{3/2}}{4a^2}$$

$$= -\frac{(4a)^{3/2}\,(x+a)^{3/2}}{4a^2} = -\frac{2(x+a)^{3/2}}{\sqrt{a}}.$$

Example 6. *Find the radius of curvature at any point (r, θ) of the curve $r = f(\theta)$.*

Sol. Let $r = f(\theta)$ be the curve

We know that $x = r\cos\theta$, $y = r\sin\theta$

Then $$\frac{dx}{d\theta} = \cos\theta\,\frac{dr}{d\theta} - r\sin\theta = r_1\cos\theta - r\sin\theta$$

$$\frac{dy}{d\theta} = \sin\theta\,\frac{dr}{d\theta} + r\cos\theta = r_1\sin\theta + r\cos\theta$$

$$\therefore \qquad \frac{dy}{dx} = \frac{\left(\frac{dy}{d\theta}\right)}{\left(\frac{dy}{dx}\right)} = \frac{r_1\sin\theta + r\cos\theta}{r_1\cos\theta - r\sin\theta}$$

and $$\frac{d^2y}{dx^2} = \frac{d}{d\theta}\left(\frac{r_1\sin\theta + r\cos\theta}{r_1\cos\theta - r\sin\theta}\right)\frac{d\theta}{dx}$$

$$= \frac{r^2 + 2r_1^2 - rr_2}{(r_1\cos\theta - r\sin\theta)^3}$$

Hence, the equation of the circle of curvature at the origin is given by,

$$(x - X)^2 + (y - Y)^2 = \rho^2$$

$$\left[x + \frac{ma\,(1+m^2)}{2}\right]^2 + \left[y - \frac{a(1+m^2)}{2}\right]^2 = \frac{a^2\,(1+m^2)^3}{4}$$

$$\Rightarrow \quad x^2 + y^2 = -\,ma\,(1 + m^2)x + a(1 + m^2)\,y$$

$$= a(1 + m^2)\,(y - mx).$$

Example 3. *Find the evolute of the curve given by* $x^{2/3} + y^{2/3} = a^{2/3}$. (K.U. April 2012)

Sol. The parametric equation of the curve is

$$x = a \cos^3 \theta,\; y = \sin^3 \theta$$

$$\frac{dx}{d\theta} = -3a \cos^2 \theta \sin \theta,\; \frac{dy}{d\theta} = 3a \sin^2 \theta \cos \theta$$

$$\frac{dy}{dx} = \frac{3a \sin^2 \theta \cos \theta}{-3a \cos^2 \theta \sin \theta} = -\tan \theta$$

$$\frac{d^2y}{dx^2} = -\sec^2 \theta \frac{d\theta}{dx} = -\frac{\sec^2 \theta}{-3a \cos^2 \theta \sin \theta}$$

$$= \frac{1}{3a} \sec^4 \theta \operatorname{cosec} \theta$$

Let C(X, Y) be the centre of curvature

$$X = x - \frac{\frac{dy}{dx}\left(1+\left(\frac{dy}{dx}\right)^2\right)}{\frac{d^2y}{dx^2}}$$

$$= a \cos^3 \theta + \frac{3a \tan \theta\,(1+\tan^2 \theta)}{\sec^4 \theta \operatorname{cosec} \theta}$$

$$= a \cos^3 \theta + 3a \sin^2 \theta \cos \theta \qquad ...(1)$$

$$Y = y + \frac{1+\left(\frac{dy}{dx}\right)^2}{\frac{d^2y}{dx^2}}$$

$$= a \sin^3 \theta + \frac{3a\,(1+\tan^2 \theta)}{\sec^4 \theta \operatorname{cosec} \theta}$$

$$= a \sin^3 \theta + 3a \cos^2 \theta \sin \theta \qquad ...(2)$$

Adding equations (1) and (2),

$$X + Y = a(\cos \theta + \sin \theta)^3$$

$$(X + Y)^{1/3} = a^{1/3} (\cos \theta + \sin \theta) \qquad ...(3)$$

Subtracting equation (2) from (1),

$$X - Y = a\,(\cos \theta - \sin \theta)^3$$

$$(X - Y)^{1/3} = a^{1/3} (\cos \theta - \sin \theta) \qquad ...(4)$$

Squaring and adding equations (3) and (4),

$$(X + Y)^{2/3} + (X - Y)^{2/3} = 2a^{2/3}$$

$\therefore$ Equation of the evolute is $(x + y)^{2/3} + (x - y)^{2/3} = 2a^{2/3}$.

Example 4. *Show that the evolute of the cycloid $x = a(\theta - \sin\theta)$, $y = a(1 - \cos\theta)$ is another cycloid.* (K.U. May 2010)

Sol. Here $\quad x = a(\theta - \sin\theta)\,;\quad y = a(1 - \cos\theta)$

$$\frac{dx}{d\theta} = a(1 - \cos\theta) \qquad \frac{dy}{d\theta} = a\sin\theta$$

$$\frac{dy}{dx} = \frac{a\sin\theta}{a(1-\cos\theta)} = \frac{2\sin\theta/2\cos\theta/2}{2\sin^2\theta/2} = \cot\frac{\theta}{2}$$

$$\frac{d^2y}{dx^2} = -\frac{1}{2}\operatorname{cosec}^2\frac{\theta}{2}\cdot\frac{d\theta}{dx}$$

$$= -\frac{1}{2}\operatorname{cosec}^2\frac{\theta}{2}\cdot\left[\frac{1}{a(1-\cos\theta)}\right] = -\frac{1}{4a}\operatorname{cosec}^4\left(\frac{\theta}{2}\right)$$

Let C (X, Y) be the centre of curvature

$$X = x - \frac{\dfrac{dy}{dx}\left[1+\left(\dfrac{dy}{dx}\right)^2\right]}{\dfrac{d^2y}{dx^2}} = a(\theta - \sin\theta) + \frac{\cot\dfrac{\theta}{2}\left(1+\cot^2\dfrac{\theta}{2}\right)4a}{\operatorname{cosec}^4\dfrac{\theta}{2}}$$

$$= a(\theta - \sin\theta) + 4a\cot\theta/2\sin^2\theta/2$$

$$= a(\theta - \sin\theta) + 4a\sin\theta/2\cos\theta/2$$

$$= a(\theta - \sin\theta) + 2a\sin\theta$$

$$= a(\theta + \sin\theta)$$

$$Y = y + \frac{1+\left(\dfrac{dy}{dx}\right)^2}{\dfrac{d^2y}{dx^2}} = a(1 - \cos\theta) - \frac{\left(1+\cot^2\dfrac{\theta}{2}\right)4a}{\operatorname{cosec}^4\dfrac{\theta}{2}}$$

$$= a(1 - \cos\theta) - 4a\sin^2\theta/2$$

$$= a(1 - \cos\theta) - 2a(1 - \cos\theta)$$

$$= -a + a\cos\theta$$

$$= -a(1 - \cos\theta)$$

Hence, the required evolute of the cycloid is $x = a(\theta + \sin\theta)$, $y = -a(1 - \cos\theta)$.

Example 5. *Show that the evolute of the tractrix $x = a\left(\cos t + \log\tan\frac{t}{2}\right)$, $y = a\sin t$ is the catenary $y = a\cosh\frac{x}{a}$.*

Sol. Here $\quad x = a\left(\cos t + \log\tan\dfrac{t}{2}\right),\ y = a\sin t$

$$\frac{dx}{dt} = a\left(-\sin t + \frac{1}{\tan\dfrac{t}{2}}\cdot\frac{1}{2}\sec^2\frac{t}{2}\right) \qquad \frac{dy}{dt} = a\cos t$$

$$= a\left(-\sin t + \frac{1}{2\sin\frac{t}{2}\cos\frac{t}{2}}\right)$$

$$= a\left(-\sin t + \frac{1}{\sin t}\right) = \frac{a\cos^2 t}{\sin t}$$

$\therefore$ $$\frac{dy}{dx} = \tan t$$

$$\frac{d^2y}{dx^2} = \sec^2 t \cdot \frac{dt}{dx} = \frac{\sec^2 t.\sin t}{a\cos^2 t} = \frac{\sin t}{a\cos^4 t}$$

Let C(X, Y) be the centre of curvature

$$X = x - \frac{\frac{dy}{dx}\left(1+\frac{dy}{dx}\right)^2}{\frac{d^2y}{dx^2}} = x - \frac{\tan t\,(1+\tan^2 t)}{\frac{\sin t}{a\cos^4 t}}$$

$$= a\,(\cos t + \log\tan t/2) - a\cos t$$

$$= a\log\tan t/2 \qquad \text{...(1)}$$

$$Y = y + \frac{1+\left(\frac{dy}{dx}\right)^2}{\frac{d^2y}{dx^2}} = y + \frac{1+\tan^2 t}{\frac{\sin t}{a\cos^4 t}}$$

$$= a\sin t + \frac{a\cos^2 t}{\sin t} = \frac{a}{\sin t} \qquad \text{...(2)}$$

From equations (1) and (2),

$$\tan t/2 = e^{X/a} \text{ and } \sin t = \frac{a}{Y}$$

But $$\sin t = \frac{2\tan t/2}{1+\tan^2 t/2}$$

$$\frac{a}{Y} = \frac{2e^{X/a}}{1+e^{2X/a}}$$

$$Y = \frac{a}{2}\left(\frac{1+e^{2X/a}}{e^{X/a}}\right) = \frac{a}{2}\left(e^{-X/a} + e^{X/a}\right)$$

$$= a\cosh\frac{X}{a}$$

Hence, the required evolute of the tractrix is $y = a\cosh\dfrac{x}{a}$.

Example 6. *Show that the evolute of the parabola* $y^2 = 4ax$ *is* $27ay^2 = 4\,(x - 2a)^3$.

Sol. The parametric equation of the parabola is $x = at^2$, $y = 2at$

$$\frac{dx}{dt} = 2at,\ \frac{dy}{dt} = 2a$$

$$\therefore \qquad \frac{dy}{dx} = \frac{1}{t}$$

$$\frac{d^2y}{dx^2} = -\frac{1}{t^2}\cdot\frac{dt}{dx} = \frac{-1}{t^2}\times\frac{1}{2at} = \frac{-1}{2at^3}$$

Let C(X, Y) be the centre of curvature

$$X = x - \frac{\frac{dy}{dx}\left[1+\left(\frac{dy}{dx}\right)^2\right]}{\frac{d^2y}{dx^2}} = at^2 - \frac{\frac{1}{t}\left(1+\frac{1}{t^2}\right)}{\frac{-1}{2at^3}} = 2a + 3at^2 \qquad ...(1)$$

$$Y = y + \frac{\left[1+\left(\frac{dy}{dx}\right)^2\right]}{\frac{d^2y}{dx^2}} = 2at + \frac{1+\frac{1}{t^2}}{\frac{-1}{2at^3}} = -2at^3 \qquad ...(2)$$

Eliminating t from equations (1) and (2)

$$\left(\frac{X-2a}{3a}\right)^3 = \frac{Y^2}{4a^2} \Rightarrow 27aY^2 = 4(X-2a)^3.$$

$\therefore$ Locus of (X, Y) is $27ay^2 = 4(x - 2a)^3$

This is the required evolute of the parabola $y^2 = 4ax$.

EXERCISE 1.3

1. Find the centre of curvature of the following curves:

(i) $y = x^3$ at $\left(\frac{1}{2}, \frac{1}{8}\right)$ (ii) $x^3 + y^3 = 2a^3$ at (a, a)

(iii) $\frac{x^2}{9} + \frac{y^2}{4} = 2$ at (3, 2) (iv) $xy = c^2$ at (c, c)

(v) $x = (1 - bt)\cos t + b \sin t$

$y = (1 - bt)\sin t - b\cos t$

2. Find the circle of curvature of the following curves:

(i) $y = \tan x$ at $\left(\frac{\pi}{4}, 1\right)$ (ii) $xy(x + y) = 2$ at (1, 1)

(iii) $y^2 = 4ax$ at $(a, 2a)$ (iv) $2xy + x + y = 4$ at (1, 1)

3. Find the evolute of the following curves:

(i) $x^2 = 4ay$ (ii) $x = a\cot^2\theta,\ y = 2a\cot\theta$

(iii) $x = a(\cos t + t\sin t),\ y = a(\sin t - t\cos t)$, *(K.U. April 2001)*

4. Find the evolute of $xy = c^2$.

5. Show that the equation of the evolute of the hyperbola $\frac{x^2}{a^2} - \frac{y^2}{b^2} = 1$ is

$$(ax)^{2/3} - (by)^{2/3} = (a^2 + b^2)^{2/3}.$$ *(K.U. November 2000)*

6. Find the evolute of $\frac{x^2}{a^2} + \frac{y^2}{b^2} = 1$. *(K.U. April 2013)*

Answers

1. (*i*) $\left(\frac{7}{64}, \frac{31}{48}\right)$ (*ii*) $\left(\frac{a}{2}, \frac{a}{2}\right)$ (*iii*) $\left(\frac{5}{6}, \frac{-5}{4}\right)$ (*iv*) $(2c, 2c)$

(*v*) $(b \sin t, -b \cos t)$

2. (*i*) $\left(x - \frac{\pi - 10}{4}\right)^2 + \left(y - \frac{9}{4}\right)^2 = \frac{125}{16}$ (*ii*) $x^2 + y^2 + 5x - 5y + 8 = 0$

(*iii*) $(x - 5a)^2 + (y + 3a)^2 = (4\sqrt{2}\, a)^2$ (*iv*) $\left(x - \frac{5}{2}\right)^2 + \left(y - \frac{5}{2}\right)^2 = \frac{9}{2}$

3. (*i*) $4(y - 2a)^3 = 27ax^2$ (*ii*) $27ay^2 = 4(x - 2a)^3$ (*iii*) $x^2 + y^2 = a^2$

4. $(x + y)^{2/3} - (x - y)^{2/3} = 2c^{2/3}$. **6.** $(ax)^{2/3}\,(by)^{2/3} = (a^2 - b^2)^{2/3}$.

1.7. INDETERMINATE FORMS

The limit of $\frac{f(x)}{g(x)}$ as $x \to a$ is, in general, equal to the limit of the numerator divided by the limit of the denominator. But what happens when both these limits are zero? The division $\left(\frac{0}{0}\right)$ then becomes meaningless. A case like this is known as indeterminate form. Other such forms are $\frac{\infty}{\infty}$, $0 \times \infty$, $\infty - \infty$, 0°, 1^∞, ∞°.

Note: The forms $\frac{0}{0}$ and $\frac{\infty}{\infty}$ are regarded as fundamental forms and all other forms are dealt with by converting them to one of these two forms.

L-Hospital's Rule

If the function $f(x)$, $g(x)$ both vanish at $x = a$ $\left(\underset{x \to a}{\text{Lt}} \frac{f(x)}{g(x)} \text{ assumes the form } \frac{0}{0}\right)$, then

$\underset{x \to a}{\text{Lt}} \frac{f(x)}{g(x)} = \underset{x \to a}{\text{Lt}} \frac{f'(x)}{g'(x)}$, provided the limit exists as x approaches a.

In general, if $f(x)$ and $g(x)$ vanish at $x = a$ and have their first $(n - 1)$ derivatives all zero at $x = a$ while their n^{th} derivatives are finite and non-zero, then

$$\underset{x \to a}{\text{Lt}} \frac{f(x)}{g(x)} = \underset{x \to a}{\text{Lt}} \frac{f^n(x)}{g^n(x)} = \frac{f^n(a)}{g^n(a)},$$ provided this limit exists finitely or infinitely.

Note: If a is infinite that is, if the fraction $\frac{f(x)}{g(x)}$ takes the form $\frac{0}{0}$ when x approaches to infinity, then $\lim\limits_{x \to \infty} \frac{f(x)}{g(x)} = \lim\limits_{x \to \infty} \frac{f'(x)}{g'(x)}$.

If the fractions $\frac{f(x)}{g(x)}$ assumes the indeterminate form $\frac{\infty}{\infty}$, as x approaches to a (finite or infinite) then, $\underset{x \to a}{\text{Lt}} \frac{f(x)}{g(x)} = \underset{x \to a}{\text{Lt}} \frac{f'(x)}{g'(x)}$.

ILLUSTRATIVE EXAMPLES

Example 1. *Evaluate* $\underset{x \to a}{Lt} \dfrac{x - \tan x}{x^3}$.

Sol. It is a $\left(\dfrac{0}{0}\right)$ form and therefore,

$$\underset{x \to 0}{\text{Lt}} \frac{x - \tan x}{x^3} = \underset{x \to 0}{\text{Lt}} \frac{1 - \sec^2 x}{3x^2} \qquad \left(\text{form } \frac{0}{0}\right)$$

$$= \underset{x \to 0}{\text{Lt}} - \frac{2 \sec x \cdot \sec x \tan x}{6x}$$

$$= \underset{x \to 0}{\text{Lt}} - \frac{\sec^2 x \tan x}{6x} \qquad \left(\text{form } \frac{0}{0}\right)$$

$$= \underset{x \to 0}{\text{Lt}} \left(\frac{-1}{3} \sec^2 x\right) \times \underset{x \to 0}{\text{Lt}} \frac{\tan x}{x}$$

$$= \frac{-1}{3} \times 1 = \frac{-1}{3}.$$

Example 2. *Evaluate* $\underset{x \to 0}{Lt} \dfrac{xe^x - \log(1 + x)}{x^2}$.

Sol. It is a $\left(\dfrac{0}{0}\right)$ form and therefore,

$$\underset{x \to 0}{\text{Lt}} \frac{xe^x - \log(1 + x)}{x^2} = \underset{x \to 0}{\text{Lt}} \frac{xe^x - e^x \cdot 1 - \dfrac{1}{1 + x}}{2x} \qquad \left(\text{form } \frac{0}{0}\right)$$

$$= \underset{x \to 0}{\text{Lt}} \frac{x \cdot e^x + e^x \cdot 1 + e^x + \dfrac{1}{(1 + x)^2}}{2 \cdot 1} = \frac{0 + 1 + 1 + 1}{2} = \frac{3}{2}.$$

Example 3. *Evaluate* $\underset{x \to 0}{Lt} \dfrac{a^x - b^x}{x}$.

Sol. It is a $\left(\dfrac{0}{0}\right)$ form and therefore,

$$\underset{x \to 0}{\text{Lt}} \frac{a^x - b^x}{x} = \underset{x \to 0}{\text{Lt}} \frac{a^x \log a - b^x \log b}{1} = \log a - \log b = \log\left(\frac{a}{b}\right).$$

Example 4. *Evaluate* $\underset{x \to 1}{Lt} \dfrac{\log(1 - x)}{\cot(\pi x)}$.

Sol. It is a $\dfrac{\infty}{\infty}$ form and therefore

$$\underset{x \to 1}{\text{Lt}} \frac{\log(1 - x)}{\cot(\pi x)} = \underset{x \to 1}{\text{Lt}} \frac{\dfrac{-1}{1 - x}}{-\pi \operatorname{cosec}^2(\pi x)}$$

$$= \underset{x \to 1}{\text{Lt}} \frac{\sin^2 (\pi x)}{\pi(1-x)} \qquad \left(\text{form } \frac{0}{0}\right)$$

$$= \underset{x \to 1}{\text{Lt}} \frac{\pi \sin (2\pi x)}{-\pi} = 0.$$

Example 5. *Evaluate* $\underset{x \to 0}{Lt} \left(\frac{1}{x^2} - \frac{1}{\sin^2 x} \right)$.

Sol. It is a $(\infty - \infty)$ form and therefore,

$$\underset{x \to 0}{\text{Lt}} \left(\frac{1}{x^2} - \frac{1}{\sin^2 x} \right) = \underset{x \to 0}{\text{Lt}} \frac{\sin^2 x - x^2}{x^2 \sin^2 x} \qquad \left(\text{form } \frac{0}{0}\right)$$

$$= \underset{x \to 0}{\text{Lt}} \frac{\sin 2x - 2x}{2x \sin^2 x + x^2 \sin 2x} \qquad \left(\text{form } \frac{0}{0}\right)$$

$$= \underset{x \to 0}{\text{Lt}} \frac{\cos 2x - 2}{(\sin^2 x + 2x \sin 2x + x^2 \cos 2x)} \qquad \left(\text{form } \frac{0}{0}\right)$$

$$= \underset{x \to 0}{\text{Lt}} \frac{-2 \sin 2x}{3 \sin 2x + 6x \cos 2x - 2x^2 \sin 2x} \qquad \left(\text{form } \frac{0}{0}\right)$$

$$= \underset{x \to 0}{\text{Lt}} \frac{-\cos 2x}{3 \cos 2x - 4x \sin 4x - x^2 \cos 2x} = -\frac{1}{3}.$$

Example 6. *Evaluate* $\underset{x \to 0}{Lt}$ *(sinx log x).*

Sol. It is a $(0 \times \infty)$ form and therefore,

$$\underset{x \to 0}{\text{Lt}} (\sin x \log x) = \underset{x \to 0}{\text{Lt}} \frac{\log x}{\text{cosec } x} \qquad \left(\text{form } \frac{\infty}{\infty}\right)$$

$$= \underset{x \to 0}{\text{Lt}} \frac{1/x}{-\text{cosec } x \cot x}$$

$$= -\underset{x \to 0}{\text{Lt}} \left(\frac{\sin x}{x} \right) \tan x$$

$$= -\underset{x \to 0}{\text{Lt}} \left(\frac{\sin x}{x} \right) \times \underset{x \to 0}{\text{Lt}} \tan x = -1.0 = 0.$$

Example 7. *Evaluate* $\underset{x \to 0}{Lt} (\cot x)^{\sin 2x}$.

Sol. It is a (∞°) form and therefore,

Let $\quad K = \underset{x \to 0}{\text{Lt}} (\cot x)^{\sin 2x}$

$\therefore \quad \log K = \underset{x \to 0}{\text{Lt}} \sin 2x \log \cot x \qquad$ | Taking logs on both sides

(form $0 \times \infty$)

$$= \underset{x \to 0}{\text{Lt}} \frac{\log \cot 2x}{\text{cosec } 2x} \qquad \left(\text{form } \frac{\infty}{\infty}\right)$$

$$= \underset{x \to 0}{\text{Lt}} \frac{-2 \text{ cosec } 2x}{-2 \text{ cosec } 2x \cot 2x} \qquad \left(\text{form } \frac{\infty}{\infty}\right)$$

$$= \underset{x \to 0}{\text{Lt}} \frac{1}{\cot 2x} = \underset{x \to 0}{\text{Lt}} \tan 2x = 0$$

$\therefore$ $$K = e^{\circ} = 1.$$

Example 8. *Evaluate* $\underset{x \to 0}{Lt} \left(\frac{\tan x}{x}\right)^{1/x^2}$.

Sol. It is a (1^{∞}) form and therefore,

Let $$K = \underset{x \to 0}{\text{Lt}} \left(\frac{\tan x}{x}\right)^{1/x^2}$$

$\therefore$ $$\log K = \underset{x \to 0}{\text{Lt}} \frac{1}{x^2} \log\left(\frac{\tan x}{x}\right)$$ | Taking logs on both sides.

$$= \underset{x \to 0}{\text{Lt}} \frac{\log\left(\frac{\tan x}{x}\right)}{x^2} \qquad \left(\text{form } \frac{0}{0}\right)$$

$$= \underset{x \to 0}{\text{Lt}} \frac{\frac{\sec^2 x}{\tan x} - \frac{1}{x}}{2x}$$

$$= \underset{x \to 0}{\text{Lt}} \frac{x \sec^2 x - \tan x}{2x^2 \tan x} \qquad \left(\text{form } \frac{0}{0}\right)$$

$$= \underset{x \to 0}{\text{Lt}} \frac{\sec^2 x \tan x}{2 \tan x + x \sec^2 x}$$

$$= \underset{x \to 0}{\text{Lt}} \frac{\tan x}{\sin 2x + x} \qquad \left(\text{form } \frac{0}{0}\right)$$

$$= \underset{x \to 0}{\text{Lt}} \frac{\sec^2 x}{2 \cos 2x + 1} = \frac{1}{3}$$

i.e., $$\underset{x \to 0}{\text{Lt}} \log K = \frac{1}{3} \Rightarrow K = e^{1/3}.$$

Example 9. *Evaluate* $\underset{x \to 1}{Lt} (1 - x^2)^{\frac{1}{\log(1-x)}}$.

Sol. It is a (0^0) form and therefore,

Let $$K = \underset{x \to 1}{\text{Lt}} (1 - x^2)^{\frac{1}{\log(1-x)}}$$

$\therefore$ $$\log K = \underset{x \to 1}{\text{Lt}} \frac{1}{\log(1-x)} \log(1 - x^2) \qquad \left(\text{form } \frac{\infty}{\infty}\right)$$

| Taking logs on both sides

$$= \underset{x \to 1}{\text{Lt}} \frac{2x(1-x)}{1-x^2} = \underset{x \to 1}{\text{Lt}} \frac{2x}{1+x} = 1$$

$\Rightarrow$ $$\underset{x \to 1}{\text{Lt}} \log K = 1 \Rightarrow K = e.$$

EXERCISE 1.4

1. Evaluate the following:

(*i*) $\underset{x \to 0}{\text{Lt}} \dfrac{\log x}{\cot x}$

(*ii*) $\underset{\theta \to \frac{\pi}{2}}{\text{Lt}} \dfrac{\log (\theta - \pi/2)}{\tan \theta}$

(*iii*) $\underset{x \to 0}{\text{Lt}} \dfrac{e^x - 2\cos x + e^{-x}}{x \sin x}$

(*iv*) $\underset{x \to 0}{\text{Lt}} \left(\dfrac{1}{x} - \cot x\right)$

(*v*) $\underset{x \to \frac{\pi}{4}}{\text{Lt}} \dfrac{\sec^2 x - 2\tan x}{1 + \cos 4x}$

(*vi*) $\underset{x \to 0}{\text{Lt}} \dfrac{e^x - e^{-x} + 2\sin x + x^3 - 4x}{10x^3}$

(*vii*) $\underset{x \to 0}{\text{Lt}} \left(\dfrac{1}{x^2} - \cot^2 x\right)$

(*viii*) $\underset{\theta \to \frac{\pi}{2}}{\text{Lt}} (\sec\theta - \tan\theta)$

(*ix*) $\underset{x \to \frac{\pi}{2}}{\text{Lt}} (\sec x)^{\cot x}$

(*x*) $\underset{x \to 0}{\text{Lt}} (1 + \sin x)^{\cot x}$

(*xi*) $\underset{x \to 0}{\text{Lt}} (\sin x)^{\tan x}$

(*xii*) $\underset{x \to 0}{\text{Lt}} (\cot x)^{\sin 2x}$.

Answers

1. (*i*) 0 (*ii*) 0 (*iii*) 2 (*iv*) 0

(*v*) $\dfrac{1}{2}$ (*vi*) $\dfrac{1}{10}$ (*vii*) $\dfrac{2}{3}$ (*viii*) 0

(*ix*) 1 (*x*) e (*xi*) 1 (*xii*) 1.

2

Partial Differentiation

2.1. FUNCTIONS OF TWO VARIABLES

If three variables x, y, z are so related that the value of z depends upon the values of x and y, then z is called a function of two variables x and y, and this is denoted by $z = f(x, y)$.

z is called the dependent variable while x and y are called independent variables.

For example, the area of a triangle is determined when its base and altitude are known. Thus, area of a triangle is a function of two variables, base and altitude.

In a similar way, a function of more than two variables can be defined.

Geometrical Interpretation:

Let $z = f(x, y)$ be a function of two independent variables x and y defined for all pairs of values of x and y which belong to an area A of the xy-plane. Then to each point (x, y) of this area corresponds a value of z given by the relation $z = f(x, y)$. Representing all these values (x, y, z) by points in space, we get a surface.

Hence the function z = f(x, y) represents a surface.

2.2. CONTINUITY

A function $f(x, y)$ is said to be continuous at a point (a, b) if, for any arbitrarily chosen positive number ε, however small, we can find a corresponding number δ such that
$| f(x, y) - f(a, b) | < \varepsilon$ for every point (x, y) within the circle with its centre at (a, b) and radius δ.
i.e., whenever $0 < (x - a)^2 + (y - b)^2 < \delta^2$.

Alternatively, $f(x, y)$ is said to be continuous at (a, b) if $\lim\limits_{\substack{x \to a \\ y \to b}} f(x, y) = f(a, b)$

irrespective of the path along which $x \to a, y \to b$.

It should not be assumed that the path along which the point (x, y) tends to (a, b) is immaterial, because $\lim\limits_{x \to a} \{\lim\limits_{y \to b} f(x, y)\}$ is not always equal to $\lim\limits_{y \to b} \{\lim\limits_{x \to a} f(x, y)\}$.

However, *usually*, the limit is the same irrespective of the path of approach.

In what follows, we shall assume that the functions considered are continuous and their partial differential co-efficients as defined in the next section, exist.

2.3. PARTIAL DERIVATIVES OF FIRST ORDER

Let $z = f(x, y)$ be a function of two independent variables x and y. If y is kept constant and x alone is allowed to vary, then z becomes a function of x only. The derivative of z, with respect to x, treating y as constant, is called partial derivative of z w.r.t. x and is denoted by

$$\frac{\partial z}{\partial x} \quad \text{or} \quad \frac{\partial f}{\partial x} \quad \text{or} \quad f_x.$$

Thus, $$\frac{\partial z}{\partial x} = \lim_{h \to 0} \frac{f(x+h, y) - f(x, y)}{h}$$

Similarly, the derivative of z, with respect to y, treating x as constant, is called partial derivative of z w.r.t. y and is denoted by $\frac{\partial z}{\partial y}$ or $\frac{\partial f}{\partial y}$ or f_y

Thus, $$\frac{\partial z}{\partial y} = \lim_{k \to 0} \frac{f(x, y+k) - f(x, y)}{k}$$

$\frac{\partial z}{\partial x}$ and $\frac{\partial z}{\partial y}$ are called **first order partial derivatives of z.**

[**In general,** if z is a function of two or more independent variables, then the **partial derivative of z** w.r.t. **any one** of the independent variables is the **ordinary derivative of z** w.r.t. that variable, **treating all other variables as constant.**]

Geometrical Interpretation:

Let $z = f(x, y)$ be a function of two variables x and y. Then by Art. 2.1, it represents a surface S. If $y = k$, a constant, then $y = k$ represents a plane parallel to the zx-plane.

$\therefore$ $z = f(x, y)$ and $y = k$ represent a plane curve C which is the section of S by $y = k$.

$\frac{\partial z}{\partial x}$ represents the slope of tangent to C at (x, k, z).

Thus, $\frac{\partial z}{\partial x}$ gives the slope of the tangent drawn to the curve of intersection of the surface $z = f(x, y)$ and a plane parallel to zx-plane.

Similarly, $\frac{\partial z}{\partial y}$ gives the slope of the tangent drawn to the curve of intersection of the surface $z = f(x, y)$ and a plane parallel to yz-plane.

2.4. PARTIAL DERIVATIVES OF HIGHER ORDER

Since the first order partial derivatives $\frac{\partial z}{\partial x}$ and $\frac{\partial z}{\partial y}$ are themselves functions of x and y, they can be further differentiated partially w.r.t. x as well as y. These are called second order partial derivatives of z. The usual notations for these second order partial derivatives are:

$$\frac{\partial}{\partial x}\left(\frac{\partial z}{\partial x}\right) = \frac{\partial^2 z}{\partial x^2} \text{ or } f_{xx}; \qquad \frac{\partial}{\partial y}\left(\frac{\partial z}{\partial y}\right) = \frac{\partial^2 z}{\partial y^2} \text{ or } f_{yy}$$

$$\frac{\partial}{\partial x}\left(\frac{\partial z}{\partial y}\right) = \frac{\partial^2 z}{\partial x \partial y} \text{ or } f_{xy}; \qquad \frac{\partial}{\partial y}\left(\frac{\partial z}{\partial x}\right) = \frac{\partial^2 z}{\partial y \partial x} \text{ or } f_{yx}$$

In general, $$\frac{\partial^2 z}{\partial x \partial y} = \frac{\partial^2 z}{\partial y \partial x} \text{ or } f_{xy} = f_{yx}.$$

Note 1. If $z = f(x)$, a function of single independent variable x, we get $\frac{dz}{dx}$.

If $z = f(x, y)$, a function of two independent variables x and y, we get $\frac{\partial z}{\partial x}$ and $\frac{\partial z}{\partial y}$.

Similarly, for a function of more than two independent variables $x_1, x_2, ..., x_n$, we get $\frac{\partial z}{\partial x_1}, \frac{\partial z}{\partial x_2}, ..., \frac{\partial z}{\partial x_n}$.

Note 2. *(i)* If $z = u + v$, where $u = f(x, y)$, $v = \phi(x, y)$ then z is a function of x and y.

$$\therefore \qquad \frac{\partial z}{\partial x} = \frac{\partial u}{\partial x} + \frac{\partial v}{\partial x}; \quad \frac{\partial z}{\partial y} = \frac{\partial u}{\partial y} + \frac{\partial v}{\partial y}.$$

(ii) If $z = uv$, where $u = f(x, y)$, $v = \phi(x, y)$ then

$$\frac{\partial z}{\partial x} = \frac{\partial}{\partial x}(uv) = u\frac{\partial v}{\partial x} + v\frac{\partial u}{\partial x}; \quad \frac{\partial z}{\partial y} = \frac{\partial}{\partial y}(uv) = u\frac{\partial v}{\partial y} + v\frac{\partial u}{\partial y}$$

(iii) If $z = \dfrac{u}{v}$, where $u = f(x, y)$, $v = \phi(x, y)$ then

$$\frac{\partial z}{\partial x} = \frac{\partial}{\partial x}\left(\frac{u}{v}\right) = \frac{v\dfrac{\partial u}{\partial x} - u\dfrac{\partial v}{\partial x}}{v^2}; \quad \frac{\partial z}{\partial y} = \frac{\partial}{\partial y}\left(\frac{u}{v}\right) = \frac{v\dfrac{\partial u}{\partial y} - u\dfrac{\partial v}{\partial y}}{v^2}$$

(iv) If $z = f(u)$, where $u = \phi(x, y)$ then $\dfrac{\partial z}{\partial x} = \dfrac{dz}{du}\cdot\dfrac{\partial u}{\partial x}$; $\dfrac{\partial z}{\partial y} = \dfrac{dz}{du}\cdot\dfrac{\partial u}{\partial y}$.

ILLUSTRATIVE EXAMPLES

Example 1. *Find the first order partial derivatives of the following*:

(i) $u = \tan^{-1}\dfrac{x^2 + y^2}{x + y}$ *(ii)* $u = \cos^{-1}\left(\dfrac{x}{y}\right)$.

Sol. *(i)*

$$u = \tan^{-1}\frac{x^2 + y^2}{x + y}$$

$$\frac{\partial u}{\partial x} = \frac{1}{1 + \left(\dfrac{x^2 + y^2}{x + y}\right)^2}\cdot\frac{\partial}{\partial x}\left(\frac{x^2 + y^2}{x + y}\right)$$

$$= \frac{(x + y)^2}{(x + y)^2 + (x^2 + y^2)^2}\cdot\frac{(x + y)\dfrac{\partial}{\partial x}(x^2 + y^2) - (x^2 + y^2)\dfrac{\partial}{\partial x}(x + y)}{(x + y)^2}$$

$$= \frac{(x + y)\cdot 2x - (x^2 + y^2)\cdot 1}{(x + y)^2 + (x^2 + y^2)^2}$$

$$\Rightarrow \qquad \frac{\partial u}{\partial x} = \frac{x^2 + 2xy - y^2}{(x + y)^2 + (x^2 + y^2)^2} \qquad \text{...(1)}$$

Since u remains the same if we interchange x and y, u is symmetrical w.r.t. x and y. Interchanging x and y in (1), we have

Similarly,
$$\frac{\partial u}{\partial y} = \frac{y^2 + 2xy - x^2}{(x + y)^2 + (x^2 + y^2)^2}$$

(ii)
$$u = \cos^{-1}\left(\frac{x}{y}\right)$$

$$\frac{\partial u}{\partial x} = \frac{-1}{\sqrt{1 - \left(\dfrac{x}{y}\right)^2}}\cdot\frac{\partial}{\partial x}\left(\frac{x}{y}\right) = \frac{-y}{\sqrt{y^2 - x^2}}\cdot\frac{1}{y} = \frac{-1}{\sqrt{y^2 - x^2}}$$

$$\frac{\partial u}{\partial y} = \frac{-1}{\sqrt{1-\left(\frac{x}{y}\right)^2}} \cdot \frac{\partial}{\partial y}\left(\frac{x}{y}\right) = \frac{-y}{\sqrt{y^2 - x^2}}\left(-\frac{x}{y^2}\right) = \frac{x}{y\sqrt{y^2 - x^2}}.$$

Example 2. *If $z(x + y) = x^2 + y^2$, show that* $\left(\frac{\partial z}{\partial x} - \frac{\partial z}{\partial y}\right)^2 = 4\left(1 - \frac{\partial z}{\partial x} - \frac{\partial z}{\partial y}\right)$.

Sol. $$z = \frac{x^2 + y^2}{x + y}$$ [z is symmetrical w.r.t. x and y]

$$\frac{\partial z}{\partial x} = \frac{(x + y)\frac{\partial}{\partial x}(x^2 + y^2) - (x^2 + y^2)\frac{\partial}{\partial x}(x + y)}{(x + y)^2}$$

$$= \frac{(x + y) \cdot 2x - (x^2 + y^2) \cdot 1}{(x + y)^2} = \frac{x^2 + 2xy - y^2}{(x + y)^2}$$

Similarly, $$\frac{\partial z}{\partial y} = \frac{y^2 + 2xy - x^2}{(x + y)^2}$$

Now $$\left(\frac{\partial z}{\partial x} - \frac{\partial z}{\partial y}\right)^2 = \left[\frac{2x^2 - 2y^2}{(x + y)^2}\right]^2 = \frac{4(x + y)^2 (x - y)^2}{(x + y)^4} = \frac{4(x - y)^2}{(x + y)^2}$$

$$4\left(1 - \frac{\partial z}{\partial x} - \frac{\partial z}{\partial y}\right) = 4\left[1 - \frac{x^2 + 2xy - y^2}{(x + y)^2} - \frac{y^2 + 2xy - x^2}{(x + y)^2}\right]$$

$$= 4\left[\frac{x^2 + 2xy + y^2 - x^2 - 2xy + y^2 - y^2 - 2xy + x^2}{(x + y)^2}\right]$$

$$= \frac{4(x^2 - 2xy + y^2)}{(x + y)^2} = \frac{4(x - y)^2}{(x + y)^2}$$

$$\therefore \quad \left(\frac{\partial z}{\partial x} - \frac{\partial z}{\partial y}\right)^2 = 4\left(1 - \frac{\partial z}{\partial x} - \frac{\partial z}{\partial y}\right).$$

Example 3. *Prove that if* $f(x, y) = \frac{1}{\sqrt{y}} \cdot e^{-\frac{(x-a)^2}{4y}}$, *then* $f_{xy} = f_{yx}$.

Sol. $$f(x, y) = \frac{1}{\sqrt{y}} \cdot e^{-\frac{(x-a)^2}{4y}} = y^{-\frac{1}{2}} e^{-\frac{(x-a)^2}{4y}}$$

$$f_x = \frac{\partial f}{\partial x} = y^{-\frac{1}{2}} \cdot e^{-\frac{(x-a)^2}{4y}} \cdot \frac{\partial}{\partial x}\left[-\frac{(x-a)^2}{4y}\right]$$

$$= y^{-\frac{1}{2}} \cdot e^{-\frac{(x-a)^2}{4y}} \cdot \left[-\frac{2(x-a)}{4y}\right] = -\frac{1}{2} y^{-\frac{3}{2}} (x - a) e^{-\frac{(x-a)^2}{4y}}$$

$$f_y = \frac{\partial f}{\partial y} = -\frac{1}{2} y^{-\frac{3}{2}} \cdot e^{-\frac{(x-a)^2}{4y}} + y^{-\frac{1}{2}} \cdot e^{-\frac{(x-a)^2}{4y}} \cdot \frac{\partial}{\partial y}\left[-\frac{(x-a)^2}{4y}\right]$$

$$= e^{-\frac{(x-a)^2}{4y}} \left[-\frac{1}{2} y^{-\frac{3}{2}} + y^{-\frac{1}{2}} \cdot \frac{(x-a)^2}{4y^2}\right] = \frac{1}{4} y^{-\frac{3}{2}} e^{-\frac{(x-a)^2}{4y}} [-2 + y^{-1}(x-a)^2]$$

$$f_{xy} = \frac{\partial}{\partial x}\left(\frac{\partial f}{\partial y}\right)$$

$$= \frac{1}{4} y^{-\frac{3}{2}} \left\{ e^{-\frac{(x-a)^2}{4y}} \cdot \frac{\partial}{\partial x}\left[-\frac{(x-a)^2}{4y}\right] \cdot [-2 + y^{-1}(x-a)^2] + e^{-\frac{(x-a)^2}{4y}} \cdot 2y^{-1}(x-a) \right\}$$

$$= \frac{1}{4} y^{-\frac{3}{2}} \cdot e^{-\frac{(x-a)^2}{4y}} \left\{ -\frac{2(x-a)}{4y} [-2 + y^{-1}(x-a)^2] + 2y^{-1}(x-a) \right\}$$

$$= \frac{1}{4} y^{-\frac{3}{2}} \cdot e^{-\frac{(x-a)^2}{4y}} \cdot \frac{x-a}{y} \left\{ -\frac{1}{2} [-2 + y^{-1}(x-a)^2] + 2 \right\}$$

$$= \frac{1}{4} y^{-\frac{5}{2}} (x-a) e^{-\frac{(x-a)^2}{4y}} \left[3 - \frac{(x-a)^2}{2y}\right]$$

$$f_{yx} = \frac{\partial}{\partial y}\left(\frac{\partial f}{\partial x}\right) = -\frac{1}{2}(x-a)\left[-\frac{3}{2} y^{-\frac{5}{2}} \cdot e^{-\frac{(x-a)^2}{4y}} + y^{-\frac{3}{2}} \cdot e^{-\frac{(x-a)^2}{4y}} \cdot \frac{(x-a)^2}{4y^2}\right]$$

$$= -\frac{1}{4}(x-a) y^{-\frac{5}{2}} \cdot e^{-\frac{(x-a)^2}{4y}} \left[-3 + \frac{(x-a)^2}{2y}\right]$$

$$= \frac{1}{4} y^{-\frac{5}{2}} (x-a) e^{-\frac{(x-a)^2}{4y}} \left[3 - \frac{(x-a)^2}{2y}\right]$$

$\therefore \quad f_{xy} = f_{yx}$.

Example 4. *If* $u = x^y$, *show that* $\dfrac{\partial^3 u}{\partial x^2 \partial y} = \dfrac{\partial^3 u}{\partial x \partial y \partial x}$.

Sol.

$$u = x^y$$

$$\frac{\partial u}{\partial y} = x^y \log x$$

$$\frac{\partial^2 u}{\partial x \partial y} = \frac{\partial}{\partial x}\left(\frac{\partial u}{\partial y}\right) = yx^{y-1} \log x + x^y \cdot \frac{1}{x} = x^{y-1}(y \log x + 1)$$

$$\frac{\partial^3 u}{\partial x^2 \partial y} = \frac{\partial}{\partial x}\left(\frac{\partial^2 u}{\partial x \partial y}\right) = \frac{\partial}{\partial x}[x^{y-1}(y \log x + 1)] \quad \ldots(1)$$

$$\frac{\partial u}{\partial x} = yx^{y-1}$$

$$\frac{\partial^2 u}{\partial y \partial x} = \frac{\partial}{\partial y}\left(\frac{\partial u}{\partial x}\right) = x^{y-1} + yx^{y-1} \log x = x^{y-1}(y \log x + 1)$$

$$\frac{\partial^3 u}{\partial x \partial y \partial x} = \frac{\partial}{\partial x}\left(\frac{\partial^2 u}{\partial y \partial x}\right) = \frac{\partial}{\partial x}\ [x^{y-1}(y \log x + 1)] \qquad ...(2)$$

From (1) and (2), $\dfrac{\partial^3 u}{\partial x^2 \partial y} = \dfrac{\partial^3 u}{\partial x \partial y \partial x}$.

Example 5. *If* $\theta = t^n e^{-\frac{r^2}{4t}}$, *find the value of n which will make* $\dfrac{1}{r^2}\dfrac{\partial}{\partial r}\left(r^2 \dfrac{\partial \theta}{\partial r}\right) = \dfrac{\partial \theta}{\partial t}$.

Sol.

$$\theta = t^n e^{-\frac{r^2}{4t}}$$

$$\frac{\partial \theta}{\partial r} = t^n \cdot e^{-\frac{r^2}{4t}} \cdot \left(-\frac{2r}{4t}\right) = -\frac{1}{2} r t^{n-1} e^{-\frac{r^2}{4t}}$$

$\therefore$
$$r^2 \frac{\partial \theta}{\partial r} = -\frac{1}{2} r^3 \cdot t^{n-1} e^{-\frac{r^2}{4t}}$$

$$\frac{\partial}{\partial r}\left(r^2 \frac{\partial \theta}{\partial r}\right) = -\frac{1}{2} t^{n-1}\left[3r^2 e^{-\frac{r^2}{4t}} + r^3 e^{-\frac{r^2}{4t}}\left(-\frac{2r}{4t}\right)\right] = -\frac{1}{2} t^{n-1} r^2 e^{-\frac{r^2}{4t}}\left[3 - \frac{r^2}{2t}\right]$$

$\therefore$
$$\frac{1}{r^2}\frac{\partial}{\partial r}\left(r^2 \frac{\partial \theta}{\partial r}\right) = \frac{1}{2} t^{n-1} e^{-\frac{r^2}{4t}}\left(\frac{r^2}{2t} - 3\right)$$

Also
$$\frac{\partial \theta}{\partial t} = n t^{n-1} e^{-\frac{r^2}{4t}} + t^n e^{-\frac{r^2}{4t}} \cdot \left(\frac{r^2}{4t^2}\right) = t^{n-1} e^{-\frac{r^2}{4t}}\left(n + \frac{r^2}{4t}\right)$$

Since
$$\frac{1}{r^2}\frac{\partial}{\partial r}\left(r^2 \frac{\partial \theta}{\partial r}\right) = \frac{\partial \theta}{\partial t} \qquad \text{(given)}$$

$\therefore$
$$\frac{1}{2} t^{n-1} e^{-\frac{r^2}{4t}}\left(\frac{r^2}{2t} - 3\right) = t^{n-1} e^{-\frac{r^2}{4t}}\left(n + \frac{r^2}{4t}\right)$$

$\Rightarrow$
$$\frac{r^2}{4t} - \frac{3}{2} = n + \frac{r^2}{4t} \quad \therefore \quad n = -\frac{3}{2}.$$

Example 6. *If* $u = (1 - 2xy + y^2)^{-1/2}$, *prove that* $\dfrac{\partial}{\partial x}\left\{(1 - x^2)\dfrac{\partial u}{\partial x}\right\} + \dfrac{\partial}{\partial y}\left\{y^2 \dfrac{\partial u}{\partial y}\right\} = 0$.

Sol.

$$u = (1 - 2xy + y^2)^{-1/2} = V^{-1/2}, \quad \text{where } V = 1 - 2xy + y^2$$

$$\frac{\partial u}{\partial x} = -\frac{1}{2} V^{-3/2} \cdot \frac{\partial V}{\partial x} = -\frac{1}{2} V^{-3/2}(-2y) = yV^{-3/2}$$

$$\frac{\partial^2 u}{\partial x^2} = y \cdot \frac{\partial}{\partial x}(V^{-3/2}) = y \cdot \left(-\frac{3}{2}\right) V^{-5/2} \cdot \frac{\partial V}{\partial x} = -\frac{3}{2} yV^{-5/2}(-2y) = 3y^2 V^{-5/2}$$

$\therefore$
$$\frac{\partial}{\partial x}\left\{(1 - x^2)\frac{\partial u}{\partial x}\right\} = (1 - x^2) \cdot \frac{\partial^2 u}{\partial x^2} + \frac{\partial u}{\partial x} \cdot \frac{\partial}{\partial x}(1 - x^2)$$

$$= (1 - x^2) \cdot 3y^2V^{-5/2} + yV^{-3/2}(-2x) = yV^{-3/2}[3yV^{-1}(1 - x^2) - 2x] \quad ...(1)$$

Also
$$\frac{\partial u}{\partial y} = -\frac{1}{2}V^{-3/2}\frac{\partial V}{\partial y} = -\frac{1}{2}V^{-3/2} \cdot (-2x + 2y) = V^{-3/2} \cdot (x - y)$$

$$\frac{\partial^2 u}{\partial y^2} = V^{-3/2} \cdot \frac{\partial}{\partial y}(x - y) + (x - y) \cdot \frac{y}{\partial y}(V^{-3/2})$$

$$= V^{-3/2} \cdot (-1) + (x - y) \cdot \left(-\frac{3}{2}V^{-5/2}\right) \cdot \frac{\partial V}{\partial y}$$

$$= -V^{-3/2} - \frac{3}{2}(x - y)V^{-5/2} \cdot (-2x + 2y) = -V^{-3/2} + 3(x - y)^2 V^{-5/2}$$

$\therefore$
$$\frac{\partial}{\partial y}\left\{y^2\frac{\partial u}{\partial y}\right\} = y^2 \cdot \frac{\partial^2 u}{\partial y^2} + \frac{\partial u}{\partial y} \cdot \frac{\partial}{\partial y}(y^2)$$

$$= y^2[-V^{-3/2} + 3(x - y)^2 V^{-5/2}] + V^{-3/2}(x - y) \cdot 2y$$

$$= yV^{-3/2}[-y + 3y(x - y)^2 V^{-1} + 2(x - y)]$$

$$= yV^{-3/2}[3y(x - y)^2 V^{-1} + (2x - 3y)] \quad ...(2)$$

Adding (1) and (2), we have

$$\frac{\partial}{\partial x}\left\{(1 - x^2)\frac{\partial u}{\partial x}\right\} + \frac{\partial}{\partial y}\left\{y^2\frac{\partial u}{\partial y}\right\} = yV^{-3/2}[3yV^{-1}(1 - x^2) - 2x + 3y(x - y)^2V^{-1} + 2x - 3y]$$

$$= yV^{-3/2}[3yV^{-1}(1 - x^2 + x^2 - 2xy + y^2) - 3y]$$

$$= yV^{-3/2}[3yV^{-1}(1 - 2xy + y^2) - 3y]$$

$$= yV^{-3/2}[3y - 3y] \quad | \because \; V = 1 - 2xy + y^2$$

$$= 0.$$

Example 7. *If* $u = \log(x^3 + y^3 + z^3 - 3xyz)$, *show that*

(*i*) $\left(\frac{\partial}{\partial x} + \frac{\partial}{\partial y} + \frac{\partial}{\partial z}\right)^2 u = -\frac{9}{(x + y + z)^2}$

(*ii*) $\frac{\partial^2 u}{\partial x^2} + \frac{\partial^2 u}{\partial y^2} + \frac{\partial^2 u}{\partial z^2} + 2\frac{\partial^2 u}{\partial y \partial z} + 2\frac{\partial^2 u}{\partial z \partial x} + 2\frac{\partial^2 u}{\partial x \partial y} = \frac{-9}{(x + y + z)^2}$.

Sol. (*i*)
$$u = \log(x^3 + y^3 + z^3 - 3xyz)$$

$$\frac{\partial u}{\partial x} = \frac{3x^2 - 3yz}{x^3 + y^3 + z^3 - 3xyz}; \quad \frac{\partial u}{\partial y} = \frac{3y^2 - 3zx}{x^3 + y^3 + z^3 - 3xyz};$$

$$\frac{\partial u}{\partial z} = \frac{3z^2 - 3xy}{x^3 + y^3 + z^3 - 3xyz}$$

Adding,
$$\frac{\partial u}{\partial x} + \frac{\partial u}{\partial y} + \frac{\partial u}{\partial z} = \frac{3(x^2 + y^2 + z^2 - xy - yz - zx)}{x^3 + y^3 + z^3 - 3xyz} = \frac{3}{x + y + z}$$

$[\because \; x^3 + y^3 + z^3 - 3xyz = (x + y + z)(x^2 + y^2 + z^2 - xy - yz - zx)]$

Now
$$\left(\frac{\partial}{\partial x} + \frac{\partial}{\partial y} + \frac{\partial}{\partial z}\right)^2 u = \left(\frac{\partial}{\partial x} + \frac{\partial}{\partial y} + \frac{\partial}{\partial z}\right)\left(\frac{\partial}{\partial x} + \frac{\partial}{\partial y} + \frac{\partial}{\partial z}\right)u$$

$$= \left(\frac{\partial}{\partial x} + \frac{\partial}{\partial y} + \frac{\partial}{\partial z}\right)\left(\frac{\partial u}{\partial x} + \frac{\partial u}{\partial y} + \frac{\partial u}{\partial z}\right) = \left(\frac{\partial}{\partial x} + \frac{\partial}{\partial y} + \frac{\partial}{\partial z}\right)\left(\frac{3}{x + y + z}\right)$$

$$= -\frac{3}{(x+y+z)^2} - \frac{3}{(x+y+z)^2} - \frac{3}{(x+y+z)^2} = -\frac{9}{(x+y+z)^2} \quad ...(1)$$

(ii) $$\left(\frac{\partial}{\partial x}+\frac{\partial}{\partial y}+\frac{\partial}{\partial z}\right)^2 u = \left(\frac{\partial}{\partial x}+\frac{\partial}{\partial y}+\frac{\partial}{\partial z}\right)\left(\frac{\partial}{\partial x}+\frac{\partial}{\partial y}+\frac{\partial}{\partial z}\right) u$$

$$= \left(\frac{\partial}{\partial x}+\frac{\partial}{\partial y}+\frac{\partial}{\partial z}\right)\left(\frac{\partial u}{\partial x}+\frac{\partial u}{\partial y}+\frac{\partial u}{\partial z}\right)$$

$$= \frac{\partial}{\partial x}\left(\frac{\partial u}{\partial x}+\frac{\partial u}{\partial y}+\frac{\partial u}{\partial z}\right)+\frac{\partial}{\partial y}\left(\frac{\partial u}{\partial x}+\frac{\partial u}{\partial y}+\frac{\partial u}{\partial z}\right)+\frac{\partial}{\partial z}\left(\frac{\partial u}{\partial x}+\frac{\partial u}{\partial y}+\frac{\partial u}{\partial z}\right)$$

$$= \frac{\partial^2 u}{\partial x^2}+\frac{\partial^2 u}{\partial x\partial y}+\frac{\partial^2 u}{\partial x\partial z}+\frac{\partial^2 u}{\partial y\partial x}+\frac{\partial^2 u}{\partial y^2}+\frac{\partial^2 u}{\partial y\partial z}+\frac{\partial^2 u}{\partial z\partial x}+\frac{\partial^2 u}{\partial z\partial y}+\frac{\partial^2 u}{\partial z^2}$$

$$= \frac{\partial^2 u}{\partial x^2}+\frac{\partial^2 u}{\partial y^2}+\frac{\partial^2 u}{\partial z^2}+2\frac{\partial^2 u}{\partial y\partial z}+2\frac{\partial^2 u}{\partial z\partial x}+2\frac{\partial^2 u}{\partial x\partial y}$$

$$\left[\because \frac{\partial^2 u}{\partial y\partial x}=\frac{\partial^2 u}{\partial x\partial y}, \frac{\partial^2 u}{\partial z\partial y}=\frac{\partial^2 u}{\partial y\partial z}, \frac{\partial^2 u}{\partial x\partial z}=\frac{\partial^2 u}{\partial z\partial x}\right]$$

$$\therefore \quad \frac{\partial^2 u}{\partial x^2}+\frac{\partial^2 u}{\partial y^2}+\frac{\partial^2 u}{\partial z^2}+2\frac{\partial^2 u}{\partial y\partial z}+2\frac{\partial^2 u}{\partial z\partial x}+2\frac{\partial^2 u}{\partial x\partial y} = \left(\frac{\partial}{\partial x}+\frac{\partial}{\partial y}+\frac{\partial}{\partial z}\right)^2 u = \frac{-9}{(x+y+z)^2}.$$

[from (1)]

Example 8. *If* $x^x y^y z^z = c$, *show that at* $x = y = z$, $\frac{\partial^2 z}{\partial x\partial y} = -(x \log ex)^{-1}$.

Sol. $x^x y^y z^z = c$ defines z as a function of x and y.

Taking log, $x \log x + y \log y + z \log z = \log c$

Differentiating partially w.r.t. y, we have

$$y \cdot \frac{1}{y} + 1 \cdot \log y + z \cdot \frac{1}{z}\frac{\partial z}{\partial y} + \log z \cdot \frac{\partial z}{\partial y} = 0$$

or $$1 + \log y + (1 + \log z)\frac{\partial z}{\partial y} = 0 \quad ...(1)$$

or $$\left.\begin{aligned} \frac{\partial z}{\partial y} &= -\frac{1+\log y}{1+\log z} \\ \text{Similarly,} \quad \frac{\partial z}{\partial x} &= -\frac{1+\log x}{1+\log z} \end{aligned}\right\} \quad ...(2)$$

Differentiating (1) partially w.r.t. x, we have

$$\left(\frac{1}{z}\frac{\partial z}{\partial x}\right)\frac{\partial z}{\partial y} + (1+\log z)\frac{\partial^2 z}{\partial x\partial y} = 0 \quad \text{or} \quad \frac{\partial^2 z}{\partial x\partial y} = -\frac{1}{z(1+\log z)}\frac{\partial z}{\partial x}\cdot\frac{\partial z}{\partial y} \quad ...(3)$$

When $x = y = z$

From (2), $\frac{\partial z}{\partial y} = -1, \frac{\partial z}{\partial x} = -1$

From (3), $$\frac{\partial^2 z}{\partial x\partial y} = -\frac{1}{x(1+\log x)}(-1)(-1) = -\frac{1}{x(\log e + \log x)} = -\frac{1}{x(\log ex)} = -(x \log ex)^{-1}.$$

Example 9. *If $u = f(r)$, where $r^2 = x^2 + y^2$, prove that* $\dfrac{\partial^2 u}{\partial x^2} + \dfrac{\partial^2 u}{\partial y^2} = f''(r) + \dfrac{1}{r} f'(r)$.

Sol. $r^2 = x^2 + y^2$ ...(1)

Differentiating partially w.r.t. x, we get $2r \dfrac{\partial r}{\partial x} = 2x$ or $\dfrac{\partial r}{\partial x} = \dfrac{x}{r}$

Similarly, $\dfrac{\partial r}{\partial y} = \dfrac{y}{r}$

Now $u = f(r)$

$$\therefore \quad \frac{\partial u}{\partial x} = f'(r) \cdot \frac{\partial r}{\partial x} = \frac{x}{r} f'(r)$$

Differentiating again w.r.t. x, we get

$$\frac{\partial^2 u}{\partial x^2} = \frac{1}{r} f'(r) + x \cdot \left(-\frac{1}{r^2} \frac{\partial r}{\partial x} \right) f'(r) + \frac{x}{r} f''(r) \cdot \frac{\partial r}{\partial x}$$

$$\left[\because \frac{\partial}{\partial x}(uvw) = vw \frac{\partial}{\partial x}(u) + uw \frac{\partial}{\partial x}(v) + uv \frac{\partial}{\partial x}(w) \right]$$

$$= \frac{1}{r} f'(r) - \frac{x}{r^2} \cdot \frac{x}{r} f'(r) + \frac{x}{r} \cdot f''(r) \cdot \frac{x}{r} = \frac{1}{r} f'(r) - \frac{x^2}{r^3} f'(r) + \frac{x^2}{r^2} f''(r)$$

$$= \frac{r^2 - x^2}{r^3} f'(r) + \frac{x^2}{r^2} f''(r) = \frac{y^2}{r^3} f'(r) + \frac{x^2}{r^2} f''(r) \quad | \text{ Using (1)}$$

Similarly, $\dfrac{\partial^2 u}{\partial y^2} = \dfrac{x^2}{r^3} f'(r) + \dfrac{y^2}{r^2} f''(r)$

$$\therefore \quad \frac{\partial^2 u}{\partial x^2} + \frac{\partial^2 u}{\partial y^2} = \frac{x^2 + y^2}{r^3} f'(r) + \frac{x^2 + y^2}{r^2} f''(r) = \frac{r^2}{r^3} f'(r) + \frac{r^2}{r^2} f''(r) = f''(r) + \frac{1}{r} f'(r).$$

Example 10. *If $x = r \cos \theta$, $y = r \sin \theta$, prove that*

(*i*) $\dfrac{\partial r}{\partial x} = \dfrac{\partial x}{\partial r}$ (*ii*) $\dfrac{1}{r} \cdot \dfrac{\partial x}{\partial \theta} = r \dfrac{\partial \theta}{\partial x}$ (*iii*) $\dfrac{\partial^2 \theta}{\partial x^2} + \dfrac{\partial^2 \theta}{\partial y^2} = 0$.

Sol. (*i*) $\dfrac{\partial r}{\partial x}$ means $\left(\dfrac{\partial r}{\partial x} \right)_y$ = The partial derivative of r w.r.t. x, treating y as constant.

$\therefore$ We express r in terms of x and y.

Squaring and adding the given relations, $r^2 = x^2 + y^2$

Differentiating partially w.r.t. x, we get $2r \dfrac{\partial r}{\partial x} = 2x$ or $\dfrac{\partial r}{\partial x} = \dfrac{x}{r}$

$\dfrac{\partial x}{\partial r}$ means $\left(\dfrac{\partial x}{\partial r} \right)_\theta$ = The partial derivative of x w.r.t. r treating θ as constant.

$\therefore$ We express x in terms of r and θ.

Thus, $x = r \cos \theta$ (given)

$$\frac{\partial x}{\partial r} = \cos \theta = \frac{x}{r} \qquad \left(\because \cos \theta = \frac{x}{r} \right)$$

$$\therefore \qquad \frac{\partial r}{\partial x} = \frac{\partial x}{\partial r}$$

(*ii*) Expressing x in terms of r and θ, we have $x = r \cos \theta$

$$\Rightarrow \qquad \frac{\partial x}{\partial \theta} = -r \sin \theta = -y \quad \Rightarrow \quad \frac{1}{r}\frac{\partial x}{\partial \theta} = -\frac{y}{r}$$

Expressing θ in terms of x and y, we have $\tan \theta = \frac{y}{x}$ or $\theta = \tan^{-1} \frac{y}{x}$

$$\frac{\partial \theta}{\partial x} = \frac{1}{1+\frac{y^2}{x^2}} \cdot \left(-\frac{y}{x^2}\right) = -\frac{y}{x^2+y^2} = \frac{-y}{r^2(\cos^2\theta + \sin^2\theta)} = -\frac{y}{r^2}$$

$$\Rightarrow \qquad r\frac{\partial \theta}{\partial x} = -\frac{y}{r} \qquad \therefore \quad \frac{1}{r} \cdot \frac{\partial x}{\partial \theta} = r\frac{\partial \theta}{\partial x}.$$

(*iii*) Expressing θ in terms of x and y, we have $\tan \theta = \frac{y}{x}$ or $\theta = \tan^{-1} \frac{y}{x}$

$$\frac{\partial \theta}{\partial x} = \frac{1}{1+\frac{y^2}{x^2}} \cdot \left(-\frac{y}{x^2}\right) = -\frac{y}{x^2+y^2} = -y(x^2+y^2)^{-1}$$

$$\frac{\partial^2 \theta}{\partial x^2} = y(x^2+y^2)^{-2} \cdot 2x = \frac{2xy}{(x^2+y^2)^2}$$

$$\frac{\partial \theta}{\partial y} = \frac{1}{1+\frac{y^2}{x^2}} \cdot \frac{1}{x} = \frac{x}{x^2+y^2} = x(x^2+y^2)^{-1}$$

$$\frac{\partial^2 \theta}{\partial y^2} = -x(x^2+y^2)^{-2} \cdot 2y = -\frac{2xy}{(x^2+y^2)^2}$$

$$\therefore \qquad \frac{\partial^2 \theta}{\partial x^2} + \frac{\partial^2 \theta}{\partial y^2} = 0.$$

Example 11. *If* $x = e^{r\cos\theta} \cos(r \sin \theta)$ *and* $y = e^{r\cos\theta} \sin(r\sin\theta)$, *prove that*

$$\frac{\partial x}{\partial r} = \frac{1}{r} \cdot \frac{\partial y}{\partial \theta}, \quad \frac{\partial y}{\partial r} = -\frac{1}{r} \cdot \frac{\partial x}{\partial \theta}.$$

Hence deduce that $\frac{\partial^2 x}{\partial r^2} + \frac{1}{r}\frac{\partial x}{\partial r} + \frac{1}{r^2}\frac{\partial^2 x}{\partial \theta^2} = 0.$

Sol. $\qquad x = e^{r\cos\theta} \cos(r\sin\theta)$

$$\therefore \qquad \frac{\partial x}{\partial r} = e^{r\cos\theta} \cdot \cos\theta \cdot \cos(r\sin\theta) - e^{r\cos\theta} \sin(r\sin\theta) \cdot \sin\theta$$

$$= e^{r\cos\theta} [\cos\theta \cos(r\sin\theta) - \sin\theta \sin(r\sin\theta)]$$

$$= e^{r\cos\theta} \cos(\theta + r\sin\theta) \qquad \text{...(1)}$$

$$\frac{\partial x}{\partial \theta} = e^{r\cos\theta} \cdot (-r\sin\theta) \cdot \cos(r\sin\theta) - e^{r\cos\theta} \sin(r\sin\theta) \cdot r\cos\theta$$

$$= -re^{r\cos\theta} [\sin\theta \cos(r\sin\theta) + \cos\theta \sin(r\sin\theta)]$$

$$= -re^{r\cos\theta} \sin(\theta + r\sin\theta) \qquad \text{...(2)}$$

Also $\qquad y = e^{r\cos\theta} \sin(r\sin\theta)$

$$\therefore \quad \frac{\partial y}{\partial r} = e^{r\cos\theta}\cdot \cos\theta \cdot \sin(r\sin\theta) + e^{r\cos\theta}\cdot \cos(r\sin\theta)\sin\theta$$

$$= e^{r\cos\theta}[\sin\theta\cos(r\sin\theta) + \cos\theta\sin(r\sin\theta)]$$

$$= e^{r\cos\theta}\sin(\theta + r\sin\theta) \qquad ...(3)$$

$$\frac{\partial y}{\partial \theta} = e^{r\cos\theta}(-r\sin\theta)\sin(r\sin\theta) + e^{r\cos\theta}\cos(r\sin\theta)\times r\cos\theta$$

$$= re^{r\cos\theta}[\cos\theta\cos(r\sin\theta) - \sin\theta\sin(r\sin\theta)]$$

$$= re^{r\cos\theta}\cos(\theta + r\sin\theta) \qquad ...(4)$$

From (1) and (4), $\quad \dfrac{\partial x}{\partial r} = \dfrac{1}{r}\cdot\dfrac{\partial y}{\partial \theta} \qquad ...(5)$

From (2) and (3), $\quad \dfrac{\partial y}{\partial r} = -\dfrac{1}{r}\cdot\dfrac{\partial x}{\partial \theta} \qquad ...(6)$

From (5), $\quad \dfrac{\partial^2 x}{\partial r^2} = -\dfrac{1}{r^2}\cdot\dfrac{\partial y}{\partial \theta} + \dfrac{1}{r}\cdot\dfrac{\partial^2 y}{\partial r\partial \theta}$

From (6), $\quad \dfrac{\partial x}{\partial \theta} = -r\dfrac{\partial y}{\partial r}$

$$\therefore \quad \frac{\partial^2 x}{\partial \theta^2} = -r\frac{\partial^2 y}{\partial \theta\partial r} = -r\frac{\partial^2 y}{\partial r\partial \theta}$$

$$\therefore \quad \frac{\partial^2 x}{\partial r^2} + \frac{1}{r}\cdot\frac{\partial x}{\partial r} + \frac{1}{r^2}\cdot\frac{\partial^2 x}{\partial \theta^2} = -\frac{1}{r^2}\cdot\frac{\partial y}{\partial \theta} + \frac{1}{r}\cdot\frac{\partial^2 y}{\partial r\partial\theta} + \frac{1}{r^2}\cdot\frac{\partial y}{\partial\theta} - \frac{1}{r}\cdot\frac{\partial^2 y}{\partial r\partial\theta} = 0.$$

Example 12. *If* $\dfrac{x^2}{a^2+u} + \dfrac{y^2}{b^2+u} + \dfrac{z^2}{c^2+u} = 1$, *prove that*

$$\left(\frac{\partial u}{\partial x}\right)^2 + \left(\frac{\partial u}{\partial y}\right)^2 + \left(\frac{\partial u}{\partial z}\right)^2 = 2\left(x\frac{\partial u}{\partial x} + y\frac{\partial u}{\partial y} + z\frac{\partial u}{\partial z}\right).$$

Sol. Given $\qquad \dfrac{x^2}{a^2+u} + \dfrac{y^2}{b^2+u} + \dfrac{z^2}{c^2+u} = 1 \qquad ...(1)$

or $\qquad x^2(a^2+u)^{-1} + y^2(b^2+u)^{-1} + z^2(c^2+u)^{-1} = 1$

Differentiating partially w.r.t. x, we have

$$2x(a^2+u)^{-1} - x^2(a^2+u)^{-2}\cdot\frac{\partial u}{\partial x} - y^2(b^2+u)^{-2}\cdot\frac{\partial u}{\partial x} - z^2(c^2+u)^{-2}\cdot\frac{\partial u}{\partial x} = 0$$

or $$\frac{2x}{a^2+u} = \left[\frac{x^2}{(a^2+u)^2} + \frac{y^2}{(b^2+u)^2} + \frac{z^2}{(c^2+u)^2}\right]\frac{\partial u}{\partial x}$$

or $$\frac{2x}{a^2+u} = V\frac{\partial u}{\partial x} \text{ where } V = \frac{x^2}{(a^2+u)^2} + \frac{y^2}{(b^2+u)^2} + \frac{z^2}{(c^2+u)^2}$$

or $$\frac{\partial u}{\partial x} = \frac{2x}{V(a^2+u)}$$

Similarly, $\quad \dfrac{\partial u}{\partial y} = \dfrac{2y}{V(b^2+u)}$ and $\dfrac{\partial u}{\partial z} = \dfrac{2z}{V(c^2+u)}$

$$\therefore \quad \left(\frac{\partial u}{\partial x}\right)^2 + \left(\frac{\partial u}{\partial y}\right)^2 + \left(\frac{\partial u}{\partial z}\right)^2 = \frac{4}{V^2}\left[\frac{x^2}{(a^2+u)^2} + \frac{y^2}{(b^2+u)^2} + \frac{z^2}{(c^2+u)^2}\right]$$

$$= \frac{4}{V^2}(V) = \frac{4}{V} \qquad ...(2)$$

$$\text{Now, } 2\left(x\frac{\partial u}{\partial x} + y\frac{\partial u}{\partial y} + z\frac{\partial u}{\partial z}\right) = 2\left[\frac{2x^2}{V(a^2+u)} + \frac{2y^2}{V(b^2+u)} + \frac{2z^2}{V(c^2+u)}\right]$$

$$= \frac{4}{V}\left[\frac{x^2}{a^2+u} + \frac{y^2}{b^2+u} + \frac{z^2}{c^2+u}\right]$$

$$= \frac{4}{V}(1) \qquad \text{[Using (1)]}$$

$$= \left(\frac{\partial u}{\partial x}\right)^2 + \left(\frac{\partial u}{\partial y}\right)^2 + \left(\frac{\partial u}{\partial z}\right)^2 \qquad \text{[Using (2)]}$$

Example 13. *If $u = lx + my$, $v = mx - ly$, show that*

$$\left(\frac{\partial u}{\partial x}\right)_y \left(\frac{\partial x}{\partial u}\right)_v = \frac{l^2}{l^2+m^2}, \left(\frac{\partial y}{\partial v}\right)_x \left(\frac{\partial v}{\partial y}\right)_u = \frac{l^2+m^2}{l^2}.$$

Sol. Given

$$u = lx + my \qquad ...(1)$$

$$v = mx - ly \qquad ...(2)$$

(*i*) $\left(\frac{\partial u}{\partial x}\right)_y$ = The partial derivative of u w.r.t. x keeping y constant.

$\therefore$ We need a relation expressing u as a function of x and y.

From (1), $\left(\frac{\partial u}{\partial x}\right)_y = l$

Also $\left(\frac{\partial x}{\partial u}\right)_v$ = The partial derivative of x w.r.t. u keeping v constant.

$\therefore$ We need a relation expressing x as a function of u and v.

Eliminating y between (1) and (2) by multiplying (1) by l, (2) by m and adding the products, we have

$$lu + mv = (l^2 + m^2)x \quad \text{or} \quad x = \frac{lu+mv}{l^2+m^2}$$

$$\therefore \quad \left(\frac{\partial x}{\partial u}\right)_v = \frac{l}{l^2+m^2}$$

Hence $\left(\frac{\partial u}{\partial x}\right)_y \left(\frac{\partial x}{\partial u}\right)_v = \frac{l^2}{l^2+m^2}$

(*ii*) $\left(\frac{\partial y}{\partial v}\right)_x$ = The partial derivative of y w.r.t. v keeping x constant.

$\therefore$ We need a relation expressing y as a function of v and x.

From (2), $y = \frac{mx - v}{l} \quad \therefore \quad \left(\frac{\partial y}{\partial v}\right)_x = -\frac{1}{l}$

Also $\left(\frac{\partial v}{\partial y}\right)_u$ = Partial derivative of v w.r.t. y keeping u constant

$\therefore$ We need a relation expressing v as a function of y and u.

Eliminating x between (1) and (2), we have $v = \frac{mu - (l^2 + m^2)y}{l}$

$$\therefore \quad \left(\frac{\partial v}{\partial y}\right)_u = -\frac{l^2 + m^2}{l}$$

$$\text{Hence} \quad \left(\frac{\partial y}{\partial v}\right)_x \left(\frac{\partial v}{\partial y}\right)_u = \left(-\frac{1}{l}\right)\left(-\frac{l^2 + m^2}{l}\right) = \frac{l^2 + m^2}{l^2}.$$

EXERCISE 2.1

1. Find the first order partial derivatives of the following functions:

   (i) $u = y^x$ (ii) $u = \log(x^2 + y^2)$

   (iii) $u = x^2 \sin \frac{y}{x}$ (iv) $u = \frac{x}{y} \tan^{-1}\left(\frac{y}{x}\right)$.

2. If $u = x^2 + y^2 + z^2$, prove that $xu_x + yu_y + zu_z = 2u$.

3. If $z = \log(x^2 + xy + y^2)$, prove that $x\frac{\partial z}{\partial x} + y\frac{\partial z}{\partial y} = 2$.

4. If $u = x^2y + y^2z + z^2x$, prove that $u_x + u_y + u_z = (x + y + z)^2$.

5. If $u = \frac{y}{z} + \frac{z}{x} + \frac{x}{y}$, show that $x\frac{\partial u}{\partial x} + y\frac{\partial u}{\partial y} + z\frac{\partial u}{\partial z} = 0$.

6. If $f(x, y) = x^3y - xy^3$, find $\left[\frac{1}{\frac{\partial f}{\partial x}} + \frac{1}{\frac{\partial f}{\partial y}}\right]_{\substack{x=1\\y=2}}$.

7. If $u = \log(\tan x + \tan y)$, prove that $\sin 2x\frac{\partial u}{\partial x} + \sin 2y\frac{\partial u}{\partial y} = 2$.

8. If $f(x, y, z) = \begin{vmatrix} x^2 & y^2 & z^2 \\ x & y & z \\ 1 & 1 & 1 \end{vmatrix}$, prove that $f_x + f_y + f_z = 0$.

9. Verify that $\frac{\partial^2 u}{\partial x \partial y} = \frac{\partial^2 u}{\partial y \partial x}$ for the following functions:

   (i) $u = ax^2 + 2hxy + by^2$ (ii) $u = \tan^{-1}\left(\frac{x}{y}\right)$ (iii) $u = \log\left(\frac{x^2 + y^2}{xy}\right)$

   (iv) $u = e^{ax} \sin by$ (v) $u = \log(x \sin y + y \sin x)$.

10. If $z = \log(e^x + e^y)$, show that $rt - s^2 = 0$; where $r = \frac{\partial^2 z}{\partial x^2}$, $t = \frac{\partial^2 z}{\partial y^2}$, $s = \frac{\partial^2 z}{\partial x \partial y}$.

11. If $u = \tan^{-1} \frac{xy}{\sqrt{1 + x^2 + y^2}}$, show that $\frac{\partial^2 u}{\partial x \partial y} = (1 + x^2 + y^2)^{-3/2}$.

12. If $u = e^{xyz}$, prove that $\dfrac{\partial^3 u}{\partial x \partial y \partial z} = (1 + 3xyz + x^2y^2z^2)e^{xyz}$.

13. If $u = \log(x^2 + y^2) + \tan^{-1}\dfrac{y}{x}$, prove that $\dfrac{\partial^2 u}{\partial x^2} + \dfrac{\partial^2 u}{\partial y^2} = 0$.

14. If $u = \tan^{-1}\left(\dfrac{2xy}{x^2 - y^2}\right)$, prove that $\dfrac{\partial^2 u}{\partial x^2} + \dfrac{\partial^2 u}{\partial y^2} = 0$.

15. Verify that $f_{xy} = f_{yx}$ when f is equal to

(*i*) $\sin^{-1}\left(\dfrac{y}{x}\right)$ (*ii*) $\log x \tan^{-1}(x^2 + y^2)$.

16. Find the value of n so that the equation $V = r^n(3\cos^2\theta - 1)$ satisfies the relation

$$\frac{\partial}{\partial r}\left(r^2 \frac{\partial V}{\partial r}\right) + \frac{1}{\sin\theta}\frac{\partial}{\partial \theta}\left(\sin\theta \frac{\partial V}{\partial \theta}\right) = 0.$$

17. If $z = \tan(y + ax) - (y - ax)^{3/2}$, show that $\dfrac{\partial^2 z}{\partial x^2} = a^2 \dfrac{\partial^2 z}{\partial y^2}$.

18. If $V = (x^2 + y^2 + z^2)^{-1/2}$, prove that $\dfrac{\partial^2 V}{\partial x^2} + \dfrac{\partial^2 V}{\partial y^2} + \dfrac{\partial^2 V}{\partial z^2} = 0$.

19. If $V = r^m$, where $r^2 = x^2 + y^2 + z^2$, show that $V_{xx} + V_{yy} + V_{zz} = m(m + 1) r^{m-2}$.

20. If $u = \log(x^2 + y^2 + z^2)$, prove that $x\dfrac{\partial^2 u}{\partial y \partial z} = y\dfrac{\partial^2 u}{\partial z \partial x} = z\dfrac{\partial^2 u}{\partial x \partial y}$.

21. If $u = \log\sqrt{x^2 + y^2 + z^2}$, prove that $(x^2 + y^2 + z^2)\left(\dfrac{\partial^2 u}{\partial x^2} + \dfrac{\partial^2 u}{\partial y^2} + \dfrac{\partial^2 u}{\partial z^2}\right) = 1$.

22. If $u = x^2 \tan^{-1}\dfrac{y}{x} - y^2 \tan^{-1}\dfrac{x}{y}$, find the value of $\dfrac{\partial^2 u}{\partial x \partial y}$.

23. If $x^2 + y^2 + z^2 = \dfrac{1}{u^2}$, prove that $\dfrac{\partial^2 u}{\partial x^2} + \dfrac{\partial^2 u}{\partial y^2} + \dfrac{\partial^2 u}{\partial z^2} = 0$.

24. If $u = \sqrt{x^2 + y^2 + z^2}$, show that

(*i*) $\left(\dfrac{\partial u}{\partial x}\right)^2 + \left(\dfrac{\partial u}{\partial y}\right)^2 + \left(\dfrac{\partial u}{\partial z}\right)^2 = 1$ (*ii*) $\dfrac{\partial^2 u}{\partial x^2} + \dfrac{\partial^2 u}{\partial y^2} + \dfrac{\partial^2 u}{\partial z^2} = \dfrac{2}{u}$.

25. If $u = e^{x-at}\cos(x - at)$, show that $\dfrac{\partial^2 u}{\partial t^2} = a^2 \dfrac{\partial^2 u}{\partial x^2}$.

26. If $v = \dfrac{1}{\sqrt{t}} e^{\frac{-x^2}{4a^2 t}}$, prove that $\dfrac{\partial v}{\partial t} = a^2 \dfrac{\partial^2 v}{\partial x^2}$.

27. If $u = (1 - 2xy + y^2)^{-1/2}$, prove that $x\dfrac{\partial u}{\partial x} - y\dfrac{\partial u}{\partial y} = y^2 u^3$.

28. If $u = e^x(x\cos y - y\sin y)$, show that $\dfrac{\partial^2 u}{\partial x^2} + \dfrac{\partial^2 u}{\partial y^2} = 0$.

29. If $x = r\cos\theta$, $y = r\sin\theta$, prove that

(*i*) $\dfrac{\partial^2 r}{\partial x^2} \cdot \dfrac{\partial^2 r}{\partial y^2} = \left(\dfrac{\partial^2 r}{\partial x \partial y}\right)^2$ (*ii*) $\dfrac{\partial^2 r}{\partial x^2} + \dfrac{\partial^2 r}{\partial y^2} = \dfrac{1}{r}\left[\left(\dfrac{\partial r}{\partial x}\right)^2 + \left(\dfrac{\partial r}{\partial y}\right)^2\right]$.

30. If $x^2 = au + bv$, $y^2 = au - bv$, prove that $\left(\frac{\partial u}{\partial x}\right)_y \cdot \left(\frac{\partial x}{\partial u}\right)_v = \frac{1}{2} = \left(\frac{\partial v}{\partial y}\right)_x \cdot \left(\frac{\partial y}{\partial v}\right)_u$.

31. Show that $\frac{\partial^2 z}{\partial x^2} - 2\frac{\partial^2 z}{\partial x \partial y} + \frac{\partial^2 z}{\partial y^2} = 0$, where $z = xf(x + y) + yg(x + y)$.

32. If $u = f(ax^2 + 2hxy + by^2)$, $v = \phi(ax^2 + 2hxy + by^2)$, prove that $\frac{\partial}{\partial y}\left(u \frac{\partial v}{\partial x}\right) = \frac{\partial}{\partial x}\left(u \frac{\partial v}{\partial y}\right)$.

[**Hint.** Given $u = f(z)$, $v = \phi(z)$, where $z = ax^2 + 2hxy + by^2$.

We have to prove that $\frac{\partial u}{\partial y} \cdot \frac{\partial v}{\partial x} + u\frac{\partial^2 v}{\partial y \partial x} = \frac{\partial u}{\partial x} \cdot \frac{\partial v}{\partial y} + u\frac{\partial^2 v}{\partial x \partial y}$ or $\frac{\partial u}{\partial y} \cdot \frac{\partial v}{\partial x} = \frac{\partial u}{\partial x} \cdot \frac{\partial v}{\partial y}$]

$\left(\because \frac{\partial^2 v}{\partial y \partial x} = \frac{\partial^2 v}{\partial x \partial y}\right)$

33. If $u = \log(x^3 + y^3 - x^2y - xy^2)$, show that $\frac{\partial^2 u}{\partial x^2} + 2\frac{\partial^2 u}{\partial x \partial y} + \frac{\partial^2 u}{\partial y^2} = -4(x + y)^{-2}$.

[**Hint.** $u = \log\{x^2(x - y) - y^2(x - y)\} = \log(x - y)(x^2 - y^2) = \log(x - y)^2(x + y)$

$= 2\log(x - y) + \log(x + y)$].

34. If $V = f(r)$ and $r^2 = x^2 + y^2 + z^2$, prove that $V_{xx} + V_{yy} + V_{zz} = f''(r) + \frac{2}{r}f'(r)$.

35. If $z = f(x + ay) + \phi(x - ay)$, prove that $\frac{\partial^2 z}{\partial y^2} = a^2 \frac{\partial^2 z}{\partial x^2}$.

36. Find p and q, if $x = \sqrt{a}(\sin u + \cos v)$, $y = \sqrt{a}(\cos u - \sin v)$, $z = 1 + \sin(u - v)$ where p and q mean $\frac{\partial z}{\partial x}$ and $\frac{\partial z}{\partial y}$ respectively.

[**Hint.** $x^2 + y^2 = 2az$, $\therefore$ $z = \frac{x^2 + y^2}{2a}$]

Answers

1. (*i*) $y^x \log y, xy^{x-1}$; (*ii*) $\frac{2x}{x^2 + y^2}, \frac{2y}{x^2 + y^2}$ (*iii*) $2x \sin\frac{y}{x} - y\cos\frac{y}{x}, x\cos\frac{y}{x}$

(*iv*) $\frac{-x}{x^2 + y^2} + \frac{1}{y}\tan^{-1}\frac{y}{x}, \frac{x^2}{y(x^2 + y^2)} - \frac{x}{y^2}\tan^{-1}\frac{y}{x}$ **6.** $-\frac{13}{22}$

16. $n = 2, -3$ **22.** $\frac{x^2 - y^2}{x^2 + y^2}$ **36.** $p = \frac{x}{a}$, $q = \frac{y}{a}$.

2.5. HOMOGENEOUS FUNCTIONS

A function $f(x, y)$ is said to be homogeneous of degree (or order) n in the variables x and y if it can be expressed in the form $x^n \phi\left(\frac{y}{x}\right)$ or $y^n \phi\left(\frac{x}{y}\right)$.

An alternative test for a function $f(x, y)$ to be homogeneous of degree (or order) n is that

$$f(tx, ty) = t^n f(x, y).$$

For example, if $f(x, y) = \dfrac{x+y}{\sqrt{x}+\sqrt{y}}$, then

$$(i)\ f(x, y) = \frac{x\left(1+\dfrac{y}{x}\right)}{\sqrt{x}\left(1+\sqrt{\dfrac{y}{x}}\right)} = x^{1/2}\,\phi\left(\frac{y}{x}\right)$$

$\Rightarrow$ $f(x, y)$ is a homogeneous function of degree $\dfrac{1}{2}$ in x and y.

$$(ii)\ f(x, y) = \frac{y\left(\dfrac{x}{y}+1\right)}{\sqrt{y}\left(\sqrt{\dfrac{x}{y}}+1\right)} = y^{1/2}\,\phi\left(\frac{x}{y}\right)$$

$\Rightarrow$ $f(x, y)$ is a homogeneous function of degree $\dfrac{1}{2}$ in x and y.

$$(iii)\ f(tx, ty) = \frac{tx+ty}{\sqrt{tx}+\sqrt{ty}} = \frac{t(x+y)}{\sqrt{t}\,(\sqrt{x}+\sqrt{y})} = t^{1/2} f(x, y)$$

$\Rightarrow$ $f(x, y)$ is a homogeneous function of degree $\dfrac{1}{2}$ in x and y.

Similarly, a function $f(x, y, z)$ is said to be homogeneous of degree (or order) n in the variables x, y, z if

$$f(x, y, z) = x^n\,\phi\left(\frac{y}{x}, \frac{z}{x}\right) \quad \text{or} \quad y^n\phi\left(\frac{x}{y}, \frac{z}{y}\right) \quad \text{or} \quad z^n\,\phi\left(\frac{x}{z}, \frac{y}{z}\right).$$

Alternative test is $f(tx, ty, tz) = t^n f(x, y, z)$.

2.6. EULER'S THEOREM ON HOMOGENEOUS FUNCTIONS

If u is a homogeneous function of degree n in x and y, then $\mathbf{x\dfrac{\partial u}{\partial x} + y\dfrac{\partial u}{\partial y} = nu}$.

Since u is a homogeneous function of degree n in x and y, it can be expressed as

$$u = x^n f\left(\frac{y}{x}\right)$$

$$\therefore \quad \frac{\partial u}{\partial x} = nx^{n-1} f\left(\frac{y}{x}\right) + x^n f'\left(\frac{y}{x}\right).\left(-\frac{y}{x^2}\right)$$

$$\Rightarrow \quad x\frac{\partial u}{\partial x} = nx^n f\left(\frac{y}{x}\right) - x^{n-1} yf'\left(\frac{y}{x}\right) \qquad \ldots(1)$$

$$\text{Also} \quad \frac{\partial u}{\partial y} = x^n f'\left(\frac{y}{x}\right).\frac{1}{x} = x^{n-1} f'\left(\frac{y}{x}\right)$$

$$\Rightarrow \quad y\frac{\partial u}{\partial y} = x^{n-1} yf'\left(\frac{y}{x}\right) \qquad \ldots(2)$$

Adding (1) and (2), we get $x\dfrac{\partial u}{\partial x} + y\dfrac{\partial u}{\partial y} = nx^n f\left(\dfrac{y}{x}\right) = nu.$

Note. Euler's theorem can be extended to a homogeneous function of any number of variables. Thus, if u is a homogeneous function of degree n in x, y and z, then $x\dfrac{\partial u}{\partial x} + y\dfrac{\partial u}{\partial y} + z\dfrac{\partial u}{\partial z} = nu.$

2.7. IF u IS A HOMOGENEOUS FUNCTION OF DEGREE n IN x AND y, THEN

$$\mathbf{x^2 \frac{\partial^2 u}{\partial x^2} + 2xy \frac{\partial^2 u}{\partial x \partial y} + y^2 \frac{\partial^2 u}{\partial y^2} = n(n-1)u}$$

Since u is a homogeneous function of degree n in x and y

$\therefore$ By Euler's theorem, we have $x \frac{\partial u}{\partial x} + y \frac{\partial u}{\partial y} = nu$ ...(1)

Differentiating (1) partially w.r.t. x, we have $1 . \frac{\partial u}{\partial x} + x \frac{\partial^2 u}{\partial x^2} + y . \frac{\partial^2 u}{\partial x \partial y} = n \frac{\partial u}{\partial x}$ ...(2)

Differentiating (1) partially, w.r.t. y, we have $x \frac{\partial^2 u}{\partial y \partial x} + 1 . \frac{\partial u}{\partial y} + y . \frac{\partial^2 u}{\partial y^2} = n . \frac{\partial u}{\partial y}$

But $$\frac{\partial^2 u}{\partial y \partial x} = \frac{\partial^2 u}{\partial x \partial y}$$

$$\therefore \quad x \frac{\partial^2 u}{\partial x \partial y} + \frac{\partial u}{\partial y} + y \frac{\partial^2 u}{\partial y^2} = n \frac{\partial u}{\partial y} \quad ...(3)$$

Multiplying (2) by x, (3) by y and adding,

$$x^2 \frac{\partial^2 u}{\partial x^2} + 2xy \frac{\partial^2 u}{\partial x \partial y} + y^2 \frac{\partial^2 u}{\partial y^2} + \left(x \frac{\partial u}{\partial x} + y \frac{\partial u}{\partial y} \right) = n \left(x \frac{\partial u}{\partial x} + y \frac{\partial u}{\partial y} \right)$$

or $$x^2 \frac{\partial^2 u}{\partial x^2} + 2xy \frac{\partial^2 u}{\partial x \partial y} + y^2 \frac{\partial^2 u}{\partial y^2} + nu = n . nu \quad \text{[Using (1)]}$$

$$\Rightarrow \quad x^2 \frac{\partial^2 u}{\partial x^2} + 2xy \frac{\partial^2 u}{\partial x \partial y} + y^2 \frac{\partial^2 u}{\partial y^2} = n^2 u - nu = n(n-1)u.$$

ILLUSTRATIVE EXAMPLES

Example 1. *Verify Euler's theorem for the functions:*

(*i*) $u = (x^{1/2} + y^{1/2})(x^n + y^n)$ (*ii*) $u = \sin^{-1} \frac{x}{y} + \tan^{-1} \frac{y}{x}$.

Sol. (*i*) $u = (x^{1/2} + y^{1/2})(x^n + y^n)$ ...(1)

$$= x^{1/2}\left(1 + \frac{y^{1/2}}{x^{1/2}}\right) x^n \left(1 + \frac{y^n}{x^n}\right) = x^{n+1/2}\left[1 + \left(\frac{y}{x}\right)^{1/2}\right]\left[1 + \left(\frac{y}{x}\right)^n\right] = x^{n+1/2} f\left(\frac{y}{x}\right)$$

[OR $f(tx, ty) = t^{n+1/2} f(x, y)$]

$\Rightarrow$ u is a homogeneous function of degree $(n + \frac{1}{2})$ in x and y

$\therefore$ By Euler's theorem, we should have $x \frac{\partial u}{\partial x} + y \frac{\partial u}{\partial y} = \left(n + \frac{1}{2}\right) u$ ...(2)

From (1), $$\frac{\partial u}{\partial x} = \frac{1}{2} x^{-1/2} (x^n + y^n) + nx^{n-1}(x^{1/2} + y^{1/2})$$

$$x \frac{\partial u}{\partial x} = \frac{1}{2} x^{1/2} (x^n + y^n) + nx^n(x^{1/2} + y^{1/2})$$

$$\frac{\partial u}{\partial y} = \frac{1}{2} y^{-1/2} (x^n + y^n) + ny^{n-1} (x^{1/2} + y^{1/2})$$

$$y\frac{\partial u}{\partial y}=\frac{1}{2}y^{1/2}(x^n+y^n)+ny^n(x^{1/2}+y^{1/2})$$

$$\therefore \quad x\frac{\partial u}{\partial x}+y\frac{\partial u}{\partial y}=\frac{1}{2}(x^{1/2}+y^{1/2})(x^n+y^n)+n(x^n+y^n)(x^{1/2}+y^{1/2})$$

$= \frac{1}{2}u + nu = (n+\frac{1}{2})u$ which is the same as (2). Hence the verification.

(*ii*) $$u=\sin^{-1}\frac{x}{y}+\tan^{-1}\frac{y}{x} \qquad \text{...(1)}$$

$$=\text{cosec}^{-1}\frac{y}{x}+\tan^{-1}\frac{y}{x}=x^0 f\left(\frac{y}{x}\right)$$

[OR $f(tx, ty) = f(x, y) = t^0 f(x, y)$]

$\Rightarrow$ **u is a homogeneous function of degree 0 in x and y.**

$\therefore$ **By Euler's theorem, we should have** $x\frac{\partial u}{\partial x}+y\frac{\partial u}{\partial y}=0\times u=0$ ...(2)

From (1), $$\frac{\partial u}{\partial x}=\frac{1}{\sqrt{1-\frac{x^2}{y^2}}}\cdot\frac{1}{y}+\frac{1}{1+\frac{y^2}{x^2}}\left(-\frac{y}{x^2}\right)=\frac{1}{\sqrt{y^2-x^2}}-\frac{y}{x^2+y^2}$$

$$x\frac{\partial u}{\partial x}=\frac{x}{\sqrt{y^2-x^2}}-\frac{xy}{x^2+y^2}$$

$$\frac{\partial u}{\partial y}=\frac{1}{\sqrt{1-\frac{x^2}{y^2}}}\left(-\frac{x}{y^2}\right)+\frac{1}{1+\frac{y^2}{x^2}}\cdot\frac{1}{x}=-\frac{x}{y\sqrt{y^2-x^2}}+\frac{x}{x^2+y^2}$$

$$y\frac{\partial u}{\partial y}=-\frac{x}{\sqrt{y^2-x^2}}+\frac{xy}{x^2+y^2}$$

$\therefore \quad x\frac{\partial u}{\partial x}+y\frac{\partial u}{\partial y}=0$ **which is the same as (2). Hence the verification.**

Example 2. *If* $u=\tan^{-1}\frac{x^3+y^3}{x-y}$, *prove that* $x\frac{\partial u}{\partial x}+y\frac{\partial u}{\partial y}=\sin 2u.$

Sol. Here u is not a homogeneous function but $\tan u=\frac{x^3+y^3}{x-y}=\frac{x^3\left[1+\left(\frac{y}{x}\right)^3\right]}{x\left[1-\frac{y}{x}\right]}=x^2 f\left(\frac{y}{x}\right)$

is a homogeneous function of degree 2 in x and y.

$\therefore$ By Euler's theorem, we have

$$x\frac{\partial}{\partial x}(\tan u)+y\frac{\partial}{\partial y}(\tan u)=2\tan u \quad \text{or} \quad x\sec^2 u\frac{\partial u}{\partial x}+y\sec^2 u\frac{\partial u}{\partial y}=2\tan u$$

or $$x\frac{\partial u}{\partial x}+y\frac{\partial u}{\partial y}=\frac{2\sin u}{\cos u}\cdot\cos^2 u=2\sin u\cos u=\sin 2u.$$

Example 3. *If* $u = \sin^{-1}\left(\dfrac{x+2y+3z}{\sqrt{x^8+y^8+z^8}}\right)$, *show that* $x\dfrac{\partial u}{\partial x} + y\dfrac{\partial u}{\partial y} + z\dfrac{\partial u}{\partial z} + 3\tan u = 0.$

Sol. Here u is not a homogeneous function.

$$\sin u = f(x, y, z) = \frac{x+2y+3z}{\sqrt{x^8+y^8+z^8}}$$

$$f(tx, ty, tz) = \frac{t\,(x+2y+3z)}{t^4\sqrt{x^8+y^8+z^8}} = t^{-3} f(x, y, z)$$

$\Rightarrow$ $\sin u$ is a homogeneous function of degree -3 in x, y, z.

$\therefore$ By Euler's theorem, we have

$$x\frac{\partial}{\partial x}(\sin u) + y\frac{\partial}{\partial y}(\sin u) + z\frac{\partial}{\partial z}(\sin u) = -3\sin u$$

or $$x\cos u\frac{\partial u}{\partial x} + y\cos u\frac{\partial u}{\partial y} + z\cos u\frac{\partial u}{\partial z} + 3\sin u = 0$$

or $$x\frac{\partial u}{\partial x} + y\frac{\partial u}{\partial y} + z\frac{\partial u}{\partial z} + 3\tan u = 0.$$

Example 4. *If* $u = \log\dfrac{x^4+y^4}{x+y}$, *show that* $x\dfrac{\partial u}{\partial x} + y\dfrac{\partial u}{\partial y} = 3.$

Sol. Here u is not a homogeneous function

$\therefore$ $$u = \log_e\left(\frac{x^4+y^4}{x+y}\right) \Rightarrow e^u = \frac{x^4+y^4}{x+y}$$

which is a homogeneous function of degree 3 in x, y.

$\therefore$ By Euler's theorem, we have $x\dfrac{\partial}{\partial x}(e^u) + y\dfrac{\partial}{\partial y}(e^u) = 3e^u$

or $$xe^u\frac{\partial u}{\partial x} + ye^u\frac{\partial u}{\partial y} = 3e^u \quad\text{or}\quad x\frac{\partial u}{\partial x} + y\frac{\partial u}{\partial y} = 3.$$

Example 5. *If* $u = \sin^{-1}\dfrac{x+y}{\sqrt{x}+\sqrt{y}}$, *prove that* $x^2\dfrac{\partial^2 u}{\partial x^2} + 2xy\dfrac{\partial^2 u}{\partial x\partial y} + y^2\dfrac{\partial^2 u}{\partial y^2} = -\dfrac{\sin u\cos 2u}{4\cos^3 u}.$

(K.U. April 2013)

Sol. $u = \sin^{-1}\dfrac{x+y}{\sqrt{x}+\sqrt{y}}$ is not a homogeneous function but

$\sin u = \dfrac{x+y}{\sqrt{x}+\sqrt{y}}$ is a homogeneous function of degree $\frac{1}{2}$ in x and y.

$\therefore$ By Euler's theorem, we have

$$x\frac{\partial}{\partial x}(\sin u) + y\frac{\partial}{\partial y}(\sin u) = \frac{1}{2}\sin u$$

or $$x\cos u\frac{\partial u}{\partial x} + y\cos u\frac{\partial u}{\partial y} = \frac{1}{2}\sin u \quad\text{or}\quad x\frac{\partial u}{\partial x} + y\frac{\partial u}{\partial y} = \frac{1}{2}\tan u \qquad ...(1)$$

Differentiating (1) partially w.r.t. x,

$$x\frac{\partial^2 u}{\partial x^2}+1\cdot\frac{\partial u}{\partial x}+y\frac{\partial^2 u}{\partial x\partial y}=\frac{1}{2}\sec^2 u\frac{\partial u}{\partial x}$$

or $$x\frac{\partial^2 u}{\partial x^2}+y\frac{\partial^2 u}{\partial x\partial y}=\left(\frac{1}{2}\sec^2 u-1\right)\frac{\partial u}{\partial x} \quad ...(2)$$

Differentiating (1) partially w.r.t. y,

$$x\frac{\partial^2 u}{\partial y\partial x}+y\frac{\partial^2 u}{\partial y^2}+1\cdot\frac{\partial u}{\partial y}=\frac{1}{2}\sec^2 u\frac{\partial u}{\partial y}$$

or $$x\frac{\partial^2 u}{\partial x\partial y}+y\frac{\partial^2 u}{\partial y^2}=\left(\frac{1}{2}\sec^2 u-1\right)\frac{\partial u}{\partial y} \quad ...(3) \quad \left[\because \frac{\partial^2 u}{\partial y\partial x}=\frac{\partial^2 u}{\partial x\partial y}\right]$$

Multiplying (2) by x, (3) by y and adding

$$x^2\frac{\partial^2 u}{\partial x^2}+2xy\frac{\partial^2 u}{\partial x\partial y}+y^2\frac{\partial^2 u}{\partial y^2}=\left(\frac{1}{2}\sec^2 u-1\right)\left(x\frac{\partial u}{\partial x}+y\frac{\partial u}{\partial y}\right)$$

$$=\left(\frac{1}{2\cos^2 u}-1\right)\cdot\frac{1}{2}\tan u \quad \text{[Using (1)]}$$

$$=-\frac{2\cos^2 u-1}{2\cos^2 u}\cdot\frac{1}{2}\frac{\sin u}{\cos u}=-\frac{\sin u\cos 2u}{4\cos^3 u}. \quad [\because\ 2\cos^2 u-1=\cos 2u]$$

Example 6. *If* $z=xf\left(\frac{y}{x}\right)+g\left(\frac{y}{x}\right)$, *show that* $x^2\frac{\partial^2 z}{\partial x^2}+2xy\frac{\partial^2 z}{\partial x\partial y}+y^2\frac{\partial^2 z}{\partial y^2}=0.$

Sol. Let $$u=xf\left(\frac{y}{x}\right) \text{ and } v=g\left(\frac{y}{x}\right)+x^0 g\left(\frac{y}{x}\right)$$

so that $$z=u+v \quad ...(1)$$

Since u is a homogeneous function of degree $n = 1$ in x, y

$$\therefore \quad x^2\frac{\partial^2 u}{\partial x^2}+2xy\frac{\partial^2 u}{\partial x\partial y}+y^2\frac{\partial^2 u}{\partial y^2}=n(n-1)u=0 \quad ...(2)$$

Since v is a homogeneous function of degree $n = 0$ in x, y

$$\therefore \quad x^2\frac{\partial^2 v}{\partial x^2}+2xy\frac{\partial^2 v}{\partial x\partial y}+y^2\frac{\partial^2 v}{\partial y^2}=n(n-1)v=0 \quad ...(3)$$

Adding (2) and (3), we have

$$x^2\frac{\partial^2}{\partial x^2}(u+v)+2xy\frac{\partial^2}{\partial x\partial y}(u+v)+y^2\frac{\partial^2}{\partial y^2}(u+v)=0$$

or $$x^2\frac{\partial^2 z}{\partial x^2}+2xy\frac{\partial^2 z}{\partial x\partial y}+y^2\frac{\partial^2 z}{\partial y^2}=0. \quad \text{[Using (1)]}$$

EXERCISE 2.2

1. Verify Euler's theorem for the functions:

(i) $f(x, y) = ax^2 + 2hxy + by^2$ (ii) $u = \dfrac{x^{1/4} + y^{1/4}}{x^{1/5} + y^{1/5}}$ (iii) $f(x, y) = \dfrac{x^2(x^2 - y^2)^3}{(x^2 + y^2)^3}$.

2. (i) If $u = f\left(\dfrac{y}{x}\right)$, show that $x\dfrac{\partial u}{\partial x} + y\dfrac{\partial u}{\partial y} = 0$ (ii) If $u = xf\left(\dfrac{y}{x}\right)$, prove that $x\dfrac{\partial u}{\partial x} + y\dfrac{\partial u}{\partial y} = u$.

3. If $V = \dfrac{x^3y^3}{x^3 + y^3}$, show that $x\dfrac{\partial V}{\partial x} + y\dfrac{\partial V}{\partial y} = 3V$.

4. If $u = \sin^{-1}\dfrac{x}{y} + \tan^{-1}\dfrac{y}{x}$, then find the value of $x\dfrac{\partial u}{\partial x} + y\dfrac{\partial u}{\partial y}$.

5. If $f(x, y) = \sqrt{x^2 - y^2}\,\sin^{-1}\dfrac{y}{x}$, prove that $x\dfrac{\partial f}{\partial x} + y\dfrac{\partial f}{\partial y} = f(x, y)$.

6. If $f(x, y) = \sqrt{y^2 - x^2}\,\sin^{-1}\dfrac{x}{y} + \dfrac{x^2 - y^2}{\sqrt{x^2 + y^2}}$, show that $x\dfrac{\partial f}{\partial x} + y\dfrac{\partial f}{\partial y} - f(x, y) = 0$.

7. If $f(x, y) = \dfrac{1}{x^2} + \dfrac{1}{xy} + \dfrac{\log x - \log y}{x^2 + y^2}$, show that $x\dfrac{\partial f}{\partial x} + y\dfrac{\partial f}{\partial y} + 2f(x, y) = 0$.

8. If $u = \cos\left(\dfrac{xy + yz + zx}{x^2 + y^2 + z^2}\right)$, prove that $x\dfrac{\partial u}{\partial x} + y\dfrac{\partial u}{\partial y} + z\dfrac{\partial u}{\partial z} = 0$.

9. If $u = \cos^{-1}\left(\dfrac{x + y}{\sqrt{x} + \sqrt{y}}\right)$, show that $x\dfrac{\partial u}{\partial x} + y\dfrac{\partial u}{\partial y} + \dfrac{1}{2}\cot u = 0$.

10. If $u = \sin^{-1}\left(\dfrac{x^2 + y^2}{x + y}\right)$, prove that $x\dfrac{\partial u}{\partial x} + y\dfrac{\partial u}{\partial y} = \tan u$.

11. (a) If $u = \sin^{-1}\dfrac{\sqrt{x} - \sqrt{y}}{\sqrt{x} + \sqrt{y}}$, show that $\dfrac{\partial u}{\partial x} = -\dfrac{y}{x}\dfrac{\partial u}{\partial y}$.

(b) If $\sin u = \dfrac{x^2y^2}{x + y}$, show that $x\dfrac{\partial u}{\partial x} + y\dfrac{\partial u}{\partial y} = 3\tan u$.

12. (a) If $u = \log\left(\dfrac{x^5 + y^5 + z^5}{x^2 + y^2 + z^2}\right)$, show that $x\dfrac{\partial u}{\partial x} + y\dfrac{\partial u}{\partial y} + z\dfrac{\partial u}{\partial z} = 3u$.

(b) Show that $xu_x + yu_y + zu_z = 2\tan u$, where $u = \sin^{-1}\left(\dfrac{x^3 + y^3 + z^3}{ax + by + cz}\right)$.

13. If $u = \dfrac{x^2y^2}{x + y}$, show that

(i) $x\dfrac{\partial^2 u}{\partial x^2} + y\dfrac{\partial^2 u}{\partial x\partial y} = 2\dfrac{\partial u}{\partial x}$ (ii) $x\dfrac{\partial^2 u}{\partial x\partial y} + y\dfrac{\partial^2 u}{\partial y^2} = 2\dfrac{\partial u}{\partial y}$.

14. Given $z = x^n f_1\left(\dfrac{y}{x}\right) + y^{-n} f_2\left(\dfrac{x}{y}\right)$, prove that $x^2\dfrac{\partial^2 z}{\partial x^2} + 2xy\dfrac{\partial^2 z}{\partial x\partial y} + y^2\dfrac{\partial^2 z}{\partial y^2} + x\dfrac{\partial z}{\partial x} + y\dfrac{\partial z}{\partial y} = n^2 z$.

15. If $u = (x^2 + y^2)^{1/3}$, show that $x^2 \dfrac{\partial^2 u}{\partial x^2} + 2xy \dfrac{\partial^2 u}{\partial x \partial y} + y^2 \dfrac{\partial^2 u}{\partial y^2} = -\dfrac{2u}{9}$.

16. If $u = \tan^{-1} \dfrac{x^3 + y^3}{x - y}$, prove that $x^2 \dfrac{\partial^2 u}{\partial x^2} + 2xy \dfrac{\partial^2 u}{\partial x \partial y} + y^2 \dfrac{\partial^2 u}{\partial y^2} = \sin 4u - \sin 2u = 2 \cos 3u \sin u$.

(K.U. May 2009)

17. If $u = \tan^{-1}\left(\dfrac{y^2}{x}\right)$, show that $x^2 \dfrac{\partial^2 u}{\partial x^2} + 2xy \dfrac{\partial^2 u}{\partial x \partial y} + y^2 \dfrac{\partial^2 u}{\partial y^2} = -\sin 2u \sin^2 u$.

18. If $u = x^2 \tan^{-1}\left(\dfrac{y}{x}\right) - y^2 \tan^{-1}\left(\dfrac{x}{y}\right)$, evaluate $x^2 \dfrac{\partial^2 u}{\partial x^2} + 2xy \dfrac{\partial^2 u}{\partial x \partial y} + y^2 \dfrac{\partial^2 u}{\partial y^2}$.

19. If $u = \operatorname{cosec}^{-1}\left(\dfrac{x^{1/2} + y^{1/2}}{x^{1/3} + y^{1/3}}\right)^{1/2}$, prove that $x^2 \dfrac{\partial^2 u}{\partial x^2} + 2xy \dfrac{\partial^2 u}{\partial x \partial y} + y^2 \dfrac{\partial^2 u}{\partial y^2} = \dfrac{\tan u}{144}(13 + \tan^2 u)$.

Answers

4. 0 **18.** $2u$.

2.8. COMPOSITE FUNCTIONS

(i) If $u = f(x, y)$ where $x = \phi(t)$, $y = \psi(t)$

then u is called a composite function of (the **single variable**) t and we can find $\dfrac{du}{dt}$.

(ii) If $z = f(x, y)$ where $x = \phi(u, v)$, $y = \psi(u, v)$

then z is called a composite function of (**two variables**) u and v so that we can find $\dfrac{\partial z}{\partial u}$ and $\dfrac{\partial z}{\partial v}$.

2.9. DIFFERENTIATION OF COMPOSITE FUNCTIONS

If u is composite function of t, defined by the relations $u = f(x, y)$; $x = \phi(t)$, $y = \psi(t)$, then

$$\frac{du}{dt} = \frac{\partial u}{\partial x} \cdot \frac{dx}{dt} + \frac{\partial u}{\partial y} \cdot \frac{dy}{dt}.$$

Proof. Here $u = f(x, y)$ ...(1)

Let δt be an increment in t and $\delta x, \delta y, \delta u$ the corresponding increments in x, y and u respectively. Then, we have

$$u + \delta u = f(x + \delta x, y + \delta y) \quad ...(2)$$

Subtracting (1) from (2), we get

$$\delta u = f(x + \delta x, y + \delta y) - f(x, y)$$

$$= f(x + \delta x, y + \delta y) - f(x, y + \delta y) + f(x, y + \delta y) - f(x, y)$$

$$\frac{\delta u}{\delta t} = \frac{f(x + \delta x, y + \delta y) - f(x, y + \delta y)}{\delta t} + \frac{f(x, y + \delta y) - f(x, y)}{\delta t}$$

$$= \frac{f(x + \delta x, y + \delta y) - f(x, y + \delta y)}{\delta x} \cdot \frac{\delta x}{\delta t} + \frac{f(x, y + \delta y) - f(x, y)}{\delta y} \cdot \frac{\delta y}{\delta t} \quad ...(3)$$

As $\delta t \to 0$, δx and δy both $\to 0$, so that

$$\lim_{\delta t \to 0} \frac{\delta u}{\delta t} = \frac{du}{dt}, \quad \lim_{\delta t \to 0} \frac{\delta x}{\delta t} = \frac{dx}{dt}, \quad \lim_{\delta t \to 0} \frac{\delta y}{\delta t} = \frac{dy}{dt}$$

$$\lim_{\delta x \to 0} \frac{f(x + \delta x, y + \delta y) - f(x, y + \delta y)}{\delta y} = \frac{\partial f}{\partial x} = \frac{\partial u}{\partial x}$$

$$\lim_{\delta y \to 0} \frac{f(x, y+\delta y) - f(x, y)}{\delta y} = \frac{\partial f}{\partial y} = \frac{\partial u}{\partial y}$$

$\therefore$ From (1), $$\frac{du}{dt} = \frac{\partial u}{\partial x} \cdot \frac{dx}{dt} + \frac{\partial u}{\partial y} \cdot \frac{dy}{dt}$$

$\frac{du}{dt}$ is called the **total derivative** of u to distinguish it from the partial derivatives $\frac{\partial u}{\partial x}$ and $\frac{\partial u}{\partial y}$.

Cor. 1. If $u = f(x, y, z)$ and x, y, z are function of t, then y is a composite function of t and

$$\frac{du}{dt} = \frac{\partial u}{\partial x} \cdot \frac{dx}{dt} + \frac{\partial u}{\partial y} \cdot \frac{dy}{dt} + \frac{\partial u}{\partial z} \cdot \frac{dz}{dt}.$$

Cor. 2. If $z = f(x, y)$ and x, y are functions of u and v, then

$$\frac{\partial z}{\partial u} = \frac{\partial z}{\partial x} \cdot \frac{\partial x}{\partial u} + \frac{\partial z}{\partial y} \cdot \frac{\partial y}{\partial u}; \quad \frac{\partial z}{\partial v} = \frac{\partial z}{\partial x} \cdot \frac{\partial x}{\partial v} + \frac{\partial z}{\partial y} \cdot \frac{\partial y}{\partial v}.$$

Now we can generalise the above result for any number of variables. Consider the function $w = f(x, y, z, ..., t)$, where $x, y, z, ..., t$ are functions of $p, q, r, ..., t$. *i.e.*, $x = x(p, q, r, ..., t)$, $y = y(p, q, r, ..., t), ..., v = v(p, q, r, ..., t)$. Using cor. 2, we find

$$\frac{\partial w}{\partial p} = \frac{\partial w}{\partial x} \cdot \frac{\partial x}{\partial p} + \frac{\partial w}{\partial y} \cdot \frac{\partial y}{\partial p} + \frac{\partial w}{\partial z} \cdot \frac{\partial z}{\partial p} + \cdots + \frac{\partial w}{\partial v} \cdot \frac{\partial v}{\partial p} \qquad ...(1)$$

Equation (1) is called the 'chain rule'. Similarly we can find $\frac{\partial w}{\partial q}, ..., \frac{\partial w}{\partial t}$.

Cor. 3. If $u = f(x, y)$ where $y = \phi(x)$ then since $x = \psi(x)$, u is a composite function of x.

$$\therefore \quad \frac{du}{dx} = \frac{\partial u}{\partial x} \cdot \frac{dx}{dx} + \frac{\partial u}{\partial y} \cdot \frac{dy}{dx} \Rightarrow \frac{du}{dx} = \frac{\partial u}{\partial x} + \frac{\partial u}{\partial y} \cdot \frac{dy}{dx}.$$

Cor. 4. If we are given an implicit function $f(x, y) = c$, then $u = f(x, y)$ where $u = c$

Using Cor. 3, we have $$\frac{du}{dx} = \frac{\partial u}{\partial x} + \frac{\partial u}{\partial y} \cdot \frac{dy}{dx}$$

But $\frac{du}{dx} = 0 \quad \therefore$ $$\frac{\partial u}{\partial x} + \frac{\partial u}{\partial y} \cdot \frac{dy}{dx} = 0 \quad \text{or} \quad \frac{\partial f}{\partial x} + \frac{\partial f}{\partial y} \cdot \frac{dy}{dx} = 0$$

$$\therefore \quad \frac{dy}{dx} = -\frac{\frac{\partial f}{\partial x}}{\frac{\partial f}{\partial y}} = -\frac{f_x}{f_y}.$$

Hence the differential co-efficient of $f(x, y)$ w.r.t. x is $\frac{df}{dx} = \frac{\partial f}{\partial x} + \frac{\partial f}{\partial y} \cdot \frac{dy}{dx}$.

ILLUSTRATIVE EXAMPLES

Example 1. *If $u = \sin^{-1}(x - y)$, $x = 3t$, $y = 4t^3$, show that* $\frac{du}{dt} = \frac{3}{\sqrt{1 - t^2}}$.

Sol. The given equations define u as a composite function of t.

$$\therefore \qquad \frac{du}{dt} = \frac{\partial u}{\partial x} \cdot \frac{dx}{dt} + \frac{\partial u}{\partial y} \cdot \frac{dy}{dt}$$

$$= \frac{1}{\sqrt{1-(x-y)^2}} \cdot 3 + \frac{1}{\sqrt{1-(x-y)^2}} (-1) \cdot 12\,t^2$$

$$= \frac{3\,(1-4t^2)}{\sqrt{1-(x-y)^2}} = \frac{3\,(1-4t^2)}{\sqrt{1-(3t-4t^3)^2}} = \frac{3\,(1-4t^2)}{\sqrt{1-9t^2+24t^4-16t^6}}$$

$$= \frac{3\,(1-4t^2)}{\sqrt{(1-t^2)\,(1-8t^2+16t^4)}} = \frac{3\,(1-4t^2)}{\sqrt{(1-t^2)\,(1-4t^2)^2}} = \frac{3}{\sqrt{1-t^2}}.$$

Example 2. *If $z = 2xy^2 - 3x^2y$ and if x increases at the rate of 2 cm per second when it passes through the value x = 3 cm, show that if y is passing through the value y = 1 cm, y must be decreasing at the rate of $2\frac{2}{15}$ cm per second, in order that z shall remain constant.*

Sol. Given: $z = 2xy^2 - 3x^2y$ and $\frac{dx}{dt} = 2$ cm/sec when $x = 3$ cm, we have to find $\frac{dy}{dt}$ when $y = 1$ cm.

Now $$\frac{dz}{dt} = \frac{\partial z}{\partial x} \cdot \frac{dx}{dt} + \frac{\partial z}{\partial y} \cdot \frac{dy}{dt} = (2y^2 - 6xy)\frac{dx}{dt} + (4xy - 3x^2)\frac{dy}{dt}$$

Since z remains constant, $\frac{dz}{dt} = 0$

$$\therefore \qquad 0 = (2y^2 - 6 \times 3 \times y) \times 2 + (4 \times 3 \times y - 3 \times 3^2)\frac{dy}{dt}$$

$$\Rightarrow \qquad 0 = (4y^2 - 36y) + (12y - 27)\frac{dy}{dt}$$

When $y = 1$ cm, we have

$$0 = (4 - 36) + (12 - 27)\frac{dy}{dt}$$

$$\Rightarrow \qquad \frac{dy}{dt} = -\frac{32}{15} \text{ which is negative}$$

$\therefore$ When $y = 1$ cm, y is decreasing at the rate of $2\frac{2}{15}$ cm/sec.

Example 3. *If z is a function of x and y, where $x = e^u + e^{-v}$ and $y = e^{-u} - e^v$, show that*

$$\frac{\partial z}{\partial u} - \frac{\partial z}{\partial v} = x\frac{\partial z}{\partial x} - y\frac{\partial z}{\partial y}.$$

Sol. Here z is a composite function of u and v.

$$\therefore \qquad \frac{\partial z}{\partial u} = \frac{\partial z}{\partial x} \cdot \frac{\partial x}{\partial u} + \frac{\partial z}{\partial y} \cdot \frac{\partial y}{\partial u} = \frac{\partial z}{\partial x} \cdot e^u + \frac{\partial z}{\partial y}(-e^{-u})$$

and $$\frac{\partial z}{\partial v} = \frac{\partial z}{\partial x} \cdot \frac{\partial x}{\partial v} + \frac{\partial z}{\partial y} \cdot \frac{\partial y}{\partial v} = \frac{\partial z}{\partial x}(-e^{-v}) + \frac{\partial z}{\partial y}(-e^v)$$

Subtracting, $$\frac{\partial z}{\partial u} - \frac{\partial z}{\partial v} = (e^u + e^{-v})\frac{\partial z}{\partial x} - (e^{-u} - e^v)\frac{\partial z}{\partial y} = x\frac{\partial z}{\partial x} - y\frac{\partial z}{\partial y}.$$

Example 4. *If* $u = f(y - z, z - x, x - y)$, *prove that* $\frac{\partial u}{\partial x} + \frac{\partial u}{\partial y} + \frac{\partial u}{\partial z} = 0.$

Sol. Here $\quad u = f(X, Y, Z)$ where $X = y - z, Y = z - x, Z = x - y$

$\therefore \quad u$ is a composite function of x, y and z.

$$\frac{\partial u}{\partial x} = \frac{\partial u}{\partial X} \cdot \frac{\partial X}{\partial x} + \frac{\partial u}{\partial Y} \cdot \frac{\partial Y}{\partial x} + \frac{\partial u}{\partial Z} \cdot \frac{\partial Z}{\partial x} = \frac{\partial u}{\partial X}(0) + \frac{\partial u}{\partial Y}(-1) + \frac{\partial u}{\partial Z}(1)$$

$$\frac{\partial u}{\partial y} = \frac{\partial u}{\partial X} \cdot \frac{\partial X}{\partial y} + \frac{\partial u}{\partial Y} \cdot \frac{\partial Y}{\partial y} + \frac{\partial u}{\partial Z} \cdot \frac{\partial Z}{\partial y} = \frac{\partial u}{\partial X}(1) + \frac{\partial u}{\partial Y}(0) + \frac{\partial u}{\partial Z}(-1)$$

$$\frac{\partial u}{\partial z} = \frac{\partial u}{\partial X} \cdot \frac{\partial X}{\partial z} + \frac{\partial u}{\partial Y} \cdot \frac{\partial Y}{\partial z} + \frac{\partial u}{\partial Z} \cdot \frac{\partial Z}{\partial z} = \frac{\partial u}{\partial X}(-1) + \frac{\partial u}{\partial Y}(1) + \frac{\partial u}{\partial Z}(0)$$

Adding, $\frac{\partial u}{\partial x} + \frac{\partial u}{\partial y} + \frac{\partial u}{\partial z} = 0.$

Example 5. *If* $w = f(x, y), x = r \cos \theta, y = r \sin \theta$, *show that*

$$\left(\frac{\partial w}{\partial r}\right)^2 + \frac{1}{r^2}\left(\frac{\partial w}{\partial q}\right)^2 = \left(\frac{\partial f}{\partial x}\right)^2 + \left(\frac{\partial f}{\partial y}\right)^2.$$

Sol. The given equations define w as a composite function of r and θ.

$$\frac{\partial w}{\partial r} = \frac{\partial w}{\partial x} \cdot \frac{\partial x}{\partial r} + \frac{\partial w}{\partial y} \cdot \frac{\partial y}{\partial r} = \frac{\partial w}{\partial x} \cdot \cos \theta + \frac{\partial w}{\partial y} \cdot \sin \theta$$

or $$\frac{\partial w}{\partial r} = \frac{\partial f}{\partial x} \cos \theta + \frac{\partial f}{\partial y} \sin \theta \qquad ...(1) \quad [\because \quad w = f(x, y)]$$

Also $$\frac{\partial w}{\partial \theta} = \frac{\partial w}{\partial x} \cdot \frac{\partial x}{\partial \theta} + \frac{\partial w}{\partial y} \cdot \frac{\partial y}{\partial \theta} = \frac{\partial w}{\partial x}(-r \sin \theta) + \frac{\partial w}{\partial y}(r \cos \theta)$$

or $$\frac{1}{r}\frac{\partial w}{\partial \theta} = -\frac{\partial f}{\partial x} \sin \theta + \frac{\partial f}{\partial y} \cos \theta \qquad ...(2)$$

Squaring and adding (1) and (2), we get

$$\left(\frac{\partial w}{\partial r}\right)^2 + \frac{1}{r^2}\left(\frac{\partial w}{\partial \theta}\right)^2 = \left(\frac{\partial f}{\partial x}\right)^2 + \left(\frac{\partial f}{\partial y}\right)^2.$$

Example 6. *If u is a homogeneous function of nth degree in x, y, z, prove that*

$$x\frac{\partial u}{\partial x} + y\frac{\partial u}{\partial y} + z\frac{\partial u}{\partial z} = nu.$$

Sol. Since u is a homogeneous function of degree n in x, y, z, let

$$u = x^n f\left(\frac{y}{x}, \frac{z}{x}\right)$$

or $$u = x^n f(t, s) \quad \text{where} \quad t = \frac{y}{x}, \quad s = \frac{z}{x}$$

Here f is a composite function of x, y, z.

$$\therefore \quad \frac{\partial u}{\partial x} = nx^{n-1} f(t, s) + x^n \left(\frac{\partial f}{\partial t} \cdot \frac{\partial t}{\partial x} + \frac{\partial f}{\partial s} \cdot \frac{\partial s}{\partial x}\right)$$

$$= nx^{n-1} f(t, s) + x^n + \left[\frac{\partial f}{\partial t} \cdot \left(-\frac{y}{x^2}\right) + \frac{\partial f}{\partial s}\left(-\frac{z}{x^2}\right)\right]$$

$$\Rightarrow \quad x\frac{\partial u}{\partial x} = nx^n f(t,s) - yx^{n-1}\frac{\partial f}{\partial t} - zx^{n-1}\frac{\partial f}{\partial s} \quad \text{...(1)}$$

$$\frac{\partial u}{\partial y} = x^n\left(\frac{\partial f}{\partial t}\cdot\frac{\partial t}{\partial y} + \frac{\partial f}{\partial s}\cdot\frac{\partial s}{\partial y}\right) = x^n\left[\frac{\partial f}{\partial t}\cdot\frac{1}{x} + \frac{\partial f}{\partial s}\cdot 0\right]$$

$$\Rightarrow \quad y\frac{\partial u}{\partial y} = yx^{n-1}\frac{\partial f}{\partial t} \quad \text{...(2)}$$

$$\frac{\partial u}{\partial z} = x^n\left[\frac{\partial f}{\partial t}\cdot\frac{\partial t}{\partial z} + \frac{\partial f}{\partial s}\cdot\frac{\partial s}{\partial z}\right] = x^n\left[\frac{\partial f}{\partial t}\cdot 0 + \frac{\partial f}{\partial s}\cdot\frac{1}{x}\right]$$

$$\Rightarrow \quad z\frac{\partial u}{\partial z} = zx^{n-1}\frac{\partial f}{\partial s} \quad \text{...(3)}$$

Adding (1), (2) and (3), we have

$$x\frac{\partial u}{\partial x} + y\frac{\partial u}{\partial y} + z\frac{\partial u}{\partial z} = nx^n f(t, s) = nu.$$

Example 7. *If by the substitution $u = x^2 - y^2$, $v = 2xy$, $f(x, y) = \theta(u, v)$, show that*

$$\frac{\partial^2 f}{\partial x^2} + \frac{\partial^2 f}{\partial y^2} = 4(x^2 + y^2)\left(\frac{\partial^2\theta}{\partial u^2} + \frac{\partial^2\theta}{\partial v^2}\right).$$

Sol. Here $u = x^2 - y^2$, $v = 2xy$ and $f(x, y) = \theta(u, v)$

$$\therefore \quad \frac{\partial f}{\partial x} = \frac{\partial f}{\partial u}\cdot\frac{\partial u}{\partial x} + \frac{\partial f}{\partial v}\cdot\frac{\partial v}{\partial x}$$

$$= \frac{\partial\theta}{\partial u}\cdot 2x + \frac{\partial\theta}{\partial v}\cdot 2y$$

$$= 2\left(x\frac{\partial}{\partial u} + y\frac{\partial}{\partial v}\right)\theta \quad \text{or} \quad \frac{\partial}{\partial x} \equiv 2\left(x\frac{\partial}{\partial u} + y\frac{\partial}{\partial v}\right)$$

$$\Rightarrow \quad \frac{\partial^2 f}{\partial x^2} = \frac{\partial}{\partial x}\left(\frac{\partial f}{\partial x}\right) = 2\left(x\frac{\partial}{\partial u} + y\frac{\partial}{\partial v}\right)\left[2\left(x\frac{\partial\theta}{\partial u} + y\frac{\partial\theta}{\partial v}\right)\right]$$

$$= 4\left[x^2\frac{\partial^2\theta}{\partial u^2} + xy\frac{\partial^2\theta}{\partial u\partial v} + yx\frac{\partial^2\theta}{\partial v\partial u} + y^2\frac{\partial^2\theta}{\partial v^2}\right]$$

$$= 4\left[x^2\frac{\partial^2\theta}{\partial u^2} + 2xy\frac{\partial^2\theta}{\partial u\partial v} + y^2\frac{\partial^2\theta}{\partial v^2}\right] \quad \text{...(1)} \quad \left(\because \frac{\partial^2\theta}{\partial u\partial v} = \frac{\partial^2\theta}{\partial v\partial u}\right)$$

Also

$$\frac{\partial f}{\partial y} = \frac{\partial f}{\partial u}\cdot\frac{\partial u}{\partial y} + \frac{\partial f}{\partial v}\cdot\frac{\partial v}{\partial y}$$

$$= \frac{\partial\theta}{\partial u}\cdot(-2y) + \frac{\partial\theta}{\partial v}\cdot 2x$$

$$= -2\left(y\frac{\partial}{\partial u} - x\frac{\partial}{\partial v}\right)\theta \quad \text{or} \quad \frac{\partial}{\partial y} \equiv -2\left(y\frac{\partial}{\partial u} - x\frac{\partial}{\partial v}\right)$$

$$\Rightarrow \quad \frac{\partial^2 f}{\partial y^2} = \frac{\partial}{\partial y}\left(\frac{\partial f}{\partial y}\right) = -2\left(y\frac{\partial}{\partial u} - x\frac{\partial}{\partial v}\right)\left[-2\left(y\frac{\partial \theta}{\partial u} - x\frac{\partial \theta}{\partial v}\right)\right]$$

$$= 4\left[y^2\frac{\partial^2 \theta}{\partial u^2} - yx\frac{\partial^2 \theta}{\partial u \partial v} - xy\frac{\partial^2 \theta}{\partial v \partial u} + x^2\frac{\partial^2 \theta}{\partial v^2}\right]$$

$$= 4\left[y^2\frac{\partial^2 \theta}{\partial u^2} - 2xy\frac{\partial^2 \theta}{\partial u \partial v} + x^2\frac{\partial^2 \theta}{\partial v^2}\right] \quad \ldots(2)$$

Adding (1) and (2), we have

$$\frac{\partial^2 f}{\partial x^2} + \frac{\partial^2 f}{\partial y^2} = 4\left[(x^2 + y^2)\frac{\partial^2 \theta}{\partial u^2} + (y^2 + x^2)\frac{\partial^2 \theta}{\partial v^2}\right]$$

$$= 4(x^2 + y^2)\left(\frac{\partial^2 \theta}{\partial u^2} + \frac{\partial^2 \theta}{\partial v^2}\right)$$

Example 8. *If $x + y = 2e^{\theta} \cos \phi$ and $x - y = 2ie^{\theta} \sin \phi$, show that*

$$\frac{\partial^2 u}{\partial \theta^2} + \frac{\partial^2 u}{\partial \phi^2} = 4xy\frac{\partial^2 u}{\partial x \partial y}.$$

Sol. Here $\quad x + y = 2e^{\theta} \cos \phi \quad$ and $\quad x - y = 2ie^{\theta} \sin \phi$

Adding $\quad x = e^{\theta}(\cos \phi + i \sin \phi) = e^{\theta} \cdot e^{i\phi}$

Subtracting $\quad y = e^{\theta}(\cos \phi - i \sin \phi) = e^{\theta} \cdot e^{-i\phi}$ (By Euler's Theorem)

Clearly, u is a composite function of θ and ϕ.

$$\therefore \quad \frac{\partial u}{\partial \theta} = \frac{\partial u}{\partial x}\cdot\frac{\partial x}{\partial \theta} + \frac{\partial u}{\partial y}\cdot\frac{\partial y}{\partial \theta} = \frac{\partial u}{\partial x}\cdot(e^{\theta}\cdot e^{i\phi}) + \frac{\partial u}{\partial y}\cdot(e^{\theta}\cdot e^{-i\phi})$$

$$= e^{\theta}\left(e^{i\phi}\frac{\partial u}{\partial x} + e^{-i\phi}\frac{\partial u}{\partial y}\right) = e^{\theta}\left(e^{i\phi}\frac{\partial}{\partial x} + e^{-i\phi}\frac{\partial}{\partial y}\right)u$$

or

$$\frac{\partial}{\partial \theta} \equiv e^{\theta}\left(e^{i\phi}\frac{\partial}{\partial x} + e^{-i\phi}\frac{\partial}{\partial y}\right)$$

$$\Rightarrow \quad \frac{\partial^2 u}{\partial \theta^2} = \frac{\partial}{\partial \theta}\left(\frac{\partial u}{\partial \theta}\right) = e^{\theta}\left(e^{i\phi}\frac{\partial}{\partial x} + e^{-i\phi}\frac{\partial}{\partial y}\right)\left[e^{\theta}\left(e^{i\phi}\frac{\partial u}{\partial x} + e^{-i\phi}\frac{\partial u}{\partial y}\right)\right]$$

$$= e^{2\theta}\left[e^{2i\phi}\frac{\partial u}{\partial x^2} + \frac{\partial^2 u}{\partial x \partial y} + \frac{\partial^2 u}{\partial y \partial x} + e^{-2i\phi}\frac{\partial^2 u}{\partial y^2}\right]$$

$$= e^{2\theta}\left[e^{2i\phi}\frac{\partial^2 u}{\partial x^2} + 2\frac{\partial^2 u}{\partial x \partial y} + e^{-2i\phi}\frac{\partial^2 u}{\partial y^2}\right] \quad \left(\because \frac{\partial^2 u}{\partial x \partial y} = \frac{\partial^2 u}{\partial y \partial x}\right) \quad \ldots(1)$$

Also

$$\frac{\partial u}{\partial \phi} = \frac{\partial u}{\partial x}\cdot\frac{\partial x}{\partial \phi} + \frac{\partial u}{\partial y}\cdot\frac{\partial y}{\partial \phi}$$

$$= \frac{\partial u}{\partial x}\cdot(e^{\theta}\cdot ie^{i\phi}) + \frac{\partial u}{\partial y}\left[e^{\theta}\cdot(-ie^{-i\phi})\right]$$

$$= ie^{\theta}\left(e^{i\phi}\frac{\partial u}{\partial x} - e^{-i\phi}\frac{\partial u}{\partial y}\right) = ie^{\theta}\left(e^{i\phi}\frac{\partial}{\partial x} - e^{-i\phi}\frac{\partial}{\partial y}\right)u$$

or $$\frac{\partial}{\partial \phi} \equiv ie^{\theta}\left(e^{i\phi}\frac{\partial}{\partial x} - e^{-i\phi}\frac{\partial}{\partial y}\right)$$

$\Rightarrow$ $$\frac{\partial^2 u}{\partial \phi^2} = \frac{\partial}{\partial \phi}\left(\frac{\partial u}{\partial \phi}\right) = ie^{\theta}\left(e^{i\phi}\frac{\partial}{\partial x} - e^{-i\phi}\frac{\partial}{\partial y}\right)\left[ie^{\theta}\left(e^{i\phi}\frac{\partial u}{\partial x} - e^{-i\phi}\frac{\partial u}{\partial y}\right)\right]$$

$$= i^2 e^{2\theta}\left[e^{2i\phi}\frac{\partial^2 u}{\partial x^2} - \frac{\partial^2 u}{\partial x \partial y} - \frac{\partial^2 u}{\partial y \partial x} + e^{-2i\phi}\frac{\partial^2 u}{\partial y^2}\right]$$

$$= -e^{2\theta}\left[e^{2i\phi}\frac{\partial^2 u}{\partial x^2} - 2\frac{\partial^2 u}{\partial x \partial y} + e^{-2i\phi}\frac{\partial^2 u}{\partial y^2}\right] \quad ...(2)$$

Adding (1) and (2), we have

$$\frac{\partial^2 u}{\partial \theta^2} + \frac{\partial^2 u}{\partial \phi^2} = e^{2\theta}\left(4\frac{\partial^2 u}{\partial x \partial y}\right) = 4xy\frac{\partial^2 u}{\partial x \partial y} \quad [\because \ xy = e^{2\theta}]$$

Example 9. *If* $z = \sqrt{x^2 + y^2}$ *and* $x^3 + y^3 + 3axy = 5a^2$, *find the value of* $\frac{dz}{dx}$, *when* $x = y = a$.

Sol. The given equations are of the form $z = f(x, y)$ and $\phi(x, y) = c$

$\therefore$ z is composite function of x.

$\Rightarrow$ $$\frac{dz}{dx} = \frac{\partial z}{\partial x}\cdot\frac{dx}{dx} + \frac{\partial z}{\partial y}\cdot\frac{dy}{dx} = \frac{\partial z}{\partial x} + \frac{\partial z}{\partial y}\cdot\frac{dy}{dx} \quad ...(1)$$

Now, $$\frac{\partial z}{\partial x} = \frac{1}{2}(x^2 + y^2)^{-1/2}\cdot 2x = \frac{x}{\sqrt{x^2 + y^2}}$$

Similarly, $$\frac{\partial z}{\partial y} = \frac{y}{\sqrt{x^2 + y^2}}$$

Also, differentiating $x^3 + y^3 + 3axy = 5a^2$ w.r.t. x, we have

$$3x^2 + 3y^2\,.\,\frac{dy}{dx} + 3ay + 3ax\,.\,\frac{dy}{dx} = 0 \quad \text{or} \quad (y^2 + ax)\frac{dy}{dx} = -(x^2 + ay)$$

$\therefore$ $$\frac{dy}{dx} = -\frac{x^2 + ay}{y^2 + ax}$$

$\therefore$ From (1), $$\frac{dz}{dx} = \frac{x}{\sqrt{x^2 + y^2}} + \frac{y}{\sqrt{x^2 + y^2}}\left(-\frac{x^2 + ay}{y^2 + ax}\right)$$

$$\left[\frac{dz}{dx}\right]_{\substack{x=a\\y=a}} = \frac{a}{\sqrt{a^2 + a^2}} + \frac{a}{\sqrt{a^2 + a^2}}\cdot\frac{a^2 + a^2}{a^2 + a^2} = 0.$$

Example 10. *If* $u = xe^y z$, *where* $y = \sqrt{a^2 - x^2}$, $z = \sin^2 x$, *find* $\frac{du}{dx}$.

Sol. Here u is a function of x, y and z while y and z are functions of x.

$$\therefore \quad \frac{du}{dx} = \frac{\partial u}{dx} \cdot \frac{dx}{dx} + \frac{\partial u}{\partial y} \cdot \frac{dy}{dx} + \frac{\partial u}{\partial z} \cdot \frac{dz}{dx}$$

$$= e^y z \cdot 1 + xe^y z \cdot \tfrac{1}{2}(a^2 - x^2)^{-1/2}(-2x) + xe^y \cdot 2 \sin x \cos x$$

$$= e^y \left[z - \frac{x^2 z}{\sqrt{a^2 - x^2}} + x \sin 2x \right].$$

Example 11. *Find* $\frac{du}{dx}$ *if* $u = \sin (x^2 + y^2)$, *where* $a^2x^2 + b^2y^2 = c^2$.

Sol. The given equations are the form $u = f(x, y)$ and $\phi(x, y) = k$

$\therefore$ u is a composite function of x.

$$\frac{du}{dx} = \frac{\partial u}{\partial x} \cdot \frac{dx}{dx} + \frac{\partial u}{\partial y} \cdot \frac{dy}{dx} = \frac{\partial u}{\partial x} + \frac{\partial u}{\partial y} \cdot \frac{dy}{dx} \quad \text{...(1)}$$

Now, $\frac{\partial u}{\partial x} = 2x \cos (x^2 + y^2)$, $\frac{\partial u}{\partial y} = 2y \cos (x^2 + y^2)$

Also, differentiating $a^2x^2 + b^2y^2 = c$ w.r.t. x, we have

$$2a^2x + 2b^2y \cdot \frac{dy}{dx} = 0 \quad \text{or} \quad \frac{dy}{dx} = -\frac{a^2x}{b^2y}$$

$\therefore$ From (1),

$$\frac{du}{dx} = 2x \cos (x^2 + y^2) + 2y \cos (x^2 + y^2) \cdot \left[-\frac{a^2x}{b^2y} \right]$$

$$= 2 \left[x - \frac{a^2x}{b^2} \right] \cos (x^2 + y^2) = \frac{2(b^2 - a^2)x}{b^2} \cdot \cos (x^2 + y^2).$$

Example 12. *Find* $\frac{dy}{dx}$, *when*

(*i*) $x^y + y^x = c$ (*ii*) $(\cos x)^y = (\sin y)^x$.

Sol. (*i*) Let $f(x, y) = x^y + y^x$, then $f(x, y) = c$

[Using Cor. 4] $\therefore$ $\frac{dy}{dx} = -\frac{f_x}{f_y} \Rightarrow \frac{dy}{dx} = -\frac{yx^{y-1} + y^x \log y}{x^y \log x + xy^{x-1}}$.

(*ii*) Let $f(x, y) = (\cos x)^y - (\sin y)^x = 0$

$$\frac{dy}{dx} = -\frac{f_x}{f_y} = -\frac{y(\cos x)^{y-1} \cdot (-\sin x) - (\sin y)^x \log (\sin y)}{(\cos x)^y \log (\cos x) - x (\sin y)^{x-1} \cdot \cos y}$$

$$= \frac{y(\cos x)^{y-1} \sin x + (\cos x)^y \log (\sin y)}{(\cos x)^y \log (\cos x) - x (\cos x)^y (\sin y)^{-1} \cos y}$$

$[\because \quad (\sin y)^x = (\cos x)^y]$

$$= \frac{(\cos x)^y \left[y \cdot \frac{\sin x}{\cos x} + \log \sin y \right]}{(\cos x)^y [\log \cos x - x \cot y]} = \frac{y \tan x + \log \sin y}{\log \cos x - x \cot y}.$$

Example 13. *If* $f(x, y) = 0$, $\phi (y, z) = 0$, *show that* $\frac{\partial f}{\partial y} \cdot \frac{\partial \phi}{\partial z} \cdot \frac{dz}{dx} = \frac{\partial f}{\partial x} \cdot \frac{\partial \phi}{\partial y}$.

Sol. $$\frac{dz}{dx} = \frac{dz}{dy} \cdot \frac{dy}{dx} \qquad ...(1)$$

$f(x, y) = 0$ gives $$\frac{dy}{dx} = -\frac{\frac{\partial f}{\partial x}}{\frac{\partial f}{\partial y}};$$ $\phi(y, z) = 0$ gives $$\frac{dz}{dy} = -\frac{\frac{\partial \phi}{\partial y}}{\frac{\partial \phi}{\partial z}}$$

$\therefore$ From (1), $$\frac{dz}{dx} = \frac{\frac{\partial f}{\partial x} \cdot \frac{\partial \phi}{\partial y}}{\frac{\partial f}{\partial y} \cdot \frac{\partial \phi}{\partial z}} \Rightarrow \frac{\partial f}{\partial y} \cdot \frac{\partial \phi}{\partial z} \cdot \frac{dz}{dx} = \frac{\partial f}{\partial x} \cdot \frac{\partial \phi}{\partial y}.$$

Example 14. *If $\phi(x, y, z) = 0$, show that* $\left(\frac{\partial y}{\partial z}\right)_x \left(\frac{\partial z}{\partial x}\right)_y \left(\frac{\partial x}{\partial y}\right)_z = -1.$

Sol. The given relation defines y as a function of x and z. Treating x as constant

$$\left(\frac{\partial y}{\partial z}\right)_x = -\frac{\frac{\partial \phi}{\partial z}}{\frac{\partial \phi}{\partial y}}$$

The given relation defines z as a function of x and y. Treating y as constant $\left(\frac{\partial z}{\partial x}\right)_y = -\frac{\frac{\partial \phi}{\partial x}}{\frac{\partial \phi}{\partial z}}$

Similarly, $$\left(\frac{\partial x}{\partial y}\right)_z = -\frac{\frac{\partial \phi}{\partial y}}{\frac{\partial \phi}{\partial x}}$$

Multiplying, we get the desired result.

Example 15. *Prove that* $\frac{\partial^2 z}{\partial x^2} + \frac{\partial^2 z}{\partial y^2} = \frac{\partial^2 z}{\partial u^2} + \frac{\partial^2 z}{\partial v^2}$

where $x = u \cos \alpha - v \sin \alpha$, $y = u \sin \alpha + v \cos \alpha$

Or

By changing the independent variables u and v to x and y by means of the relations $x = u \cos \alpha - v \sin \alpha$, $y = u \sin \alpha + v \cos \alpha$, show that $\frac{\partial^2 z}{\partial u^2} + \frac{\partial^2 z}{\partial v^2}$ transforms into $\frac{\partial^2 z}{\partial x^2} + \frac{\partial^2 z}{\partial y^2}$.

Sol. Here z is a composite function of u and v

$$\frac{\partial z}{\partial u} = \frac{\partial z}{\partial x} \cdot \frac{\partial x}{\partial u} + \frac{\partial z}{\partial y} \cdot \frac{\partial y}{\partial u} = \cos \alpha \frac{\partial z}{\partial x} + \sin \alpha \frac{\partial z}{\partial y}$$

or $$\frac{\partial}{\partial u}(z) = \left(\cos \alpha \frac{\partial}{\partial x} + \sin \alpha \frac{\partial}{\partial y}\right) z \Rightarrow \frac{\partial}{\partial u} \equiv \cos \alpha \frac{\partial}{\partial x} + \sin \alpha \frac{\partial}{\partial y} \qquad ...(1)$$

Also $$\frac{\partial z}{\partial v} = \frac{\partial z}{\partial x} \cdot \frac{\partial x}{\partial v} + \frac{\partial z}{\partial y} \cdot \frac{\partial y}{\partial v} = -\sin \alpha \frac{\partial z}{\partial x} + \cos \alpha \frac{\partial z}{\partial y}$$

or $$\frac{\partial}{\partial v}(z) = \left(-\sin\alpha\frac{\partial}{\partial x} + \cos\alpha\frac{\partial}{\partial y}\right)z$$

$$\Rightarrow \qquad \frac{\partial}{\partial v} \equiv -\sin\alpha\frac{\partial}{\partial x} + \cos\alpha\frac{\partial}{\partial y} \qquad ...(2)$$

Now we shall make use of the equivalence of operators as given by (1) and (2).

$$\frac{\partial^2 z}{\partial u^2} = \frac{\partial}{\partial u}\left(\frac{\partial z}{\partial u}\right) = \left(\cos\alpha\frac{\partial}{\partial x} + \sin\alpha\frac{\partial}{\partial y}\right)\left(\cos\alpha\frac{\partial z}{\partial x} + \sin\alpha\frac{\partial z}{\partial y}\right)$$

$$= \cos^2\alpha\frac{\partial^2 z}{\partial x^2} + \cos\alpha\sin\alpha\frac{\partial^2 z}{\partial x\partial y} + \sin\alpha\cos\alpha\frac{\partial^2 z}{\partial y\partial x} + \sin^2\alpha\frac{\partial^2 z}{\partial y^2}$$

$$= \cos^2\alpha\frac{\partial^2 z}{\partial x^2} + 2\cos\alpha\sin\alpha\frac{\partial^2 z}{\partial x\partial y} + \sin^2\alpha\frac{\partial^2 z}{\partial y^2} \qquad ...(3)$$

$$\frac{\partial^2 z}{\partial v^2} = \frac{\partial}{\partial v}\left(\frac{\partial z}{\partial v}\right) = \left(-\sin\alpha\frac{\partial}{\partial x} + \cos\alpha\frac{\partial}{\partial y}\right)\left(-\sin\alpha\frac{\partial z}{\partial x} + \cos\alpha\frac{\partial z}{\partial y}\right)$$

$$= \sin^2\alpha\frac{\partial^2 z}{\partial x^2} - \sin\alpha\cos\alpha\frac{\partial^2 z}{\partial x\partial y} - \cos\alpha\sin\alpha\frac{\partial^2 z}{\partial y\partial x} + \cos^2\alpha\frac{\partial^2 z}{\partial y^2}$$

$$= \sin^2\alpha\frac{\partial^2 z}{\partial x^2} - 2\cos\alpha\sin\alpha\frac{\partial^2 z}{\partial x\partial y} + \cos^2\alpha\frac{\partial^2 z}{\partial y^2} \qquad ...(4)$$

Adding (3) and (4), $\dfrac{\partial^2 z}{\partial u^2} + \dfrac{\partial^2 z}{\partial v^2} = \dfrac{\partial^2 z}{\partial x^2} + \dfrac{\partial^2 z}{\partial y^2}$.

Example 16. *Transform the equation* $\dfrac{\partial^2 u}{\partial x^2} + \dfrac{\partial^2 u}{\partial y^2} = 0$ *into polar co-ordinates.*

Sol. The relations connecting cartesian co-ordinates, (x, y) with polar co-ordinates (r, θ) are

$$x = r\cos\theta,\ y = r\sin\theta$$

Squaring and adding, $r^2 = x^2 + y^2$

Dividing, $\tan\theta = \dfrac{y}{x}$

$$r = \sqrt{x^2 + y^2},\ \theta = \tan^{-1}\left(\frac{y}{x}\right)$$

$$\frac{\partial r}{\partial x} = \frac{x}{\sqrt{x^2+y^2}} = \frac{r\cos\theta}{r} = \cos\theta$$

$$\frac{\partial r}{\partial y} = \frac{y}{\sqrt{x^2+y^2}} = \frac{r\sin\theta}{r} = \sin\theta$$

$$\frac{\partial\theta}{\partial x} = \frac{1}{1+\dfrac{y^2}{x^2}}\cdot\left(-\frac{y}{x^2}\right) = -\frac{y}{x^2+y^2} = -\frac{r\sin\theta}{r^2} = -\frac{\sin\theta}{r}$$

$$\frac{\partial\theta}{\partial y} = \frac{1}{1+\dfrac{y^2}{x^2}}\cdot\frac{1}{x} = \frac{x}{x^2+y^2} = \frac{r\cos\theta}{r^2} = \frac{\cos\theta}{r}$$

Here u is a composite function of x and y

$$\frac{\partial u}{\partial x}=\frac{\partial u}{\partial r}\cdot\frac{\partial r}{\partial x}+\frac{\partial u}{\partial \theta}\cdot\frac{\partial \theta}{\partial x}=\cos\theta\,\frac{\partial u}{\partial r}-\frac{\sin\theta}{r}\cdot\frac{\partial u}{\partial \theta}$$

or
$$\frac{\partial}{\partial x}(u)=\left(\cos\theta\,\frac{\partial}{\partial r}-\frac{\sin\theta}{r}\cdot\frac{\partial}{\partial\theta}\right)u$$

$$\Rightarrow\qquad \frac{\partial}{\partial x}\equiv\cos\theta\,\frac{\partial}{\partial r}-\frac{\sin\theta}{r}\cdot\frac{\partial}{\partial\theta}\qquad\text{...(1)}$$

Also
$$\frac{\partial u}{\partial y}=\frac{\partial u}{\partial r}\cdot\frac{\partial r}{\partial y}+\frac{\partial u}{\partial \theta}\cdot\frac{\partial \theta}{\partial y}=\sin\theta\,\frac{\partial u}{\partial r}+\frac{\cos\theta}{r}\cdot\frac{\partial u}{\partial \theta}$$

or
$$\frac{\partial}{\partial y}(u)=\left(\sin\theta\,\frac{\partial}{\partial r}+\frac{\cos\theta}{r}\cdot\frac{\partial}{\partial\theta}\right)u\ \Rightarrow\ \frac{\partial}{\partial y}\equiv\sin\theta\,\frac{\partial}{\partial r}+\frac{\cos\theta}{r}\cdot\frac{\partial}{\partial\theta}\qquad\text{...(2)}$$

Now we shall make use of the equivalence of cartesian and polar operators as given by (1) and (2),

$$\frac{\partial^2u}{\partial x^2}=\frac{\partial}{\partial x}\left(\frac{\partial u}{\partial x}\right)=\left(\cos\theta\,\frac{\partial}{\partial r}-\frac{\sin\theta}{r}\cdot\frac{\partial}{\partial\theta}\right)\left(\cos\theta\,\frac{\partial u}{\partial r}-\frac{\sin\theta}{r}\cdot\frac{\partial u}{\partial\theta}\right)$$

$$=\cos\theta\,\frac{\partial}{\partial r}\left(\cos\theta\,\frac{\partial u}{\partial r}-\frac{\sin\theta}{r}\cdot\frac{\partial u}{\partial\theta}\right)-\frac{\sin\theta}{r}\cdot\frac{\partial}{\partial\theta}\left(\cos\theta\,\frac{\partial u}{\partial r}-\frac{\sin\theta}{r}\cdot\frac{\partial u}{\partial\theta}\right)$$

$$=\cos\theta\left[\cos\theta\,\frac{\partial^2u}{\partial r^2}-\sin\theta\,\frac{\partial u}{\partial\theta}\left(-\frac{1}{r^2}\right)-\frac{\sin\theta}{r}\cdot\frac{\partial^2u}{\partial\theta\,\partial r}\right]$$

$$-\frac{\sin\theta}{r}\left[-\sin\theta\,\frac{\partial u}{\partial r}+\cos\theta\,\frac{\partial^2u}{\partial\theta\,\partial r}-\frac{\cos\theta}{r}\frac{\partial u}{\partial\theta}-\frac{\sin\theta}{r}\cdot\frac{\partial^2u}{\partial\theta^2}\right]$$

$$=\cos^2\theta\,\frac{\partial^2u}{\partial r^2}+\frac{2\cos\theta\sin\theta}{r^2}\cdot\frac{\partial u}{\partial\theta}+\frac{\sin^2\theta}{r}\cdot\frac{\partial u}{\partial r}-\frac{2\cos\theta\sin\theta}{r^2}\cdot\frac{\partial^2u}{\partial r\,\partial\theta}+\frac{\sin^2\theta}{r^2}\cdot\frac{\partial^2u}{\partial\theta^2}\qquad\text{...(3)}$$

$$\frac{\partial^2u}{\partial y^2}=\frac{\partial}{\partial y}\left(\frac{\partial u}{\partial y}\right)=\left(\sin\theta\,\frac{\partial}{\partial r}+\frac{\cos\theta}{r}\cdot\frac{\partial}{\partial\theta}\right)\left(\sin\theta\,\frac{\partial u}{\partial r}+\frac{\cos\theta}{r}\cdot\frac{\partial u}{\partial\theta}\right)$$

$$=\sin\theta\,\frac{\partial}{\partial r}\left(\sin\theta\,\frac{\partial u}{\partial r}+\frac{\cos\theta}{r}\cdot\frac{\partial u}{\partial\theta}\right)+\frac{\cos\theta}{r}\cdot\frac{\partial}{\partial\theta}\left(\sin\theta\,\frac{\partial u}{\partial r}+\frac{\cos\theta}{r}\cdot\frac{\partial u}{\partial\theta}\right)$$

$$=\sin\theta\left[\sin\theta\,\frac{\partial^2u}{\partial r^2}-\frac{\cos\theta}{r^2}\cdot\frac{\partial u}{\partial\theta}+\frac{\cos\theta}{r}\cdot\frac{\partial^2u}{\partial r\,\partial\theta}\right]$$

$$+\frac{\cos\theta}{r}\left[\cos\theta\,\frac{\partial u}{\partial r}+\sin\theta\,\frac{\partial^2u}{\partial\theta\,\partial r}-\frac{\sin\theta}{r}\cdot\frac{\partial u}{\partial\theta}+\frac{\cos\theta}{r}\cdot\frac{\partial^2u}{\partial\theta^2}\right]$$

$$=\sin^2\theta\,\frac{\partial^2u}{\partial r^2}-\frac{2\cos\theta\sin\theta}{r^2}\cdot\frac{\partial u}{\partial\theta}+\frac{\cos^2\theta}{r}\cdot\frac{\partial u}{\partial r}+\frac{2\cos\theta\sin\theta}{r}\cdot\frac{\partial^2u}{\partial r\,\partial\theta}+\frac{\cos^2\theta}{r^2}\cdot\frac{\partial^2u}{\partial\theta^2}\qquad\text{...(4)}$$

Adding (3) and (4),
$$\frac{\partial^2u}{\partial x^2}+\frac{\partial^2u}{\partial y^2}=\frac{\partial^2u}{\partial r^2}+\frac{1}{r}\frac{\partial u}{\partial r}+\frac{1}{r^2}\cdot\frac{\partial^2u}{\partial\theta^2}$$

$\therefore\quad \dfrac{\partial^2u}{\partial x^2}+\dfrac{\partial^2u}{\partial y^2}=0$ transforms into $\dfrac{\partial^2u}{\partial r^2}+\dfrac{1}{r}\cdot\dfrac{\partial u}{\partial r}+\dfrac{1}{r^2}\cdot\dfrac{\partial^2u}{\partial\theta^2}=0.$

EXERCISE 2.3

1. (*a*) Find $\frac{du}{dt}$ when $u = x^2 + y^2$, $x = at^2$, $y = 2at$. Also verify by direct substitution.

(*b*) If $u = x^2 + y^2 + z^2$ and $x = e^{2t}$, $y = e^{2t} \cos 3t$, $z = e^{2t} \sin 3t$; find $\frac{du}{dt}$ as a total derivative and verify the result by direct substitution.

2. If $u = \sin \frac{x}{y}$, $x = e^t$, $y = t^2$, find $\frac{du}{dt}$ as a function of t. Verify your result by direct substitution.

3. If $u = x^3 + y^3$, where $x = a \cos t$, $y = b \sin t$, find $\frac{du}{dt}$ and verify the result.

4. (*a*) At a given instant the sides of a rectangle are 4 ft. and 3 ft. respectively and they are increasing at the rate of 1.5 ft./sec. and 0.5 ft./sec. respectively. Find the rate at which the area is increasing at that instant.

(*b*) At a given instant the major and minor axes of an ellipse are 60 cm and 40 cm respectively and they are increasing at the rate of 2 cm/sec and 1.5 cm/sec respectively. Find the rate at which the area is increasing at that instant. (Area of Ellipse = πaf)

(*c*) If $z = 2xy - 3x^2y$ and if x increases at the rate of 3 cm/sec at the instant when $x = 4$ cm and $y = 2$ cm, at what rate must y be changing in order that z shall be neither increasing nor decreasing?

5. (*a*) If $z = u^2 + v^2$, $u = r \cos \theta$, $v = r \sin \theta$, find $\frac{\partial z}{\partial r}$ and $\frac{\partial z}{\partial \theta}$.

(*b*) If $z = \log (u^2 + v)$, $u = e^{x^2 + y^2}$, $v = x^2 + y$, find $\frac{\partial z}{\partial x}$ and $\frac{\partial z}{\partial y}$.

(*c*) If $u = \tan^{-1} \left(\frac{y}{x}\right)$ where $x = e^t - e^{-t}$ and $y = e^t + e^{-t}$, find $\frac{du}{dt}$.

6. If $u = f(r, s)$, $r = x + y$, $s = x - y$, show that $\frac{\partial u}{\partial x} + \frac{\partial u}{\partial y} = 2\frac{\partial u}{\partial r}$.

7. If $z = e^{ax+by} f(ax - by)$, prove that $b \frac{\partial z}{\partial x} + a \frac{\partial z}{\partial y} = 2abz$.

8. (*i*) If $x = u + v$, $y = uv$ and z is a function of x, y; show that $u \frac{\partial z}{\partial u} + v\frac{\partial z}{\partial v} = x\frac{\partial z}{\partial x} + 2y\frac{\partial z}{\partial y}$.

(*ii*) If $u = f(r, s)$, $r = x + at$, $s = y + bt$, show that $\frac{\partial u}{\partial t} = a\frac{\partial u}{\partial x} + f\frac{\partial u}{\partial y}$.

9. If $u = x \log (xy)$, where $x^3 + y^3 + 3xy = 1$, find $\frac{du}{dx} = \frac{\partial u}{\partial x} + \frac{\partial u}{\partial y} \cdot \frac{dy}{dx}$.

10. (*a*) If $u = f(r, s, t)$ and $r = \frac{x}{y}$, $s = \frac{y}{z}$, $t = \frac{z}{x}$, prove that $x \frac{\partial u}{\partial x} + y\frac{\partial u}{\partial y} + z\frac{\partial u}{\partial z} = 0$.

(*b*) If $x = u + v + w$, $y = vw + wu + uv$, $z = uvw$ and F is a function of x, y, z, show that

$$u \frac{\partial F}{\partial u} + v\frac{\partial F}{\partial v} + w\frac{\partial F}{\partial w} = x\frac{\partial F}{\partial x} + 2y\frac{\partial F}{\partial y} + 3z\frac{\partial F}{\partial z}.$$

(c) If $u = f(2x - 3y, 3y - 4z, 4z - 2x)$, prove that $\frac{1}{2}\frac{\partial u}{\partial x} + \frac{1}{3}\frac{\partial u}{\partial y} + \frac{1}{4}\frac{\partial u}{\partial z} = 0$.

(d) If $u = f(x^2 + 2yz, y^2 + 2zx)$, prove that $(y^2 - zx)\frac{\partial u}{\partial x} + (x^2 - yz)\frac{\partial u}{\partial y} + (z^2 - xy)\frac{\partial u}{\partial z} = 0$.

11. If $z = x^2y$ and $x^2 + xy + y^2 = 1$, show that $\frac{dz}{dx} = 2xy - \frac{x^2(2x + y)}{x + 2y}$.

12. If $x^3 + 3x^2y + 6xy^2 + y^3 = 1$, find $\frac{dy}{dx}$.

13. If $x^y = y^x$, show that $\frac{dy}{dx} = \frac{y(y - x\log y)}{x(x - y\log x)}$ using partial derivative method.

14. If the curves $f(x, y) = 0$ and $\phi(x, y) = 0$ touch, show that at the point of contact, $\frac{\partial f}{\partial x}\cdot\frac{\partial \phi}{\partial y} = \frac{\partial f}{\partial y}\cdot\frac{\partial \phi}{\partial x}$.

15. (a) If $f(x, y) = 0$, show that $\frac{d^2y}{dx^2} = -\frac{q^2r - 2pqs + p^2t}{q^3}$.

(b) If $f(x, y) = 0$, show that $\left(\frac{\partial f}{\partial y}\right)^3 \cdot \frac{d^2y}{dx^2} = 2\left(\frac{\partial f}{\partial x}\right)\left(\frac{\partial f}{\partial y}\right)\left(\frac{\partial^2 f}{\partial x\partial y}\right) - \left(\frac{\partial f}{\partial y}\right)^2\left(\frac{\partial^2 f}{\partial x^2}\right) - \left(\frac{\partial f}{\partial x}\right)^2\left(\frac{\partial^2 f}{\partial y^2}\right)$.

16. (a) If $x^3 + y^3 - 3axy = 0$, find $\frac{dy}{dx}$ and $\frac{d^2y}{dx^2}$ using partial derivative method.

(b) Prove that if $y^3 - 3ax^2 + x^3 = 0$, then $\frac{d^2y}{dx^2} + \frac{2a^2x^2}{y^5} = 0$.

17. If $z = xyf\left(\frac{y}{x}\right)$ and z is a constant, show that $\frac{f'\left(\frac{y}{x}\right)}{f\left(\frac{y}{x}\right)} = \frac{x\left[y + x\frac{dy}{dx}\right]}{y\left[y - x\frac{dy}{dx}\right]}$.

18. By changing the independent variables x and y to u and v by means of the relations $u = x - ay$, $v = x + ay$, show that $a^2\frac{\partial^2 z}{\partial x^2} - \frac{\partial^2 z}{\partial y^2}$ transforms into $4a^2\frac{\partial^2 z}{\partial u\partial v}$.

19. If z is a function of x and y, and u and v be two other variables such that $u = lx + my$, $v = ly - mx$, show that $\frac{\partial^2 z}{\partial x^2} + \frac{\partial^2 z}{\partial y^2} = (l^2 + m^2)\left(\frac{\partial^2 z}{\partial u^2} + \frac{\partial^2 z}{\partial v^2}\right)$.

20. If $x = e^r\cos\theta$, $y = e^r\sin\theta$, show that $\frac{\partial^2 u}{\partial x^2} + \frac{\partial^2 u}{\partial y^2} = e^{-2r}\left(\frac{\partial^2 u}{\partial r^2} + \frac{\partial^2 u}{\partial \theta^2}\right)$.

21. If $x = \rho\cos\phi$, $y = \rho\sin\phi$, show that $\left(\frac{\partial v}{\partial x}\right)^2 + \left(\frac{\partial v}{\partial y}\right)^2 = \left(\frac{\partial v}{\partial \rho}\right)^2 + \frac{1}{\rho^2}\left(\frac{\partial v}{\partial \phi}\right)^2$.

22. Prove that $\frac{\partial^2 u}{\partial x^2} + \frac{\partial^2 u}{\partial y^2} = \frac{\partial^2 u}{\partial s^2} + \frac{\partial^2 u}{\partial t^2}$, where $x = s\cos\alpha - t\sin\alpha$ and $y = s\sin\alpha + t\cos\alpha$.

23. If $z = f(x, y)$, where $x = s + t$, and $y = s - t$, show that $\left(\frac{\partial z}{\partial x}\right)^2 - \left(\frac{\partial z}{\partial y}\right)^2 = \frac{\partial z}{\partial s}\cdot\frac{\partial z}{\partial t}$.

24. If $u = f(x, y)$, where $x = e^s \cos t, y = e^s \sin t$, show that $\left(\frac{\partial u}{\partial x}\right)^2 + \left(\frac{\partial u}{\partial y}\right)^2 = e^{-2s}\left[\left(\frac{\partial u}{\partial s}\right)^2 + \left(\frac{\partial u}{\partial t}\right)^2\right]$.

25. If f satisfies the equation, $f(tx, ty) = t^n f(x, y)$ for all t, where n is a positive integer. Then show that $f_x(tx, ty) = t^{n-1} f_x(x, y)$.

26. Suppose that the equation $F(x, y, z) = 0$ implicitly defines each of the three variables x, y and z as functions of the other two; $z = f(x, y), y = g(x, z), x = h(y, z)$. If F is differentiable and F_x, F_y, and F_z are all non-zero, show that $\frac{\partial z}{\partial x}\frac{\partial x}{\partial y}\frac{\partial y}{\partial z} = -1$.

27. If $z = y + f(x^2 - y^2)$, where f is differentiable, show that $y\frac{\partial z}{\partial x} + x\frac{\partial z}{\partial y} = x$.

28. If $z = f(u, v)$, where $u = xy$, $v = \frac{y}{x}$, show that $x^2\frac{\partial^2 z}{\partial x^2} - y^2\frac{\partial^2 z}{\partial y^2} = -4uv\frac{\partial^2 z}{\partial u\,\partial v} + 2v\frac{\partial z}{\partial v}$.

29. If $x^x y^y z^z = e$, show that at $x = y = z$, $\frac{\partial^2 z}{\partial x\,\partial y} = -[x \log ex]^{-1}$

Answers

1. (*a*) $4a^2t(t^2 + 2)$ (*b*) $8e^{4t}$ **2.** $\frac{e^t(t-2)}{t^3}\cos\left(\frac{e^t}{t^2}\right)$

3. $-3a^3\cos^2 t \sin t + 3b^3 \sin^2 t \cos t$

4. (*a*) 6.5 sq. ft./sec (*b*) 85π cm²/sec

(*c*) Decreasing at the rate of 3.3 cm/sec

5. (*a*) $2r$, 0 (*b*) $\frac{2x(2u^2+1)}{u^2+v}, \frac{4yu^2+1}{u^2+v}$ (*c*) $\frac{-2}{e^{2t}+e^{-2t}}$

9. $1 + \log(xy) - \frac{x(x^2+y)}{y(y^2+x)}$ **12.** $-\frac{x^2+2xy+2y^2}{x^2+4xy+y^2}$ **16.** (*a*) $\frac{x^2-ay}{ax-y^2}, \frac{2a^3xy}{(ax-y^2)^3}$.

2.10. JACOBIANS

If u and v are functions of two independent variables x and y, then the determinant $\begin{vmatrix} \frac{\partial u}{\partial x} & \frac{\partial u}{\partial y} \\ \frac{\partial v}{\partial x} & \frac{\partial v}{\partial y} \end{vmatrix}$ is called Jacobian of u, v with respect to x, y and is denoted by the symbol $J\left(\frac{u, v}{x, y}\right)$ or $\frac{\partial(u, v)}{\partial(x, y)}$.

Similarly, if u, v, w be functions of x, y, z, then the Jacobian of u, v, w with respect to x, y, z is

$$J\left(\frac{u, v, w}{x, y, z}\right) \quad \text{or} \quad \frac{\partial(u, v, w)}{\partial(x, y, z)} = \begin{vmatrix} \frac{\partial u}{\partial x} & \frac{\partial u}{\partial y} & \frac{\partial u}{\partial z} \\ \frac{\partial v}{\partial x} & \frac{\partial v}{\partial y} & \frac{\partial v}{\partial z} \\ \frac{\partial w}{\partial x} & \frac{\partial w}{\partial y} & \frac{\partial w}{\partial z} \end{vmatrix}.$$

2.11. PROPERTIES OF JACOBIANS

I. If u, v are functions of r, s where r, s are functions of x, y, then

$$\frac{\partial(\mathbf{u}, \mathbf{v})}{\partial(\mathbf{x}, \mathbf{y})} = \frac{\partial(\mathbf{u}, \mathbf{v})}{\partial(\mathbf{r}, \mathbf{s})} \times \frac{\partial(\mathbf{r}, \mathbf{s})}{\partial(\mathbf{x}, \mathbf{y})}. \qquad \text{[Chain Rule for Jacobians]}$$

Proof. Since u, v are composite functions of x, y

$$\therefore \quad \left.\begin{aligned} \frac{\partial u}{\partial x} &= \frac{\partial u}{\partial r}\cdot\frac{\partial r}{\partial x} + \frac{\partial u}{\partial s}\cdot\frac{\partial s}{\partial x} = u_r r_x + u_s s_x \\ \frac{\partial u}{\partial y} &= \frac{\partial u}{\partial r}\cdot\frac{\partial r}{\partial y} + \frac{\partial u}{\partial s}\cdot\frac{\partial s}{\partial y} = u_r r_y + u_s s_y \\ \frac{\partial v}{\partial x} &= \frac{\partial v}{\partial r}\cdot\frac{\partial r}{\partial x} + \frac{\partial v}{\partial s}\cdot\frac{\partial s}{\partial x} = v_r r_x + v_s s_x \\ \frac{\partial v}{\partial y} &= \frac{\partial v}{\partial r}\cdot\frac{\partial r}{\partial y} + \frac{\partial v}{\partial s}\cdot\frac{\partial s}{\partial y} = v_r r_y + v_s s_y \end{aligned}\right] \quad \text{...(A)}$$

Now $$\frac{\partial(u, v)}{\partial(r, s)} \times \frac{\partial(r, s)}{\partial(x, y)} = \begin{vmatrix} u_r & u_s \\ v_r & v_s \end{vmatrix} \cdot \begin{vmatrix} r_x & r_y \\ s_x & s_y \end{vmatrix}$$

Interchanging rows and columns in the second determinant

$$= \begin{vmatrix} u_r & u_s \\ v_r & v_s \end{vmatrix} \cdot \begin{vmatrix} r_x & s_x \\ r_y & s_y \end{vmatrix} = \begin{vmatrix} u_r r_x + u_s s_x & u_r r_y + u_s s_y \\ v_r r_x + v_s s_x & v_r r_y + v_x s_y \end{vmatrix} = \begin{vmatrix} \frac{\partial u}{\partial x} & \frac{\partial u}{\partial y} \\ \frac{\partial v}{\partial x} & \frac{\partial v}{\partial y} \end{vmatrix} \qquad \text{[Using (A)]}$$

$$= \frac{\partial(u, v)}{\partial(x, y)}.$$

II. If J_1 is the Jacobian of u, v, with respect to x, y and J_2 is the Jacobian of x, y, with respect to u, v, then $J_1J_2 = 1$ i.e., $\frac{\partial(\mathbf{u}, \mathbf{v})}{\partial(\mathbf{x}, \mathbf{y})}\cdot\frac{\partial(\mathbf{x}, \mathbf{y})}{\partial(\mathbf{u}, \mathbf{v})} = \mathbf{1}.$

Proof. Let $u = u(x, y)$ and $v = v(x, y)$, so that u and v are functions of x, y.

Differentiating partially w.r.t. u and v, we get

$$\left.\begin{aligned} 1 &= \frac{\partial u}{\partial x}\cdot\frac{\partial x}{\partial u} + \frac{\partial u}{\partial y}\cdot\frac{\partial y}{\partial u} = u_x x_u + u_y y_u \\ 0 &= \frac{\partial u}{\partial x}\cdot\frac{\partial x}{\partial v} + \frac{\partial u}{\partial y}\cdot\frac{\partial y}{\partial v} = u_x x_v + u_y y_v \\ 0 &= \frac{\partial v}{\partial x}\cdot\frac{\partial x}{\partial u} + \frac{\partial v}{\partial y}\cdot\frac{\partial y}{\partial u} = v_x x_u + v_y y_u \\ 1 &= \frac{\partial v}{\partial x}\cdot\frac{\partial x}{\partial v} + \frac{\partial v}{\partial y}\cdot\frac{\partial y}{\partial v} = v_x x_v + v_y y_v \end{aligned}\right] \quad \text{...(A)}$$

Now $$\frac{\partial(u,v)}{\partial(x,y)}\cdot\frac{\partial(x,y)}{\partial(u,v)}=\begin{vmatrix} u_x & u_y \\ v_x & v_y \end{vmatrix}\begin{vmatrix} x_u & x_v \\ y_u & y_v \end{vmatrix}$$

Interchanging rows and columns in the second determinant

$$=\begin{vmatrix} u_x & u_y \\ v_x & v_y \end{vmatrix}\begin{vmatrix} x_u & y_u \\ x_v & y_v \end{vmatrix}=\begin{vmatrix} u_x x_u + u_y y_u & u_x x_v + u_y y_v \\ v_x x_u + v_y y_u & v_x x_v + v_y y_v \end{vmatrix}$$

$$=\begin{vmatrix} 1 & 0 \\ 0 & 1 \end{vmatrix}=1.$$ [Using (A)]

ILLUSTRATIVE EXAMPLES

Example 1. *If* $r=\sqrt{x^2+y^2}$, $\theta=\tan^{-1}\frac{y}{x}$, *evaluate* $\frac{\partial(r,\theta)}{\partial(x,y)}$.

Sol. $$r=\sqrt{x^2+y^2},\ \theta=\tan^{-1}\frac{y}{x}$$

$$\frac{\partial r}{\partial x}=\frac{x}{\sqrt{x^2+y^2}},\qquad \frac{\partial r}{\partial y}=\frac{y}{\sqrt{x^2+y^2}}$$

$$\frac{\partial\theta}{\partial x}=\frac{1}{1+\frac{y^2}{x^2}}\left(-\frac{y}{x^2}\right)=-\frac{y}{x^2+y^2},\qquad \frac{\partial\theta}{\partial y}=\frac{1}{1+\frac{y^2}{x^2}}\left(\frac{1}{x}\right)=\frac{x}{x^2+y^2}$$

$$\frac{\partial(r,\theta)}{\partial(x,y)}=\begin{vmatrix} \frac{\partial r}{\partial x} & \frac{\partial r}{\partial y} \\ \frac{\partial\theta}{\partial x} & \frac{\partial\theta}{\partial y} \end{vmatrix}=\begin{vmatrix} \frac{x}{\sqrt{x^2+y^2}} & \frac{y}{\sqrt{x^2+y^2}} \\ -\frac{y}{\sqrt{x^2+y^2}} & \frac{x}{\sqrt{x^2+y^2}} \end{vmatrix}$$

$$=\frac{x^2}{(x^2+y^2)^{3/2}}+\frac{y^2}{(x^2+y^2)^{3/2}}=\frac{x^2+y^2}{(x^2+y^2)^{3/2}}=\frac{1}{\sqrt{x^2+y^2}}.$$

Example 2. *If* $x=r\sin\theta\cos\phi,\ y=r\sin\theta\sin\phi,\ z=r\cos\theta$, *show that* $\frac{\partial(x,y,z)}{\partial(r,\theta,\phi)}=r^2\sin\theta$.

(K.U. May 2010)

Sol. $$\frac{\partial(x,y,z)}{\partial(r,\theta,\phi)}=\begin{vmatrix} \frac{\partial x}{\partial r} & \frac{\partial x}{\partial\theta} & \frac{\partial x}{\partial\phi} \\ \frac{\partial y}{\partial r} & \frac{\partial y}{\partial\theta} & \frac{\partial y}{\partial\phi} \\ \frac{\partial z}{\partial r} & \frac{\partial z}{\partial\theta} & \frac{\partial z}{\partial\phi} \end{vmatrix}=\begin{vmatrix} \sin\theta\cos\phi & r\cos\theta\cos\phi & -r\sin\theta\sin\phi \\ \sin\theta\sin\phi & r\cos\theta\sin\phi & r\sin\theta\cos\phi \\ \cos\theta & -r\sin\theta & 0 \end{vmatrix}$$

Taking out common factors (r from second column and $r\sin\theta$ from third column)

$$=r^2\sin\theta\begin{vmatrix} \sin\theta\cos\phi & \cos\theta\cos\phi & -\sin\phi \\ \sin\theta\sin\phi & \cos\theta\sin\phi & \cos\phi \\ \cos\theta & -\sin\theta & 0 \end{vmatrix}$$

Expanding by third row

$$= r^2 \sin\theta \left\{\cos\theta \begin{vmatrix} \cos\theta\cos\phi & -\sin\phi \\ \cos\theta\sin\phi & \cos\phi \end{vmatrix} + \sin\theta \begin{vmatrix} \sin\theta\cos\phi & -\sin\phi \\ \sin\theta\sin\phi & \cos\phi \end{vmatrix}\right\}$$

$$= r^2 \sin\theta\,[\cos\theta\,(\cos\theta\cos^2\phi + \cos\theta\sin^2\phi) + \sin\theta\,(\sin\theta\cos^2\phi + \sin\theta\sin^2\phi)]$$

$$= r^2 \sin\theta\,(\cos^2\theta + \sin^2\theta) = r^2\sin\theta.$$

Note. Here (x, y, z) and (r, θ, ϕ) are respectively the Cartesian and spherical polar coordinates of a point.

Example 3. *If* $u = \dfrac{x}{y-z}, v = \dfrac{y}{z-x}, w = \dfrac{z}{x-y}$, *show that* $\dfrac{\partial(u, v, w)}{\partial(x, y, z)} = 0$.

Sol. Here $\qquad u = \dfrac{x}{y-z}, v = \dfrac{y}{z-x}, w = \dfrac{z}{x-y}$

$$\Rightarrow \quad \log u = \log x - \log(y-z) \qquad \text{...(1)}$$

$$\log v = \log y - \log(z-x) \qquad \text{...(2)}$$

$$\log w = \log z - \log(x-y) \qquad \text{...(3)}$$

Differentiating (1) partially w.r.t. x

$$\frac{1}{u}\cdot\frac{\partial u}{\partial x} = \frac{1}{x} \quad \Rightarrow \quad \frac{\partial u}{\partial x} = \frac{u}{x}$$

Differentiating (1) partially w.r.t. y

$$\frac{1}{u}\cdot\frac{\partial u}{\partial y} = -\frac{1}{y-z} \quad \Rightarrow \quad \frac{\partial u}{\partial y} = \frac{-u}{y-z}$$

Differentiating (1) partially w.r.t. z

$$\frac{1}{u}\cdot\frac{\partial u}{\partial z} = -\frac{1}{y-z}(-1) \quad \Rightarrow \quad \frac{\partial u}{\partial z} = \frac{u}{y-z}$$

Similarly, from (2) and (3), we have

$$\frac{\partial v}{\partial x} = \frac{v}{z-x}, \quad \frac{\partial v}{\partial y} = \frac{v}{y}, \quad \frac{\partial v}{\partial z} = \frac{-v}{z-x}$$

$$\frac{\partial w}{\partial x} = \frac{-w}{x-y}, \quad \frac{\partial w}{\partial y} = \frac{w}{x-y}, \quad \frac{\partial w}{\partial z} = \frac{w}{z}$$

$$\therefore \quad \frac{\partial(u, v, w)}{\partial(x, y, z)} = \begin{vmatrix} \dfrac{\partial u}{\partial x} & \dfrac{\partial u}{\partial y} & \dfrac{\partial u}{\partial z} \\ \dfrac{\partial v}{\partial x} & \dfrac{\partial v}{\partial y} & \dfrac{\partial v}{\partial z} \\ \dfrac{\partial w}{\partial x} & \dfrac{\partial w}{\partial y} & \dfrac{\partial w}{\partial z} \end{vmatrix}$$

$$= \begin{vmatrix} \dfrac{u}{x} & \dfrac{-u}{y-z} & \dfrac{u}{y-z} \\ \dfrac{v}{z-x} & \dfrac{v}{y} & \dfrac{-v}{z-x} \\ \dfrac{-w}{x-y} & \dfrac{w}{x-y} & \dfrac{w}{z} \end{vmatrix}$$

Taking out u, v, w from R_1, R_2, R_3 respectively

$$= uvw \begin{vmatrix} \frac{1}{x} & \frac{-1}{y-z} & \frac{1}{y-z} \\ \frac{1}{z-x} & \frac{1}{y} & \frac{-1}{z-x} \\ \frac{-1}{x-y} & \frac{1}{x-y} & \frac{1}{z} \end{vmatrix}$$ **Multiplying R_1, R_2, R_3 by $y - z, z - x, x - y$, respectively**

$$= \frac{uvw}{(y-z)(z-x)(x-y)} \begin{vmatrix} \frac{y-z}{x} & -1 & 1 \\ 1 & \frac{z-x}{y} & -1 \\ -1 & 1 & \frac{x-y}{z} \end{vmatrix}$$ **Multiplying C_1, C_2, C_3 by x, y, z, respectively**

$$= \frac{uvw}{xyz(y-z)(z-x)(x-y)} \begin{vmatrix} y-z & -y & z \\ x & z-x & -z \\ -x & y & x-y \end{vmatrix}$$ **Operating $C_1 \to C_1 + C_2 + C_3$**

$$= \frac{1}{(y-z)^2 (z-x)^2 (x-y)^2} \begin{vmatrix} 0 & -y & z \\ 0 & z-x & -z \\ 0 & y & x-y \end{vmatrix} = 0$$

$$\left[\because \frac{u}{x} = \frac{1}{y-z}, \frac{v}{y} = \frac{1}{z-x}, \frac{w}{z} = \frac{1}{x-y} \right]$$

Example 4. *If* $u = \frac{xy}{z}, v = \frac{yz}{x}, w = \frac{zx}{y}$ *find,* $\frac{\partial (u, v, w)}{\partial (x, y, z)}$.

Sol.
$$\frac{\partial (u, v, w)}{\partial (x, y, z)} = \begin{vmatrix} \frac{\partial u}{\partial x} & \frac{\partial u}{\partial y} & \frac{\partial u}{\partial z} \\ \frac{\partial v}{\partial x} & \frac{\partial v}{\partial y} & \frac{\partial v}{\partial z} \\ \frac{\partial w}{\partial x} & \frac{\partial w}{\partial y} & \frac{\partial w}{\partial z} \end{vmatrix} = \begin{vmatrix} \frac{y}{z} & \frac{x}{z} & \frac{-xy}{z^2} \\ \frac{-yz}{x^2} & \frac{z}{x} & \frac{y}{x} \\ \frac{z}{y} & \frac{-zx}{y^2} & \frac{x}{y} \end{vmatrix}$$

$$= \frac{y}{z}\left(\frac{z}{y} + \frac{z}{y}\right) - \frac{x}{z}\left(\frac{-z}{x} + \frac{-z}{x}\right) - \frac{xy}{z^2}\left(\frac{z^2}{xy} - \frac{z^2}{xy}\right)$$

$$= 2 + 2 = \mathbf{4}.$$

EXERCISE 2.4

1. If $u = x^2 - 2y$, $v = x + y$, prove that $\frac{\partial (u, v)}{\partial (x, y)} = 2x + 2$.
2. If $u = x(1 - y)$, $v = xy$; prove that $JJ' = 1$.
3. (*a*) In polar coordinates, $x = r \cos \theta$, $y = r \sin \theta$, show that $\frac{\partial (x, y)}{\partial (r, \theta)} = r$.

(*b*) If $x = r\cos\theta$, $y = r\sin\theta$, verify that $\dfrac{\partial(x, y)}{\partial(r, \theta)} \cdot \dfrac{\partial(r, \theta)}{\partial(x, y)} = 1$.

$\left[\textbf{Hint. } r = \sqrt{x^2 + y^2},\ \theta = \tan^{-1}\dfrac{y}{x}.\right]$

4. If $x = r\cos\theta$, $y = r\sin\theta$, $z = z$, evaluate $\dfrac{\partial(x, y, z)}{\partial(r, \theta, z)}$.

Note. Here (x, y, z) and (r, θ, z) are respectively the Cartesian and cylindrical coordinates of a point.

5. (*a*) If $y_1 = \dfrac{x_2x_3}{x_1}$, $y_2 = \dfrac{x_3x_1}{x_2}$, $y_3 = \dfrac{x_1x_2}{x_3}$, show that $\dfrac{\partial(y_1, y_2, y_3)}{\partial(x_1, x_2, x_3)} = 4$.

(*b*) If $u = \dfrac{yz}{x}$, $v = \dfrac{zx}{y}$, $w = \dfrac{xy}{z}$, show that $\dfrac{\partial(x, y, z)}{\partial(u, v, w)} = \dfrac{1}{4}$.

6. If $u = \dfrac{y^2}{2x}$, $v = \dfrac{x^2 + y^2}{2x}$, find $\dfrac{\partial(u, v)}{\partial(x, y)}$.

7. If $x = a\cosh\xi\cos\eta$, $y = a\sinh\xi\sin\eta$, show that $\dfrac{\partial(x, y)}{\partial(\xi, \eta)} = \dfrac{a^2}{2}(\cosh 2\xi - \cos 2\eta)$.

$\left[\textbf{Hint. } \cosh^2 x = \dfrac{\cosh 2x + 1}{2},\ \sinh^2 x = \dfrac{\cosh 2x - 1}{2}\right]$

8. If $F = xu + v - y$, $G = u^2 + vy + w$, $H = zu - v + vw$, compute $\dfrac{\partial(F, G, H)}{\partial(u, w, v)}$.

9. If $u = \dfrac{x + y}{1 - xy}$ and $v = \tan^{-1}x + \tan^{-1}y$, find $\dfrac{\partial(u, v)}{\partial(x, y)}$.

10. If $u = x^2 - 2y$, $v = x + y + z$, $w = x - 2y + 3z$, find $\dfrac{\partial(u, v, w)}{\partial(x, y, z)}$.

11. If $u = x + y + z$, $uv = y + z$, $uvw = z$, show that $\dfrac{\partial(x, y, z)}{\partial(u, v, w)} = u^2v$.

12. If $u = xyz$, $v = xy + yz + zx$, $w = x + y + z$, show that $\dfrac{\partial(u, v, w)}{\partial(x, y, z)} = (x - y)(y - z)(z - x)$.

13. If $y_1 = x_1^3$, $y_2 = e^{x_2}$ and $y_3 = x_1 + \sin x_3$, find $\dfrac{\partial(y_1, y_2, y_3)}{\partial(x_1, x_2, x_3)}$.

14. If $u = 2xy$, $v = x^2 - y^2$, $x = r\cos\theta$, $y = r\sin\theta$, evaluate $\dfrac{\partial(u, v)}{\partial(r, \theta)}$. (*K.U. May 2012*)

$\left[\textbf{Hint. } \dfrac{\partial(u, v)}{\partial(r, \theta)} = \dfrac{\partial(u, v)}{\partial(x, y)} \times \dfrac{\partial(x, y)}{\partial(r, \theta)}\right]$

15. If $u = \sin^{-1}x + \sin^{-1}y$, $v = x\sqrt{1 - y^2} + y\sqrt{1 - x^2}$, find $\dfrac{\partial(u, v)}{\partial(x, y)}$.

16. If $x = uv$ and $y = \dfrac{u + v}{u - v}$, find $\dfrac{\partial(u, v)}{\partial(x, y)}$.

Answers

4. r **6.** $-\dfrac{y}{2x}$ **8.** $xw - x - xyv + 2uv - z$

9. 0 **10.** $10x + 4$ **13.** $3x_1^2\, e^{x_2} \cos x_3$ **14.** $-4r^3$

15. 0 **16.** $\dfrac{(u-v)^2}{4uv}$.

2.12. MAXIMA AND MINIMA OF FUNCTIONS OF TWO VARIABLES

A function $f(x, y)$ is said to have a **maximum value at x = a, y = b if f(a, b) > f(a + h, b + k),** for small and independent values of h and k, positive or negative.

A function $f(x, y)$ is said to have a **minimum value at x = a, y = b** if $f(a, b) < f(a + h, b + k)$, for small and independent values of h and k, positive or negative.

Thus $f(x, y)$ has a maximum or minimum value at a point (a, b) according as

$$\Delta f = f(a + h, b + k) - f(a, b) < \text{or} > 0.$$

Geometrically, the surface $z = f(x, y)$ has a maximum at the point (a, b) if $f(a, b)$ is the greatest in the small neighbourhood of the point (a, b). As such, the surface descends in all directions at this point. The point of maxima may well be compared with the highest point of a dome. The surface has a minimum at the point (a, b) if $f(a, b)$ is the smallest in the small neighbourhood of the point (a, b). As such, the surface ascends in all directions at this point. The point of minima may well be compared with the lowest point of a bowl. However, sometimes, at the point (a, b), the tangent plane is horizontal and the surface descends in certain directions and ascends in other directions. Such a point is called a *saddle point*.

A maximum or a minimum value of a function is called its *extreme value*.

2.13. CONDITIONS FOR f(x, y) TO BE MAXIMUM OR MINIMUM

By Taylor's theorem, we have

$$\Delta f = f(a + h, b + k) - f(a, b)$$

$$= [hf_x(a, b) + kf_y(a, b)] + \frac{1}{2!}[h^2 f_{xx}(a, b) + 2hkf_{xy}(a, b) + k^2 f_{yy}(a, b)] + \cdots \quad ...(1)$$

For small values of h and k, the second and higher order terms are still smaller and may be neglected.

Thus, sign of Δf = sign of $[hf_x(a, b) + kf_y(a, b)]$

Taking $h = 0$, the sign of Δf changes with the sign of k. Similarly taking $k = 0$, the sign of Δf changes with the sign of h. Since Δf changes sign with h and k, $f(x, y)$ cannot have a maximum or a minimum value at (a, b) unless $f_x(a, b) = 0 = f_y(a, b)$.

Hence the necessary conditions for (x, y) to have a maximum or a minimum value at (a, b) are

$$f_x(a, b) = 0, f_y(a, b) = 0$$

If these conditions are satisfied, then for small values of h and k, we have, from (1),

$$\Delta f = \frac{1}{2!}[h^2 f_{xx}(a, b) + 2hkf_{xy}(a, b) + k^2 f_{yy}(a, b)] + \cdots$$

$$= \frac{1}{2!}[h^2 r + 2hks + k^2 t] + \cdots \text{ where } r = f_{xx}(a, b),\ s = f_{xy}(a, b),\ t = f_{yy}(a, b)$$

or

$$\Delta f = \frac{1}{2r}[h^2 r^2 + 2hkrs + k^2 rt] + \cdots = \frac{1}{2r}[(h^2 r^2 + 2hkrs + k^2 s^2) + k^2 rt - k^2 s^2)] + \cdots$$

$$= \frac{1}{2r}\,[(hr + ks)^2 + k^2\,(rt - s^2)] + \cdots \qquad ...(2)$$

Now $(hr + ks)^2$ is always positive and $k^2(rt - s^2)$ will be positive if $rt - s^2 > 0$. In this case, Δf will have the same sign as that of r for all values of h and k.

Hence if $rt - s^2 > 0$, then $f(x, y)$ has a maximum or a minimum at (a, b) according as $r < 0$ or $r > 0$.

If $rt - s^2 < 0$, then Δf changes sign with h and k. Hence there is neither a maximum nor a minimum value at (a, b). The point (a, b) is a saddle point in this case.

If $rt - s^2 = 0$, no conclusion can be drawn about a maximum or a minimum value at (a, b) and hence further investigation is required.

Note. The point (a, b) is called a *stationary point* if $f_x(a, b) = 0, f_y(a, b) = 0$. The value $f(a, b)$ is called a *stationary value*. Thus every extreme value is a stationary value but the converse may not be true.

2.14. RULE TO FIND THE EXTREME VALUES OF A FUNCTION z = f(x, y)

(*i*) Find $\frac{\partial z}{\partial x}$ and $\frac{\partial z}{\partial y}$.

(*ii*) Solve $\frac{\partial z}{\partial x} = 0$ and $\frac{\partial z}{\partial y} = 0$ simultaneously.

Let (a, b); (c, d) ... be the solutions of these equations.

(*iii*) For each solution in step (*ii*), find $r = \frac{\partial^2 f}{\partial x^2}$, $s = \frac{\partial^2 f}{\partial x \partial y}$, $t = \frac{\partial^2 f}{\partial y^2}$.

(*iv*) (*a*) If $rt - s^2 > 0$ and $r < 0$ for a particular solution (a, b) of step (*ii*), then z has a maximum value at (a, b).

(*b*) If $rt - s^2 > 0$ and $r > 0$ for a particular solution (a, b) of step (*ii*), then z has a minimum value at (a, b).

(*c*) If $rt - s^2 < 0$ for a particular solution (a, b) of step (*ii*), then z has no extreme value at (a, b).

(*d*) If $rt - s^2 = 0$, the case is doubtful and requires further investigation.

ILLUSTRATIVE EXAMPLES

Example 1. *Examine the function* $x^3 + y^3 - 3axy$ *for maxima and minima.*

Sol. Here $f(x, y) = x^3 + y^3 - 3axy$

$$f_x = 3x^2 - 3ay, \quad f_y = 3y^2 - 3ax, \quad r = f_{xx} = 6x, \quad s = f_{xy} = -3a, \quad t = f_{yy} = 6y$$

Now $\quad f_x = 0$ and $f_y = 0$

$\Rightarrow \quad x^2 - ay = 0 \qquad ...(1) \qquad y^2 - ax = 0 \qquad ...(2)$

From (1), $\quad y = \frac{x^2}{a}$

$\therefore$ From (2), $\quad \frac{x^4}{a^2} - ax = 0$ or $x(x^3 - a^3) = 0$ or $x = 0, a$

when $x = 0$, $y = 0$; when $x = a$, $y = a$

$\therefore$ There are two stationary points $(0, 0)$ and (a, a).

Now $\quad rt - s^2 = 36xy - 9a^2$

At (0, 0) $rt - s^2 = -9a^2 < 0$

$\Rightarrow$ There is no extreme value at (0, 0).

At (a, a) $rt - s^2 = 36a^2 - 9a^2 = 27a^2 > 0$

$\Rightarrow$ $f(x, y)$ has extreme value at (a, a)

Now $r = 6a$

If $a > 0$, $r > 0$ so that $f(x, y)$ has a minimum value at (a, a).

Minimum value $= f(a, a) = a^3 + a^3 - 3a^3 = -a^3$

If $a < 0$, $r < 0$ so that $f(x, y)$ has a maximum value at (a, a).

Maximum value $= f(a, a) = a^3 + a^3 - 3a^3 = -a^3$.

Example 2. *Discuss the maxima and minima of* $x^3y^2(1 - x - y)$.

Sol. Here $f(x, y) = x^3y^2(1 - x - y) = x^3y^2 - x^4y^2 - x^3y^3$

$$f_x = 3x^2y^2 - 4x^3y^2 - 3x^2y^3, f_y = 2x^3y - 2x^4y - 3x^3y^2$$

$$r = f_{xx} = 6xy^2 - 12x^2y^2 - 6xy^3 \;;\; s = f_{xy} = 6x^2y - 8x^3y - 9x^2y^2 \;;\; t = f_{yy} = 2x^3 - 2x^4 - 6x^3y$$

Now $f_x = 0$ and $f_y = 0$

$\Rightarrow$ $x^2y^2(3 - 4x - 3y) = 0$ ...(1) $x^3y(2 - 2x - 3y) = 0$ ...(2)

Solving (1) and (2), the stationary points are (0, 0) and $(\frac{1}{2}, \frac{1}{3})$.

Now
$$rt - s^2 = 6xy^2(1 - 2x - y)\, 2x^3(1 - x - 3y) - [x^2y(6 - 8x - 9y)]^2$$
$$= 12x^4y^2(1 - 2x - y)(1 - x - 3y) - x^4y^2(6 - 8x - 9y)^2$$

At (0, 0), $rt - s^2 = 0$

$\therefore$ At (0, 0) further investigation is needed.

Consider $f(a + h, b + k) - f(a, b) = f(h, k) - f(0, 0) = h^3k^2(1 - h - k)$

$= h^3k^2$ (neglecting $h^4k^2 + h^3k^3$ which is very small as compared to h^3k^2)

which is > 0 if $h > 0$ and < 0 if $h < 0$

$\therefore$ $f(a + h, b + k) - f(a, b)$ does not keep same sign for all small values of h and k (positive or negative).

$\therefore$ There is no extreme value at (0, 0).

At $\left(\frac{1}{2}, \frac{1}{3}\right)$, $rt - s^2 = 12 \cdot \frac{1}{16} \cdot \frac{1}{9}\left(-\frac{1}{3}\right)\left(-\frac{1}{2}\right) - \frac{1}{16} \cdot \frac{1}{9}(1) = \frac{1}{144}(2 - 1) = \frac{1}{14} > 0.$

Also
$$r = 6 \cdot \frac{1}{2} \cdot \frac{1}{9} - 12 \cdot \frac{1}{4} \cdot \frac{1}{9} - 6 \cdot \frac{1}{2} \cdot \frac{1}{8} = \frac{1}{3} - \frac{1}{3} - \frac{3}{8} = -\frac{3}{8} < 0.$$

$\therefore$ $f(x, y)$ has a maximum value at $(\frac{1}{2}, \frac{1}{3})$

Maximum value $= f\left(\frac{1}{2}, \frac{1}{3}\right) = \frac{1}{8} \cdot \frac{1}{9}\left(1 - \frac{1}{2} - \frac{1}{3}\right) = \frac{1}{432}$.

Example 3. *Locate the stationary points of* $x^4 + y^4 - 2x^2 + 4xy - 2y^2$ *and determine their nature.*

Sol. Here $f(x, y) = x^4 + y^4 - 2x^2 + 4xy - 2y^2$

$$f_x = 4x^3 - 4x + 4y, \quad f_y = 4y^3 + 4x - 4y$$
$$r = f_{xx} = 12x^2 - 4, \quad s = f_{xy} = 4, t = f_{yy} = 12y^2 - 4$$

Now $f_x = 0$ and $f_y = 0$

$\Rightarrow$ $x^3 - x + y = 0$ ...(1) $y^3 + x - y = 0$ ...(2)

Adding (1) and (2), $x^3 + y^3 = 0 \Rightarrow y = -x.$

Putting $y = -x$ in (1), we get $x^3 - 2x = 0$ or $x(x^2 - 2) = 0$ $\therefore$ $x = 0, \pm\sqrt{2}$.

Since $y = -x$, the stationary points are $(0, 0)$, $(\sqrt{2}, -\sqrt{2})$ and $(-\sqrt{2}, \sqrt{2})$.

Now $\quad rt - s^2 = (12x^2 - 4)(12y^2 - 4) - 16$

At (0, 0), $\quad rt - s^2 = 16 - 16 = 0.$

$\therefore$ At (0, 0), further investigation is needed.

Now $\quad f(0, 0) = 0$

$$f(x, y) = x^4 + y^4 - 2(x^2 - 2xy + y^2) = x^4 + y^4 - 2(x - y)^2$$

When h and k are small, $f(h, k) = 2h^4 > 0$ for $h = k$

and $\quad f(h, k) = -2(h - k)^2 < 0$ for $h \neq k$

(neglecting $h^4 + k^4$ which is very small)

$\Rightarrow \quad f(0, 0) < f(h, k)$ for $h = k$

$\quad f(0, 0) > f(h, k)$ for $h \neq k$

$\Rightarrow$ There is no extreme value at (0, 0).

At $(\sqrt{2}, -\sqrt{2})$, $\quad rt - s^2 = (12 \times 2 - 4)^2 - 16 = 384 > 0$

$$r = 12 \times 2 - 4 = 20 > 0$$

$\therefore$ $f(x, y)$ has a minimum value at $(\sqrt{2}, -\sqrt{2})$.

$$\text{Minimum Value} = (\sqrt{2})^4 + (-\sqrt{2})^4 - 2(\sqrt{2})^2 + 4\sqrt{2}(-\sqrt{2}) - 2(-\sqrt{2})^2$$
$$= 4 + 4 - 4 - 8 - 4 = -8.$$

Similarly, $f(x, y)$ has a minimum value -8 at $(-\sqrt{2}, \sqrt{2})$.

Example 4. *Examine for minimum and maximum values: sin x + sin y + sin (x + y).*

Sol. Here $\quad f(x, y) = \sin x + \sin y + \sin (x + y)$

$$f_x = \cos x + \cos (x + y),\ f_y = \cos y + \cos (x + y)$$
$$r = f_{xx} = -\sin x - \sin (x + y),\ s = f_{xy} = -\sin (x + y),$$
$$t = f_{yy} = -\sin y - \sin (x + y)$$

Now $\quad f_x = 0$ and $f_y = 0$

$\Rightarrow \quad \cos x + \cos (x + y) = 0 \quad$ ...(1) $\quad \cos y + \cos (x + y) = 0 \quad$ ...(2)

Subtracting (2) from (1), $\cos x - \cos y = 0$ or $\cos x = \cos y$ $\therefore$ $x = y$

From (1), $\quad \cos x + \cos 2x = 0$ or $\cos 2x = -\cos x = \cos (\pi - x)$ or $2x = \pi - x$ $\therefore$ $x = \dfrac{\pi}{3}$

$\therefore \quad x = y = \dfrac{\pi}{3}$ is a stationary point.

At $\left(\dfrac{\pi}{3}, \dfrac{\pi}{3}\right)$, $\quad r = -\dfrac{\sqrt{3}}{2} - \dfrac{\sqrt{3}}{2} = -\sqrt{3} < 0,\ s = \dfrac{\sqrt{3}}{2},\ t = -\dfrac{\sqrt{3}}{2} - \dfrac{\sqrt{3}}{2} = -\sqrt{3}$

or $\quad rt - s^2 = 3 - \dfrac{3}{4} = \dfrac{9}{4} > 0$

Also $r < 0$

$\therefore$ $f(x, y)$ has a maximum value at $\left(\dfrac{\pi}{3}, \dfrac{\pi}{3}\right)$.

$$\text{Maximum value} = f\left(\frac{\pi}{3}, \frac{\pi}{3}\right) = \sin \frac{\pi}{3} + \sin \frac{\pi}{3} + \sin \frac{2\pi}{3} = \frac{\sqrt{3}}{2} + \frac{\sqrt{3}}{2} + \frac{\sqrt{3}}{2} = \frac{3\sqrt{3}}{2}.$$

Example 5. *A rectangular box, open at the top, is to have a given capacity. Find the dimensions of the box requiring least material for its construction.*

Sol. Let x, y and z be the length, breadth and height respectively. Let V be the given capacity and S, the surface.

(V is given $\Rightarrow$ V is constant)

$$V = xyz \quad \text{or} \quad z = \frac{V}{xy} \qquad ...(1)$$

$$S = xy + 2xz + 2yz = xy + \frac{2V}{y} + \frac{2V}{x} = f(x, y) \qquad \text{[Using (1)]}$$

$$f_x = y - \frac{2V}{x^2}, \; f_y = x - \frac{2V}{y^2}$$

$$r = f_{xx} = \frac{4V}{x^3}, \; s = f_{xy} = 1, \; t = f_{yy} = \frac{4V}{y^3}$$

Now $\quad f_x = 0 \quad \text{and} \quad f_y = 0$

$$\Rightarrow \quad y - \frac{2V}{x^2} = 0 \quad ...(1) \qquad x - \frac{2V}{y^2} = 0 \quad ...(2)$$

From (1), $\quad y = \frac{2V}{x^2}$

$\therefore$ From (2), $x - 2V \cdot \frac{x^4}{4V^2} = 0$ or $x\left(1 - \frac{x^3}{2V}\right) = 0$ or $x = (2V)^{1/3}$ $\quad | \because \; x \neq 0$

and $\quad y = \frac{2V}{x^2} = \frac{2V}{(2V)^{3/2}} = (2V)^{1/3}$

$\therefore \; x = y = (2V)^{1/3}$ is a stationary point.

At this point, $\quad r = \frac{4V}{2V} = 2 > 0, \; s = 1, \; t = \frac{4V}{2V} = 2$

so that $\quad rt - s^2 = 4 - 1 = 3 > 0$ and $r > 0$

$\Rightarrow$ S is minimum when $x = y = (2V)^{1/3}$

Also $\quad z = \frac{V}{xy} = \frac{V}{(2V)^{2/3}} = \frac{V^{1/3}}{2^{2/3}} = \frac{(2V)^{1/3}}{2} = \frac{y}{2}$

Hence S is minimum when $x = y = 2z = (2V)^{1/3}$.

Example 6. *Prove that if the perimeter of a triangle is constant, its area is maximum when the triangle is equilateral.*

Sol. Let a, b, c be the sides of a triangle whose perimeter $2s$ is constant.

Then $\quad 2s = a + b + c$ or $c = 2s - a - b \qquad ...(1)$

$$\Delta = \sqrt{s(s-a)(s-b)(s-c)} = \sqrt{s(s-a)(s-b)(a+b-s)} \qquad | \text{ Using (1)}$$

Let $\quad z = \Delta^2 = s(s-a)(s-b)(a+b-s) = f(a, b)$

$$f_a = s(s-b)\frac{\partial}{\partial a}[(s-a)(a+b-s)]$$

$$= s(s-b)[-(a+b-s) + (s-a)] = s(s-b)(2s-2a-b)$$

$$f_b = s(s-a)\frac{\partial}{\partial b}[(s-b)(a+b-s)]$$

$$= s(s - a)[- (a + b - s) + (s - b)] = s(s - a)(2s - a - 2b)$$

$$r = f_{aa} = - 2s(s - b),\ s = f_{ab} = s\ [- (2s - a - 2b) - (s - a)] = s(2a + 2b - 3s)$$

$$t = f_{bb} = - 2s(s - a)$$

Now $\qquad f_a = 0 \quad \text{and} \quad f_b = 0$

$\Rightarrow \quad s(s - b)(2s - 2a - b) = 0$ and $s(s - a)(2s - a - 2b) = 0$

$\Rightarrow \quad (s - b)(2s - 2a - b) = 0 \qquad ...(1)$

$\qquad (s - a)(2s - a - 2b) = 0 \qquad ...(2) \qquad | \because \ s \neq 0$

From (1), $\qquad s = b \quad \text{or} \quad 2s = 2a + b$

When $s = b$, from (2), $(b - a)(- a) = 0 \quad \text{or} \quad b = a \qquad | \because \ a \neq 0$

when $\ 2s = 2a + b$, from (2), $\dfrac{b}{2}(a - b) = 0 \quad \text{or} \quad a = b \qquad | \because \ b \neq 0$

If we express z as a function of b and c, we similarly get $b = c$

$$\therefore \qquad a = b = c = \frac{2s}{3}$$

$$r = - 2s\left(\frac{s}{3}\right) = - \frac{2s^2}{3} < 0$$

$$s = s\left(\frac{4s}{3} + \frac{4s}{3} - 3s\right) = s\left(-\frac{s}{3}\right) = -\frac{s^2}{3}$$

$$t = - 2s\left(\frac{s}{3}\right) = - \frac{2s^2}{3}$$

$$rt - s^2 = \frac{4s^4}{9} - \frac{s^4}{9} = \frac{s^4}{3} > 0. \text{ Also } r < 0$$

$\therefore \quad z$ and hence Δ is maximum when $a = b = c = \dfrac{2s}{3}$ *i.e.* when the triangle is equilateral.

Example 7. *Show that the rectangular solid of maximum volume that can be inscribed in a given sphere is a cube.*

Sol. Let x, y, z be the length, breadth and height of the rectangular solid. If V is the volume of the solid, then $V = xyz \qquad ...(1)$

Since each diagonal of the solid passes through the centre of the sphere

$\therefore \qquad$ each diagonal = diameter of sphere = d (say)

$$\Rightarrow \qquad \sqrt{x^2 + y^2 + z^2} = d \text{ or } x^2 + y^2 + z^2 = d^2 \quad \text{or} \quad z = \sqrt{d^2 - x^2 - y^2}$$

$\therefore \quad$ From (1), $\qquad V = xy\sqrt{d^2 - x^2 - y^2}$

or $\qquad V^2 = x^2y^2(d^2 - x^2 - y^2) = 8^2d^2x^2y^2 - 8^2x^4y^2 - 8^2x^2y^4 = f(x, y)$

$$f_x = 2d^2xy^2 - 4x^3y^2 - 2xy^4 = 2xy^2(d^2 - 2x^2 - y^2)$$

$$f_y = 2d^2x^2y - 2x^4y - 4x^2y^3 = 2x^2y(d^2 - x^2 - 2y^2)$$

$$r = f_{xx} = 2d^2y^2 - 12x^2y^2 - 2y^4,\ s = f_{xy} = 4d^2xy - 8x^3y - 8xy^3$$

$$t = f_{yy} = 2d^2x^2 - 2x^4 - 12x^2y^2$$

Now $\qquad f_x = 0$ and $f_y = 0$

$$\Rightarrow \quad d^2 - 2x^2 - y^2 = 0 \quad ...(1) \quad [\because \; x \neq 0, y \neq 0]$$

$$d^2 - x^2 - 2y^2 = 0 \quad ...(2)$$

Subtracting (2) from (1), $\quad -x^2 + y^2 = 0 \Rightarrow y = x$

$\therefore$ From (1), $\quad x = y = \dfrac{d}{\sqrt{3}}$

$$z = \sqrt{d^2 - \frac{d^2}{3} - \frac{d^2}{3}} = \frac{d}{\sqrt{3}}$$

At $x = y = \dfrac{d}{\sqrt{3}}$,

$$r = 2d^2 \cdot \frac{d^2}{3} - 12 \cdot \frac{d^2}{3} \cdot \frac{d^2}{3} - 2 \cdot \frac{d^4}{9} = -\frac{8d^4}{9} < 0$$

$$s = 4d^2 \cdot \frac{d^2}{3} - 8 \cdot \frac{d^3}{3\sqrt{3}} \cdot \frac{d}{\sqrt{3}} - 8 \cdot \frac{d}{\sqrt{3}} \cdot \frac{d^3}{3\sqrt{3}} = -\frac{4d^4}{9}$$

$$t = 2d^2 \cdot \frac{d^2}{3} - 2 \cdot \frac{d^4}{9} - 12 \cdot \frac{d^2}{3} \cdot \frac{d^2}{3} = -\frac{8d^4}{9}$$

$$rt - s^2 = \frac{64d^8}{81} - \frac{16d^8}{81} = \frac{16d^8}{27} > 0, \text{ Also } r < 0.$$

$\therefore$ V^2 and hence, V is maximum when $x = y = z$ *i.e.*, when the rectangular solid is a cube.

Example 8. *Find the volume of the largest rectangular parallelopiped that can be inscribed in the ellipsoid,* $\dfrac{x^2}{a^2} + \dfrac{y^2}{b^2} + \dfrac{z^2}{c^2} = 1$.

Sol. Let (x, y, z) be a vertex of the parallelopiped, then it lies on the ellipsoid

$$\frac{x^2}{a^2} + \frac{y^2}{b^2} + \frac{z^2}{c^2} = 1 \quad ...(1)$$

Also its dimensions are $2x$, $2y$, $2z$ so that the volume V is given by $V = 2x \cdot 2y \cdot 2z = 8xyz$

$$\Rightarrow \quad V^2 = 64x^2y^2z^2 = 64x^2y^2c^2\left(1 - \frac{x^2}{a^2} - \frac{y^2}{b^2}\right) = 64c^2\left(x^2y^2 - \frac{x^4y^2}{a^2} - \frac{x^2y^4}{b^2}\right) = f(x, y)$$

$$f_x = 64c^2\left(2xy^2 - \frac{4x^3y^2}{a^2} - \frac{2xy^4}{b^2}\right); \quad f_y = 64c^2\left(2x^2y - \frac{2x^4y}{a^2} - \frac{4x^2y^3}{b^2}\right)$$

$$r = f_{xx} = 64c^2\left(2y^2 - \frac{12x^2y^2}{a^2} - \frac{2y^4}{b^2}\right); \quad s = f_{xy} = 64c^2\left(4xy - \frac{8x^3y}{a^2} - \frac{8xy^3}{b^2}\right)$$

$$t = f_{yy} = 64c^2\left(2x^2 - \frac{2x^4}{a^2} - \frac{12x^2y^2}{b^2}\right)$$

Now $\quad f_x = 0$ and $f_y = 0$

$$\Rightarrow \quad 128c^2xy^2\left(1 - \frac{2x^2}{a^2} - \frac{y^2}{b^2}\right) = 0 \quad \text{and} \quad 128c^2x^2y\left(1 - \frac{x^2}{a^2} - \frac{2y^2}{b^2}\right) = 0$$

$$\Rightarrow \quad 1 - \frac{2x^2}{a^2} - \frac{y^2}{b^2} = 0 \quad ...(1) \quad [\because \; x \neq 0, y \neq 0]$$

$$1 - \frac{x^2}{a^2} - \frac{2y^2}{b^2} = 0 \qquad \text{...(2)}$$

Subtracting (2) from (1), $-\frac{x^2}{a^2} + \frac{y^2}{b^2} = 0$ or $y = \frac{bx}{a}$

$\therefore$ From (1), $1 - \frac{2x^2}{a^2} - \frac{x^2}{a^2} = 0$ or $x^2 = \frac{a^2}{3}$ or $x = \frac{a}{\sqrt{3}}$

$\therefore$ $$y = \frac{b}{a} \cdot \frac{a}{\sqrt{3}} = \frac{b}{\sqrt{3}}$$

and $$z^2 = c^2\left(1 - \frac{x^2}{a^2} - \frac{y^2}{b^2}\right) = c^2\left(1 - \frac{1}{3} - \frac{1}{3}\right) = \frac{c^2}{3}$$

$\therefore$ $$z = \frac{c}{\sqrt{3}}$$

Thus $x = \frac{a}{\sqrt{3}}, y = \frac{b}{\sqrt{3}}$ is a stationary point.

At this point, $r = 64c^2\left[\frac{2b^2}{3} - \frac{12}{a^2} \cdot \frac{a^2}{3} \cdot \frac{b^2}{3} - \frac{2}{b^2} \cdot \frac{b^4}{9}\right] = -\frac{512}{9} b^2c^2 < 0$

$$s = 64c^2\left[4 \cdot \frac{a}{\sqrt{3}} \cdot \frac{b}{\sqrt{3}} - \frac{8}{a^2} \cdot \frac{a^3}{3\sqrt{3}} \cdot \frac{b}{\sqrt{3}} - \frac{8}{b^2} \cdot \frac{a}{\sqrt{3}} \cdot \frac{b^3}{3\sqrt{3}}\right] = -\frac{256}{9} abc^2$$

$$t = 64c^2\left[2 \cdot \frac{a^2}{3} - \frac{2}{a^2} \cdot \frac{a^4}{9} - \frac{12}{b^2} \cdot \frac{a^2}{3} \cdot \frac{b^2}{3}\right] = -\frac{512}{9} a^2c^2$$

$$rt - s^2 = \left(\frac{512}{9}\right)^2 a^2b^2c^4 - \left(\frac{256}{9}\right)^2 a^2b^2c^4 = \left(\frac{256}{9}\right)^2 a^2b^2c^4 (4 - 1) > 0.$$

Also $r < 0$

$\therefore$ V^2 and hence, V is maximum.

when $x = \frac{a}{\sqrt{3}}, y = \frac{b}{\sqrt{3}}, z = \frac{c}{\sqrt{3}}$

Maximum volume $= 8 \cdot \frac{a}{\sqrt{3}} \cdot \frac{b}{\sqrt{3}} \cdot \frac{c}{\sqrt{3}} = \frac{8abc}{3\sqrt{3}}$.

Example 9. *In a plane triangle ABC, find the maximum value of cos A cos B cos C.*

Sol. cos A cos B cos C = cos A cos B cos [π – (A + B)] | ∵ A + B + C = π

$= -\cos A \cos B \cos (A + B) = f(A, B)$

$$\frac{\partial f}{\partial A} = -\cos B[-\sin A \cos (A + B) - \cos A \sin (A + B)]$$
$$= \cos B \sin [A + (A + B)] = \cos B \sin (2A + B)$$

$$\frac{\partial f}{\partial B} = -\cos A[-\sin B \cos (A + B) - \cos B \sin (A + B)]$$
$$= \cos A \sin [B + (A + B)] = \cos A \sin (A + 2B)$$

$$r = 2 \cos B \cos (2A + B)$$
$$s = -\sin A \sin (A + 2B) + \cos A \cos (A + 2B)$$

$$= \cos [A + (A + 2B)] = \cos (2A + 2B)$$

$$t = 2 \cos A \cos (A + 2B)$$

Now $\quad \dfrac{\partial f}{\partial A} = 0$ and $\dfrac{\partial f}{\partial B} = 0$

$\Rightarrow \quad \cos B \sin (2A + B) = 0 \quad \text{...(1)}$

$\quad \cos A \sin (A + 2B) = 0 \quad \text{...(2)}$

If $\cos B = 0$, then $B = \dfrac{\pi}{2}$

From (2), $\cos A \sin (A + \pi) = 0$ or $\cos A(-\sin A) = 0$

$\Rightarrow$ either $\cos A = 0$ *i.e.* $A = \dfrac{\pi}{2}$ which is not possible $\quad (\because \; A + B + C = \pi \Rightarrow C = 0)$

or $\quad \sin A = 0$ *i.e.* $A = 0$ or π which is not possible.

$\therefore \quad \cos B \neq 0$. Similarly $\cos A \neq 0$

$\therefore$ From (1), $\quad \sin (2A + B) = 0 \quad$ or $\quad 2A + B = \pi$

From (2), $\quad \sin (A + 2B) = 0 \quad$ or $\quad A + 2B = \pi$

Solving these equations, $A = B = \dfrac{\pi}{3}$

At $A = B = \dfrac{\pi}{3}$

$$r = 2 \cos \frac{\pi}{3} \cos \pi = -1, \; s = \cos \frac{4\pi}{3} = \cos \left(\pi + \frac{\pi}{3}\right) = -\cos \pi = -\frac{1}{2}$$

$$t = 2 \cos \frac{\pi}{3} \cos \pi = -1$$

$$rt - s^2 = 1 - \frac{1}{4} = \frac{3}{4} > 0$$

Also $\quad r = -1 < 0$

$\Rightarrow \; f(A, B)$ is maximum at $A = B = \dfrac{\pi}{3}$

Maximum value $= f\left(\dfrac{\pi}{3}, \dfrac{\pi}{3}\right) = -\cos \dfrac{\pi}{3} \cos \dfrac{\pi}{3} \cos \dfrac{2\pi}{3} = -\left(\dfrac{1}{2}\right)\left(\dfrac{1}{2}\right)\left(-\dfrac{1}{2}\right) = \dfrac{1}{8}$.

2.15. LAGRANGE'S METHOD OF UNDETERMINED MULTIPLIERS

To find the maximum or minimum values of a function of three (or more) variables, when the variables are not independent but are connected by some given relation, we try to convert the given function to the one, having least number of independent variables with the help of the given relation. [See examples with 2.14]

When this procedure is not practicable, we use Lagrange's method.

Let $f(x, y, z)$ be a function of x, y, z which is to be examined for maximum or minimum value.

Let the variables x, y, z be connected by the relation $\phi(x, y, z) = 0 \quad \text{...(1)}$

For $f(x, y, z)$ to have a maximum or minimum value, the necessary condition is

$$\frac{\partial f}{\partial x} = 0, \; \frac{\partial f}{\partial y} = 0, \; \frac{\partial f}{\partial z} = 0$$

$$\therefore \qquad \frac{\partial f}{\partial x}\, dx + \frac{\partial f}{\partial y}\, dy + \frac{\partial f}{\partial z}\, dz = 0 \qquad ...(2)$$

Also, from (1), taking differentials, we get $\frac{\partial \phi}{\partial x}\, dx + \frac{\partial \phi}{\partial y}\, dy + \frac{\partial \phi}{\partial z}\, dz = 0$ ...(3)

Multiplying (3) by a parameter λ and adding to (2), we get

$$\left(\frac{\partial f}{\partial x} + \lambda\frac{\partial \phi}{\partial x}\right) dx + \left(\frac{\partial f}{\partial y} + \lambda\frac{\partial \phi}{\partial y}\right) dy + \left(\frac{\partial f}{\partial z} + \lambda\frac{\partial \phi}{\partial z}\right) dz = 0$$

This equation will hold good if $\frac{\partial f}{\partial x} + \lambda\frac{\partial \phi}{\partial s} = 0,\ \frac{\partial f}{\partial y} + \lambda\frac{\partial \phi}{\partial y} = 0,\ \frac{\partial f}{\partial z} + \lambda\frac{\partial \phi}{\partial z} = 0$

These equations together with equation (1), give the values of x, y, z and λ for a maximum or minimum.

Lagrange's method does not enable us to find whether there is a maximum or minimum. This fact is determined from the physical considerations of the problem.

Note. The above equations can be easily obtained by considering Lagrange's function

$$F(x, y, z) = f(x, y, z) + \lambda\phi(x, y, z)$$

and considering the stationary values of $F(x, y, z)$. For stationary values of $F(x, y, z)$, $dF = 0$

$$\Rightarrow \qquad \left(\frac{\partial f}{\partial x} + \lambda\frac{\partial \phi}{\partial x}\right) dx + \left(\frac{\partial f}{\partial y} + \lambda\frac{\partial \phi}{\partial y}\right) dy + \left(\frac{\partial f}{\partial z} + \lambda\frac{\partial \phi}{\partial z}\right) dz = 0$$

$$\Rightarrow \qquad \frac{\partial f}{\partial x} + \lambda\frac{\partial \phi}{\partial x} = 0,\ \frac{\partial f}{\partial y} + \lambda\frac{\partial \phi}{\partial y} = 0,\ \frac{\partial f}{\partial z} + \lambda\frac{\partial \phi}{\partial z} = 0.$$

Example 10. *Find the maximum and minimum distances of the point (3, 4, 12) from the sphere*

$$x^2 + y^2 + z^2 = 1.$$

Sol. Let (x, y, z) be any point on the sphere.

Distance of the point A(3, 4, 12) from (x, y, z) is given by $\sqrt{(x-3)^2 + (y-4)^2 + (z-12)^2}$

If the distance is maximum or minimum, so will be the square of the distance.

Let $\qquad f(x, y, z) = (x - 3)^2 + (y - 4)^2 + (z - 12)^2$ ...(1)

subject to the condition that $\phi(x, y, z) = x^2 + y^2 + z^2 - 1 = 0$ ...(2)

Consider Lagrange's function $F(x, y, z) = f(x, y, z) + \lambda\phi(x, y, z)$

$$= (x - 3)^2 + (y - 4)^2 + (z - 12)^2 + \lambda(x^2 + y^2 + z^2 - 1)$$

For stationary values, $dF = 0$.

$$\Rightarrow \qquad [2(x - 3) + 2\lambda x]dx + [2(y - 4) + 2\lambda y]dy + [2(z - 12) + 2\lambda z]dz = 0$$

$$\Rightarrow \qquad 2(x - 3) + 2\lambda x = 0 \qquad ...(3)$$

$$2(y - 4) + 2\lambda y = 0 \qquad ...(4)$$

$$2(z - 12) + 2\lambda z = 0 \qquad ...(5)$$

Multiplying (3) by x, (4) by y, (5) by z and adding, we get

$$2(x^2 + y^2 + z^2) - 6x - 8y - 24z + 2\lambda(x^2 + y^2 + z^2) = 0$$

or $\qquad 2 - 6x - 8y - 24z + 2\lambda = 0$ | Using (2)

or $\qquad 3x + 4y + 12z = 1 + \lambda$ ...(6)

From (3), (4) and (5), $\quad x = \frac{3}{1+\lambda},\ y = \frac{4}{1+\lambda},\ z = \frac{12}{1+\lambda}.$

Putting these values of x, y, z in (6), we have

$$\frac{9}{1+\lambda}+\frac{16}{1+\lambda}+\frac{144}{1+\lambda}=1+\lambda \quad \text{or} \quad (1+\lambda)^2=169 \text{ or } 1+\lambda=\pm 13$$

$$\therefore \qquad \lambda = 12 \quad \text{or} \quad -14$$

When $\lambda = 12$, $\quad x=\frac{3}{13},\ y=\frac{4}{13},\ z=\frac{12}{13}$

When $\lambda = -14$, $\quad x=-\frac{3}{13},\ y=-\frac{4}{13},\ z=-\frac{12}{13}$

Thus, we get two points $P\left(\frac{3}{13},\frac{4}{13},\frac{12}{13}\right)$ and $Q\left(-\frac{3}{13},-\frac{4}{13},-\frac{12}{13}\right)$ on the sphere which are at a *maximum or minimum* distance from the given point A.

Now $\quad AP=\sqrt{\left(3-\frac{3}{13}\right)^2+\left(4-\frac{4}{13}\right)^2+\left(12-\frac{12}{13}\right)^2}=12$

$$AQ=\sqrt{\left(3+\frac{3}{13}\right)^2+\left(4+\frac{4}{13}\right)^2+\left(12+\frac{12}{13}\right)^2}=14$$

$\therefore\ P\left(\frac{3}{13},\frac{4}{13},\frac{12}{13}\right)$ is at a minimum distance from A and the minimum distance = 12.

$Q\left(-\frac{3}{13},-\frac{4}{13},-\frac{12}{13}\right)$ is at a maximum distance from A and the maximum distance = 14.

Example 11. *Find the minimum value of $x^2 + y^2 + z^2$, given that $ax + by + cz = p$.*

Sol. Let $\quad u = x^2 + y^2 + z^2 \quad$ ...(1)

where $\quad \phi(x, y, z) = ax + by + cz - p = 0 \quad$ ...(2)

Consider Lagrange's function, $F(x, y, z) = (x^2 + y^2 + z^2) + \lambda(ax + by + cz - p)$

For stationary values, $dF = 0$

$$\Rightarrow \quad (2x + \lambda a)dx + (2y + \lambda b)dy + (2z + \lambda c)dz = 0$$

$\Rightarrow \quad 2x + \lambda a = 0 \quad$ ...(3) $\quad 2y + \lambda b = 0 \quad$ ...(4) $\quad 2z + \lambda c = 0 \quad$ ...(5)

Multiplying (3) by x, (4) by y, (5) by z and adding, we get

$$2(x^2 + y^2 + z^2) + \lambda(ax + by + cz) = 0$$

or $\quad 2u + \lambda p = 0 \quad$ | Using (1) and (2)

$$\therefore \qquad \lambda = -\frac{2u}{p}$$

From (3), (4) and (5), $\quad x=\frac{au}{p},\ y=\frac{bu}{p},\ z=\frac{cu}{p}$

$\therefore$ From (1), $\quad u=\frac{(a^2+b^2+c^2)u^2}{p^2} \quad$ or $\quad u=\frac{p^2}{a^2+b^2+c^2}$.

This is the maximum or minimum value of u. Now u is the square of the distance of any point $P(x, y, z)$ on the plane (2) from the origin. Also, the length of perpendicular from O on the plane is $\frac{p}{\sqrt{a^2+b^2+c^2}}$.

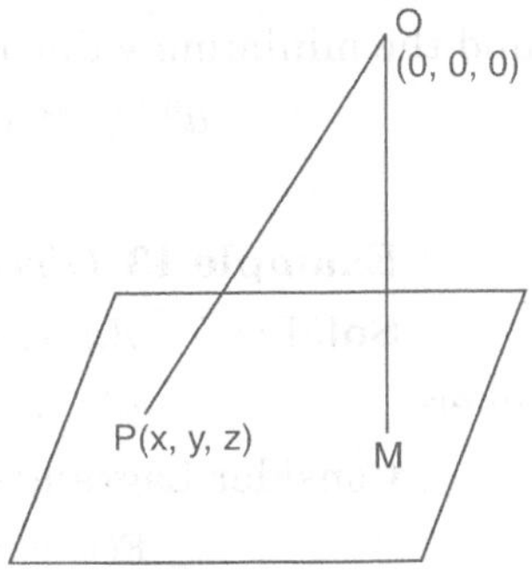

Fig. 2.1

Clearly, OP is least when P coincides with M, the foot of the perpendicular from O on the plane.

Hence the minimum value of $u = \dfrac{p^2}{a^2 + b^2 + c^2}$.

Note. Try it by eliminating z so that

$$u = x^2 + y^2 + \left(\frac{p - ax - by}{c}\right)^2.$$

Example 12. *Find the minimum value of $ax + by + cz$ subject to the condition $\dfrac{a^2}{x} + \dfrac{b^2}{y} + \dfrac{c^2}{z} = 1$.*

Sol. Let $\quad f(x, y, z) = ax + by + cz \quad$ ...(1)

where $\quad \phi(x, y, z) = \dfrac{a^2}{x} + \dfrac{b^2}{y} + \dfrac{c^2}{z} - 1 = 0 \quad$ ...(2)

Consider Lagrange's function

$$F(x, y, z) = (ax + by + cz) + \lambda\left(\frac{a^2}{x} + \frac{b^2}{y} + \frac{c^2}{z} - 1\right)$$

For stationary values, $dF = 0$

$$\Rightarrow \quad \left(a - \frac{\lambda a^2}{x^2}\right)dx + \left(b - \frac{\lambda b^2}{y^2}\right)dy + \left(c - \frac{\lambda c^2}{z^2}\right)dz = 0$$

$$\Rightarrow \quad a - \frac{\lambda a^2}{x^2} = 0 \quad \text{...(3)}$$

$$b - \frac{\lambda b^2}{y^2} = 0 \quad \text{...(4)}$$

$$c - \frac{\lambda c^2}{z^2} = 0 \quad \text{...(5)}$$

From (3), (4) and (5)

$$x = \sqrt{\lambda a}, \; y = \sqrt{\lambda b}, \; z = \sqrt{\lambda c}$$

Putting these values in (2), we have

$$\frac{a^2}{\sqrt{\lambda a}} + \frac{b^2}{\sqrt{\lambda b}} + \frac{c^2}{\sqrt{\lambda c}} = 1$$

$$\Rightarrow \quad \sqrt{\lambda} = a^{3/2} + b^{3/2} + c^{3/2} \quad \Rightarrow \quad \lambda = (a^{3/2} + b^{3/2} + c^{3/2})^2$$

$$\therefore \quad x = a^{1/2}(a^{3/2} + b^{3/2} + c^{3/2}), \quad y = b^{1/2}(a^{3/2} + b^{3/2} + c^{3/2}),$$

$$z = c^{1/2}(a^{3/2} + b^{3/2} + c^{3/2})$$

$\Rightarrow$ the stationary point is

$$(a^{1/2}(a^{3/2} + b^{3/2} + c^{3/2}), \; b^{1/2}(a^{3/2} + b^{3/2} + c^{3/2}), \; c^{1/2}(a^{3/2} + b^{3/2} + c^{3/2}))$$

and the minimum value of f is

$$a^{3/2}(a^{3/2}+b^{3/2}+c^{3/2})+b^{3/2}(a^{3/2}+b^{3/2}+c^{3/2})+c^{3/2}(a^{3/2}+b^{3/2}+c^{3/2})$$
$$=(a^{3/2}+b^{3/2}+c^{3/2})^2.$$

Example 13. *Given $x + y + z = a$, find the maximum value of $x^m y^n z^p$.*

Sol. Let $f(x, y, z) = x^m y^n z^p$ ...(1)

where $\phi(x, y, z) = x + y + z - a = 0$ ...(2)

Consider Lagrange's function

$$F(x, y, z) = x^m y^n z^p + \lambda(x + y + z - a)$$

For stationary values, $dF = 0$

$$\Rightarrow \quad (m\,x^{m-1}y^n z^p + \lambda)dx + (n\,x^m y^{n-1} z^p + \lambda)dy + (p\,x^m y^n z^{p-1} + \lambda)dz = 0$$

$$\Rightarrow \quad m\,x^{m-1}y^n z^p + \lambda = 0 \quad \text{...(3)}$$
$$n\,x^m y^{n-1} z^p + \lambda = 0 \quad \text{...(4)}$$
$$p\,x^m y^n z^{p-1} + \lambda = 0 \quad \text{...(5)}$$

Multiplying (3), (4), (5) by x, y, z respectively and adding

$$(m+n+p)x^m y^n z^p + \lambda(x+y+z) = 0$$

$$\Rightarrow \quad (m+n+p)f + \lambda a = 0 \quad \text{[by (2)]}$$

$$\Rightarrow \quad \lambda = -\frac{(m+n+p)f}{a}$$

From (3), multiplying by x, we have

$$mf + \lambda x = 0 \Rightarrow x = -\frac{mf}{\lambda}$$

$$\therefore \quad x = \frac{ma}{m+n+p}$$

Similarly, $y = \dfrac{na}{m+n+p}, \quad z = \dfrac{pa}{m+n+p}$

$\therefore$ Maximum value of f

$$= \left(\frac{ma}{m+n+p}\right)^m \left(\frac{na}{m+n+p}\right)^n \left(\frac{pa}{m+n+p}\right)^p$$

$$= \frac{m^m n^n p^p\, a^{m+n+p}}{(m+n+p)^{m+n+p}}.$$

Example 14. *Find the dimensions of the rectangular box, open at the top, of maximum capacity whose surface is 432 sq. cm.*

Sol. Let x, y, z be the edges of the box.

Capacity of box $= xyz$, and

surface area $= xy + 2yz + 2zx = 432$

Let $f(x, y, z) = xyz$ ...(1)

where $\phi(x, y, z) = xy + 2yz + 2zx - 432 = 0$ ...(2)

Consider Lagrange's function

$$F(x, y, z) = xyz + \lambda(xy + 2yz + 2zx - 432)$$

For stationary values, $dF = 0$

$$\Rightarrow \quad [yz + \lambda(y + 2z)]dx + [xz + \lambda(x + 2z)]dy + [xy + \lambda(2y + 2x)]dz = 0$$

$$\Rightarrow \quad yz + \lambda(y + 2z) = 0 \quad ...(3)$$

$$xz + \lambda(x + 2z) = 0 \quad ...(4)$$

$$xy + \lambda(2y + 2x) = 0 \quad ...(5)$$

Multiplying (3) by x, (4) by y and subtracting, we get

$$2z\lambda(x - y) = 0 \Rightarrow x = y$$

Multiplying (4) by y, (5) by z and subtracting, we get

$$x\lambda(y - 2z) = 0 \Rightarrow y = 2z$$

$$\therefore \quad x = y = 2z$$

$\therefore$ From (2) $x \cdot x + x \cdot x + x \cdot x = 432$

$$\Rightarrow \quad 3x^2 = 432 \Rightarrow x^2 = 144$$

$$\therefore \quad x = 12 \Rightarrow y = 12, \quad z = 6$$

Hence the dimensions of the box are 12 cm, 12 cm, 6 cm.

Example 15. *Prove that the stationary values of* $u = \dfrac{x^2}{a^4} + \dfrac{y^2}{b^4} + \dfrac{z^2}{c^4}$ *where* $lx + my + nz = 0$ *and* $\dfrac{x^2}{a^2} + \dfrac{y^2}{b^2} + \dfrac{z^2}{c^2} = 1$ *are the roots of the equation*

$$\frac{l^2a^4}{1-a^2u} + \frac{m^2b^4}{1-b^2u} + \frac{n^2c^4}{1-c^2u} = 0.$$

Sol. Consider Lagrange's function,

$$F(x, y, z) = \left(\frac{x^2}{a^4} + \frac{y^2}{b^4} + \frac{z^2}{c^4}\right) + \lambda\,(lx + my + nz) + \mu\left(\frac{x^2}{a^2} + \frac{y^2}{b^2} + \frac{z^2}{c^2} - 1\right)$$

For stationary values, $dF = 0$

$$\Rightarrow \quad \left(\frac{2x}{a^4} + \lambda l + \frac{2\mu x}{a^2}\right)dx + \left(\frac{2y}{b^4} + \lambda m + \frac{2\mu y}{b^2}\right)dy + \left(\frac{2z}{c^4} + \lambda n + \frac{2\mu z}{c^2}\right)dz = 0$$

$$\Rightarrow \quad \frac{2x}{a^4} + \lambda l + \frac{2\mu x}{a^2} = 0 \quad ...(1)$$

$$\Rightarrow \quad \frac{2y}{b^4} + \lambda m + \frac{2\mu y}{b^2} = 0 \quad ...(2)$$

$$\frac{2z}{c^4} + \lambda n + \frac{2\mu z}{c^2} = 0 \quad ...(3)$$

Multiplying (1), (2), (3) by x, y, z respectively and adding, we get

$$2\left(\frac{x^2}{a^4}+\frac{y^2}{b^4}+\frac{z^2}{c^4}\right)+\lambda\,(lx+my+nz)+2\mu\left(\frac{x^2}{a^2}+\frac{y^2}{b^2}+\frac{z^2}{c^2}\right)=0$$

$\Rightarrow$ $$2u+\lambda(0)+2\mu\,(1)=0$$ | from given relations

$\Rightarrow$ $$\mu=-u$$

$\therefore$ Equation (1) becomes $\dfrac{2x}{a^4}+\lambda l-\dfrac{2ux}{a^2}=0$ or $\dfrac{2x}{a^4}(1-a^2u)=-\lambda l$ or $x=-\dfrac{\lambda l a^4}{2(1-a^2u)}$

Similarly, $$y=-\frac{\lambda m b^4}{2(1-b^2u)},\ z=-\frac{\lambda n c^4}{2(1-c^2u)}$$

To eliminate λ between them, multiply these values of x, y, z by l, m, n respectively and add. Then

$$lx+my+nz=-\frac{\lambda}{2}\left[\frac{l^2a^4}{1-a^2u}+\frac{m^2b^4}{1-b^2u}+\frac{n^2c^4}{1-c^2u}\right]$$

Since $lx+my+nz=0$, we have $\dfrac{l^2a^4}{1-a^2u}+\dfrac{m^2b^4}{1-b^2u}+\dfrac{n^2c^4}{1-c^2u}=0$

which is a quadratic in u and gives two stationary values of u.

EXERCISE 2.5

1. Examine for extreme values:

   (i) $x^2+y^2+6x+12$ (ii) x^3+y^3+3xy (iii) $2(x^2-y^2)-x^4+y^4$

   (iv) $x^2y^2-5x^2-8xy-5y^2$ (v) $x^3+3xy^2-15x^2-15y^2+72x$

   (vi) $x^2-xy+y^2+3x-2y+1$ (vii) $x^3+y^3-63(x+y)+12xy$ (viii) $xy+\dfrac{a^3}{x}+\dfrac{a^3}{y}$

   (ix) $\sin x\sin y\sin(x+y)\,;\ 0<x, y<\pi$

   (x) $\cos x+\cos y+\cos(x+y)\,;\ 0<x, y<\pi$.

2. Find the stationary points of the function $z=x^3y^2(12-x-y)$ satisfying the condition $x>0, y>0$ and examine their nature.

3. Find the minimum values of $x^2+y^2+z^2$ when

   (i) $x+y+z=3a$ (ii) $xyz=a^3$.

4. Find the minimum value of the function $x+y+z$ subject to the condition $\dfrac{a}{x}+\dfrac{b}{y}+\dfrac{c}{z}=1$.

5. Find the minimum value of the function $x^2+y^2+z^2$ subject to the condition

   (i) $ax+by+cz=a+b+c$ (ii) $xy+yz+zx=3a^2$.

6. Find the points on the surface $z^2=xy+1$ nearest to the origin.

7. Given $f(x, y, z)=\dfrac{5xyz}{x+2y+4z}$, find the values of x, y, z for which $f(x, y, z)$ is a maximum, subject to the condition $xyz=8$.

8. A rectangular box, open at the top, is to have a volume of 32 c.c. Find the dimensions of the box requiring least material for its construction. *(K.U., May 2010)*

9. Try examples 5, 6, 7 and 8 by Lagrange's method.

10. A tent on a square base of side x, has its sides vertical of height y and the top is a regular pyramid of height h. Find x and y in terms of h, if the canvas required for its construction is to be minimum for the tent to have a given capacity.

11. The temperature T at any point (x, y, z) in space is $T = 400\,xyz^2$. Find the highest temperature on the surface of the unit sphere $x^2 + y^2 + z^2 = 1$.

12. Find the shortest and the longest distances from the point (1, 2, – 1) to the sphere

$$x^2 + y^2 + z^2 = 24.$$

13. Find the maximum and minimum distances from the origin to the curve $5x^2 + 6xy + 5y^2 - 8 = 0$.

14. Prove that of all the rectangular parallelopipeds of the same volume, the cube has the least surface.

15. Prove that of all the rectangulars parallelopipeds of given surface, cube has the maximum volume.

16. Divide 24 into three parts such that the continued product of the first, square of the second and the cube of the third may be maximum.

17. Find the stationary values of $x^2 + y^2 + z^2$ subject to $ax^2 + by^2 + cz^2 = 1$ and $lx + my + nz = 0$.

18. If $u = a^3x^2 + b^3y^2 + c^3z^2$ where $x^{-1} + y^{-1} + z^{-1} = 1$, show that the stationary value of u is given by

$$x = \frac{\Sigma a}{a}, y = \frac{\Sigma a}{b}, z = \frac{\Sigma a}{c}.$$

19. Find the local maximum and minimum values and saddle points of the function.

(*a*) $f(x, y) = \dfrac{\left(x^2y^2 - 8x + y\right)}{xy}$ (*b*) $f(x, y) = e^x \sin y$

(*c*) $f(x, y) = e^x \cos y$.

20. Find the absolute maximum and minimum values of f on the set D.

(*a*) $f(x, y) = x^2 + y^2 + x^2y + 4$, $D = \{(x, y)/ \mid x \mid \le 1, \quad \mid y \mid \le 1\}$

(*b*) $f(x, y) = 1 + xy - x - y$; D is the region bounded by $y = x^2$ and the line $y = 4$.

21. Find three positive numbers whose sum is 100 and whose product is a maximum.

22. Find the volume of the largest rectangular box with edges parallel to the axes that can be inscribed in the ellipsoid $9x^2 + 36y^2 + 4z^2 = 36$.

23. Find the volume of the largest rectangular box in the first octant with three faces in the co-ordinate planes and one vertex in the plane $x + 2y + 3z = 6$.

24. Show that the function $f(x, y) = -(x^2 - 1)^2 - (x^2y - x - 1)^2$ has only two critical points, but has local maximum at both of them.

25. The plane $x + y + 2z = 2$ intersects the paraboloid $z = x^2 + y^2$ in an ellipse. Find the points On this ellipse that are nearest to and farthest from the origin.

Answers

1. (*i*) Min. value = 3 at (–3, 0) (*ii*) Max. value = 1 at (–1, –1)

(*iii*) Max. value = 1 at (±1, 0) ; Min. value = –1 at (0, ±1)

(*iv*) Max. value = 0 at (0, 0) (*v*) Max. value = 112 at (4, 0); Min. value – 108 at (6, 0) (*vi*) Min. value = $-\frac{4}{3}$ at $\left(-\frac{4}{3}, \frac{1}{3}\right)$

(*vii*) Max. value = 784 at (–7, –7) ; Min. value = –216 at (3, 3)

(*viii*) Min. value = $3a^2$ at (a, a)

(*ix*) Max. value = $\frac{3\sqrt{3}}{8}$ at $\left(\frac{\pi}{3}, \frac{\pi}{3}\right)$; Min. value = $-\frac{3\sqrt{3}}{8}$ at $\left(\frac{2\pi}{3}, \frac{2\pi}{3}\right)$

(*x*) Min. value = $-\frac{3}{2}$ at $\left(\frac{2\pi}{3}, \frac{2\pi}{3}\right)$

2. Maximum at (6, 4) **3.** (*i*) $3a^2$ (*ii*) $3a^2$

4. $(\sqrt{a}+\sqrt{b}+\sqrt{c})^2$ at $(\sqrt{a}(\sqrt{a}+\sqrt{b}+\sqrt{c}), \sqrt{b}(\sqrt{a}+\sqrt{b}+\sqrt{c}), \sqrt{c}(\sqrt{a}+\sqrt{b}+\sqrt{c}))$

5. (*i*) $\frac{(a+b+c)^2}{a^2+b^2+c^2}$ (*ii*) $3a^2$ at (a, a, a) and $(-a, -a, -a)$

6. (0, 0, ±1) **7.** (4, 2, 1) **8.** 4 cm × 4 cm × 2 cm

10. $x = \sqrt{5h}, y = \frac{1}{2}h$ **11.** 50 **12.** $\sqrt{6}, 3\sqrt{6}$

13. 4, 1 **16.** 4, 8, 12

17. Two stationary values of u are roots of $\frac{l^2}{au-1} + \frac{m^2}{bu-1} + \frac{n^2}{cu-1} = 0$

19. (*a*) Max $f\left(-\frac{1}{2}, 4\right) = -6$

(*b*) Saddle points $(0, n\pi)$, n integer

(*c*) None

20. (*a*) Max $f(\pm 1, 1) = 7$, mini $f(0, 0) = 4$

(*b*) Max $f(2, 4) = 3$, mini $f(-2, 4) = -9$

21. $\frac{100}{3}, \frac{100}{3}, \frac{100}{3}$. **22.** $\frac{16}{\sqrt{3}}$

23. $\frac{4}{3}$ **24.** Max $f(-1, 0) = f(1, 2) = 0$

25. Nearest $\left(\frac{1}{2}, \frac{1}{2}, \frac{1}{2}\right)$, Farthest (–1, –1, 2).

3
Multiple Integrals

3.1. DOUBLE INTEGRALS

The definite integral $\int_a^b f(x)\,dx$ is defined as the limits of the sum $f(x_1)\delta x_1 + f(x_2)\delta x_2 + \cdots + f(x_n)\delta x_n$ when $n \to \infty$ and each of the lengths $\delta x_1, \delta x_2, \ldots, \delta x_n$ tends to zero. Here $\delta x_1, \delta x_2, \ldots, \delta x_n$ are n sub-intervals into which the range $b - a$ has been divided and $x_1, x_2, \ldots, x_n$ are values of x lying respectively in the first, second, ..., nth sub-interval.

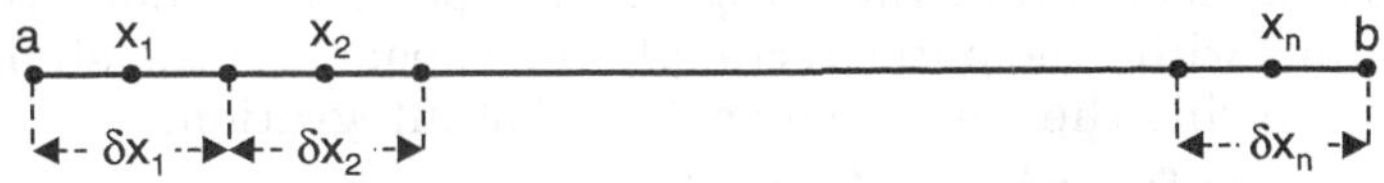

A double integral is its counterpart in two dimensions. Let a single-valued and bounded function $f(x, y)$ of two independent variables x, y be defined in a closed region R of the xy-plane. Divide the region R into sub-regions by drawing lines parallel to co-ordinate axes as shown in Fig. 3.1. Number the rectangles which lie entirely inside the region R, from 1 to n. Let (x_r, y_r) be any point inside the rth rectangle whose area is δA_r.

Consider the sum

$$f(x_1, y_1)\delta A_1 + f(x_2, y_2)\delta A_2 + \cdots + f(x_n, y_n)\delta A_n$$

$$= \sum_{r=1}^{n} f(x_r, y_r)\delta A_r \qquad \ldots(1)$$

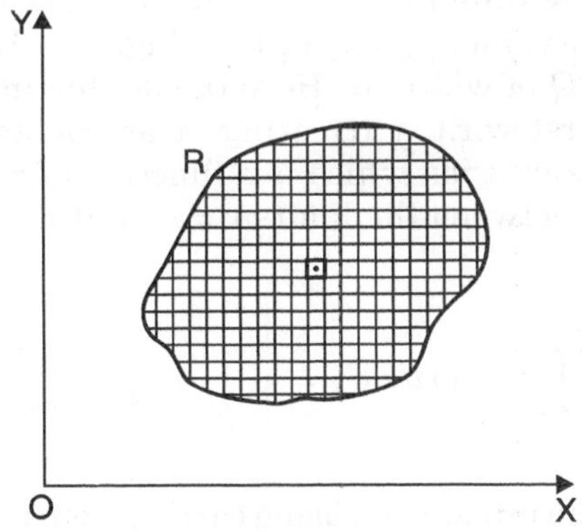

Fig. 3.1

Let the number of these sub-regions increase indefinitely, such that the largest linear dimension (*i.e.*, diagonal) of δA_r approaches zero. The limit of the sum (1), if it exists, irrespective of the mode of sub-division, is called the *double integral* of $f(x, y)$ over the region R and is denoted by $\iint_R f(x, y)\,dA$

In other words, $$\lim_{\substack{n \to \infty \\ \delta A_r \to 0}} \sum_{r=1}^{n} f(x_r, y_r)\,\delta A_r = \iint_R f(x, y)\,dA$$

which is also expressed as $$\iint_R f(x, y)\,dx\,dy \quad \text{or} \quad \iint_R f(x, y)\,dy\,dx$$

3.2. EVALUATION OF DOUBLE INTEGRALS

The methods of evaluating the double integrals depend upon the nature of the curves bounding the region R. Let the region R be bounded by the curves $x = x_1$, $x = x_2$ and $y = y_1$, $y = y_2$.

(*i*) **When x_1, x_2 are functions of y and y_1, y_2 are constants.** Let AB and CD be the curves $x_1 = \phi_1(y)$ and $x_2 = \phi_2(y)$ as shown in Fig. 3.2.

Take a horizontal strip PQ of width δy. Here the double integral is evaluated first w.r.t. x (treating y as a constant). The resulting expression which is a function of y is integrated w.r.t. y between the limits $y = y_1$ and $y = y_2$. Thus

$$\iint_R f(x, y)\, dx\, dy = \boxed{\int_{y_1}^{y_2} \boxed{\int_{x_1 = \phi_1(y)}^{x_2 = \phi_2(y)} f(x, y)\, dx}\; dy}$$

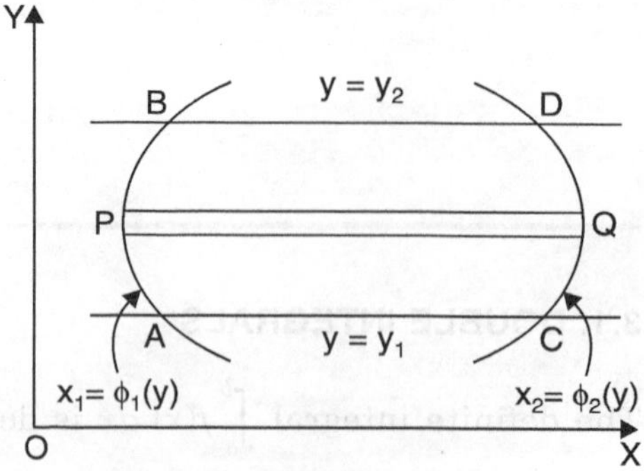

Fig. 3.2

the integration being carried from the inner to the outer rectangle. Geometrically, the integral in the inner rectangle indicates that the integration is performed along the horizontal strip PQ (keeping y constant) while the outer rectangle corresponds to the sliding of the strip PQ from AC to BD thus covering the entire region ABDC of integration.

(*ii*) **When y_1, y_2 are functions of x and x_1, x_2 are constants.** Let AB and CD be the curves $y_1 = \phi_1(x)$ and $y_2 = \phi_2(x)$. (See Fig. 3.3) Take a vertical strip PQ of width δx. Here the double integral is evaluated first w.r.t. y (treating x as constant). The resulting expression which is a function of x is integrated w.r.t. x between the limits $x = x_1$ and $x = x_2$. Thus,

$$\iint_R f(x, y)\, dx\, dy = \boxed{\int_{x_1}^{x_2} \boxed{\int_{y_1 = \phi_1(x)}^{y_2 = \phi_2(x)} f(x, y)\, dy}\; dx}$$

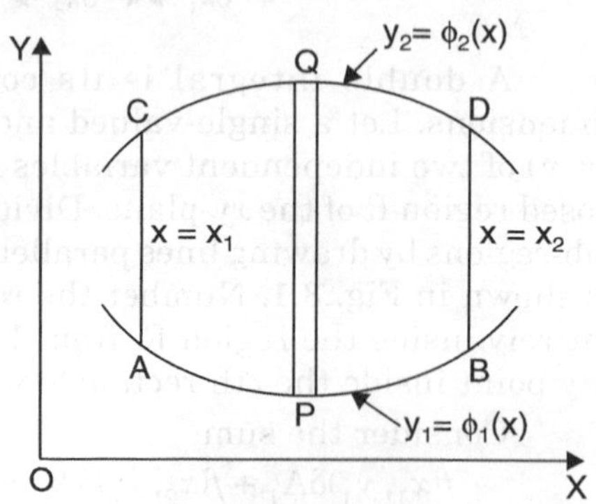

Fig. 3.3

the integration being carried from the inner to the outer rectangle. Geometrically, the integral in the inner rectangle indicates that the integration is performed along the vertical strip PQ (keeping x constant) while the outer rectangle corresponds to the sliding of the strip PQ from AC to BD thus covering the entire region ABDC of integration.

(*iii*) **When x_1, x_2, y_1, y_2 are constants.** Here the region of integration R is the rectangle ABDC. It is immaterial whether we integrate first along the horizontal strip PQ and then slide it from AB to CD; or we integrate first along the vertical strip P′Q′ and then

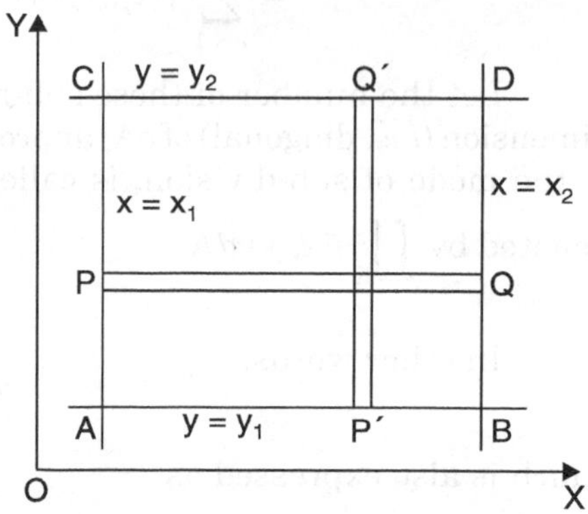

Fig. 3.4

slide it from AC to BD. (Refer Fig. 3.4.) Thus the order of integration is immaterial, provided the limits of integration are changed accordingly.

$$\iint_R f(x, y)\, dx\, dy = \boxed{\int_{y_1}^{y_2} \boxed{\int_{x_1}^{x_2} f(x, y)\, dx}\, dy} = \boxed{\int_{x_1}^{x_2} \boxed{\int_{y_1}^{y_2} f(x, y)\, dy}\, dx}$$

Note 1. From cases (*i*) and (*ii*) above, we observe that *integration is to be performed w.r.t. that variable having variable limits first and then w.r.t. the variable with constant limits.*

Note 2. If $f(x, y)$ has discontinuities within or on the boundary of the region of integration, then the change of the order of integration does not result into the same integrals.

ILLUSTRATIVE EXAMPLES

Example 1. *Prove that* $\int_1^2 \int_3^4 (xy + e^y)\, dy\, dx = \int_3^4 \int_1^2 (xy + e^y)\, dx\, dy.$

Sol. $\int_1^2 \int_3^4 (xy + e^y)\, dy\, dx = \int_1^2 \left[\int_3^4 (xy + e^y)\, dy\right] dx$

$$= \int_1^2 \left[\frac{xy^2}{2} + e^y\right]_3^4 dx = \int_1^2 \left(8x + e^4 - \frac{9}{2}x - e^3\right) dx$$

$$= \int_1^2 \left(\frac{7}{2}x + e^4 - e^3\right) dx = \left[\frac{7x^2}{4} + (e^4 - e^3)x\right]_1^2$$

$$= 7 + 2(e^4 - e^3) - \frac{7}{4} - (e^4 - e^3) = \frac{21}{4} + e^4 - e^3$$

$$\int_3^4 \int_1^2 (xy + e^y)\, dx\, dy = \int_3^4 \left[\int_1^2 (xy + e^y)\, dx\right] dy = \int_3^4 \left[\frac{yx^2}{2} + xe^y\right]_1^2 dy$$

$$= \int_3^4 \left(2y + 2e^y - \frac{y}{2} - e^y\right) dy = \int_3^4 \left(\frac{3y}{2} + e^y\right) dy$$

$$= \left[\frac{3y^2}{4} + e^y\right]_3^4 = 12 + e^4 - \frac{27}{4} - e^3 = \frac{21}{4} + e^4 - e^3$$

Hence the result.

Example 2. *Evaluate* $\int_0^1 \int_0^1 \frac{dx\, dy}{\sqrt{(1 - x^2)(1 - y^2)}}.$

Sol. $\int_0^1 \int_0^1 \frac{dx\, dy}{\sqrt{(1-x^2)(1-y^2)}} = \int_0^1 \left[\int_0^1 \frac{dx}{\sqrt{(1-x^2)(1-y^2)}}\right] dy = \int_0^1 \frac{1}{\sqrt{1-y^2}} \left[\sin^{-1} x\right]_0^1 dy$

$$= \int_0^1 \frac{1}{\sqrt{1-y^2}} \cdot \frac{\pi}{2}\, dy = \frac{\pi}{2} \left[\sin^{-1} y\right]_0^1 = \frac{\pi}{2} \cdot \frac{\pi}{2} = \frac{\pi^2}{4}.$$

Example 3. *Show that* $\int_0^1 dx \int_0^1 \frac{x-y}{(x+y)^3}\, dy \neq \int_0^1 dy \int_0^1 \frac{x-y}{(x+y)^3}\, dx.$

Sol. $$\text{LHS} = \int_0^1 dx \int_0^1 \frac{2x-(x+y)}{(x+y)^3}\, dy = \int_0^1 dx \int_0^1 \left[\frac{2x}{(x+y)^3} - \frac{1}{(x+y)^2}\right] dy$$

$$= \int_0^1 \left[2x \cdot \frac{(x+y)^{-2}}{-2} - \frac{(x+y)^{-1}}{-1}\right]_0^1 dx = \int_0^1 \left[\frac{-x}{(x+y)^2} + \frac{1}{x+y}\right]_0^1 dx$$

$$= \int_0^1 \left[\frac{-x}{(x+1)^2} + \frac{1}{x+1} + \frac{1}{x} - \frac{1}{x}\right] dx = \int_0^1 \frac{-x+x+1}{(x+1)^2}\, dx = \int_0^1 \frac{1}{(x+1)^2}\, dx$$

$$= \left[-\frac{1}{x+1}\right]_0^1 = -\frac{1}{2} + 1 = \frac{1}{2}$$

$$\text{RHS} = \int_0^1 dy \int_0^1 \frac{(x+y)-2y}{(x+y)^3}\, dx = \int_0^1 dy \int_0^1 \left[\frac{1}{(x+y)^2} - \frac{2y}{(x+y)^3}\right] dx$$

$$= \int_0^1 \left[\frac{(x+y)^{-1}}{-1} - 2y \cdot \frac{(x+y)^{-2}}{-2}\right]_0^1 dy = \int_0^1 \left[-\frac{1}{x+y} + \frac{y}{(x+y)^2}\right]_0^1 dy$$

$$= \int_0^1 \left[-\frac{1}{1+y} + \frac{y}{(1+y)^2} + \frac{1}{y} - \frac{1}{y}\right] dy = \int_0^1 \frac{-1-y+y}{(1+y)^2}\, dy = -\int_0^1 \frac{1}{(1+y)^2}\, dy$$

$$= \left[\frac{1}{1+y}\right]_0^1 = \frac{1}{2} - 1 = -\frac{1}{2}$$

$\therefore$ The two integrals are not equal.

Example 4. *Evaluate* $\int_0^1 \int_0^{\sqrt{1+x^2}} \frac{dy\, dx}{1+x^2+y^2}.$ (K.U. April 2012)

Sol. $$I = \int_0^1 \left[\int_0^{\sqrt{1+x^2}} \frac{1}{(1+x^2)+y^2}\, dy\right] dx$$

$$= \int_0^1 \frac{1}{\sqrt{1+x^2}} \left[\tan^{-1} \frac{y}{\sqrt{1+x^2}}\right]_0^{\sqrt{1+x^2}} dx$$

$$= \int_0^1 \frac{1}{\sqrt{1+x^2}} [\tan^{-1} 1 - \tan^{-1} 0]\, dx = \frac{\pi}{4} \int_0^1 \frac{dx}{\sqrt{1+x^2}}$$

$$= \frac{\pi}{4} \left[\log (x + \sqrt{1+x^2})\right]_0^1 = \frac{\pi}{4} [\log (1+\sqrt{2}) - \log 1] = \frac{\pi}{4} \log (\sqrt{2}+1).$$

Example 5. *Evaluate* $\int_0^a \int_0^{\sqrt{a^2-y^2}} \sqrt{a^2-x^2-y^2}\, dx\, dy.$

Sol. $$I = \int_0^a \left[\int_0^{\sqrt{a^2 - y^2}} \sqrt{(a^2 - y^2) - x^2}\, dx \right] dy.$$

$$= \int_0^a \left[\frac{x\sqrt{a^2 - y^2 - x^2}}{2} + \frac{a^2 - y^2}{2} \sin^{-1} \frac{x}{\sqrt{a^2 - y^2}} \right]_0^{\sqrt{a^2 - y^2}} dy$$

$$\left[\because \int \sqrt{a^2 - x^2}\, dx = \frac{x\sqrt{a^2 - x^2}}{2} + \frac{a^2}{2} \sin^{-1} \frac{x}{a} \right]$$

$$= \int_0^a \frac{a^2 - y^2}{2} \sin^{-1} 1\, dy = \frac{\pi}{4} \int_0^a (a^2 - y^2)\, dy$$

$$= \frac{\pi}{4} \left[a^2 y - \frac{y^3}{3} \right]_0^a = \frac{\pi}{4} \left[a^3 - \frac{a^3}{3} \right] = \frac{\pi a^3}{6}.$$

Example 6. *Evaluate* $\iint_R e^{2x + 3y}\, dx\, dy$ *over the triangle bounded by* $x = 0$, $y = 0$ *and* $x + y = 1$.

Sol. The region R of integration is the triangle OAB. Here x varies from 0 to 1 and y varies from x-axis upto the line $x + y = 1$ *i.e.*, from 0 to $1 - x$.

$\therefore$ The region R can be expressed as

$$0 \le x \le 1,\ 0 \le y \le 1 - x$$

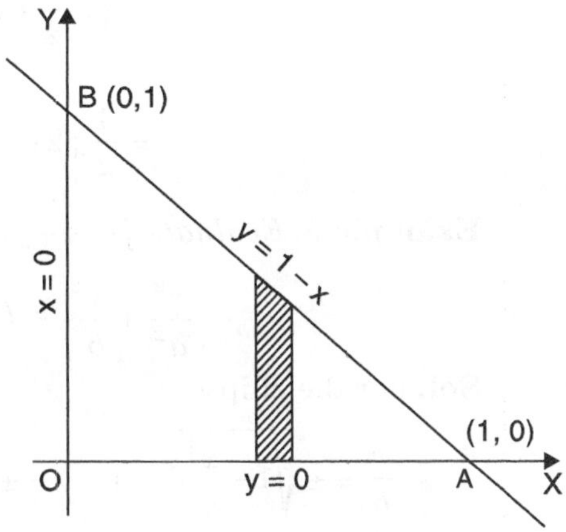

Fig. 3.5

$$\therefore \iint_R e^{2x + 3y}\, dx\, dy$$

$$= \int_0^1 \int_0^{1-x} e^{2x + 3y}\, dy\, dx$$

$$= \int_0^1 \left[\frac{1}{3} e^{2x + 3y} \right]_0^{1-x} dx$$

$$= \frac{1}{3} \int_0^1 (e^{3-x} - e^{2x})\, dx$$

$$= \frac{1}{3} \left[-e^{3-x} - \frac{1}{2} e^{2x} \right]_0^1 = -\frac{1}{3} \left[\left(e^2 + \frac{1}{2} e^2 \right) - \left(e^3 + \frac{1}{2} \right) \right]$$

$$= -\frac{1}{3} \left[-e^2(e - 1) + \frac{1}{2}(e^2 - 1) \right]$$

$$= \frac{1}{6} (e - 1)\, [2e^2 - (e + 1)] = \frac{1}{6} (e - 1)(2e^2 - e - 1)$$

$$= \frac{1}{6} (e - 1)(e - 1)(2e + 1) = \frac{1}{6} (e - 1)^2(2e + 1).$$

Example 7. *Evaluate* $\iint_R y\,dx\,dy$, *where R is the region bounded by the parabolas* $y^2 = 4x$ *and* $x^2 = 4y$.

Sol. Solving $y^2 = 4x$ and $x^2 = 4y$, we have

$$\left(\frac{x^2}{4}\right)^2 = 4x \quad \text{or} \quad x(x^3 - 64) = 0$$

$\therefore$ $x = 0, 4$

When $x = 4,\ y = 4$

$\therefore$ Co-ordinates of A are (4, 4)

The region R can be expressed as

$$0 \le x \le 4,\ \frac{x^2}{4} \le y \le 2\sqrt{x}$$

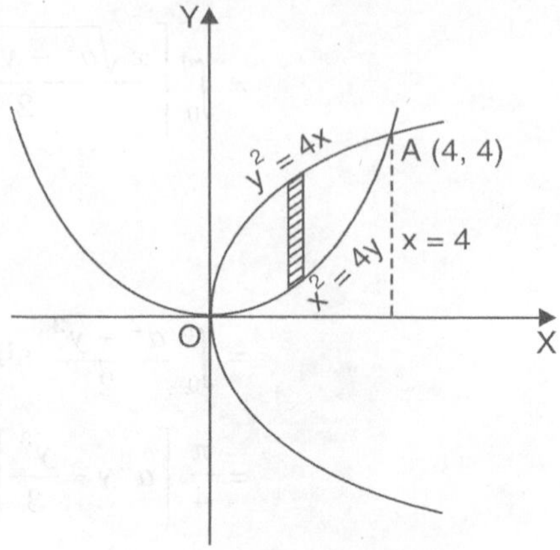

Fig. 3.6

$$\therefore \quad \iint_R y\,dx\,dy = \int_0^4 \int_{x^2/4}^{2\sqrt{x}} y\,dy\,dx$$

$$= \int_0^4 \frac{1}{2}\left[y^2\right]_{x^2/4}^{2\sqrt{x}} dx = \frac{1}{2}\int_0^4 \left(4x - \frac{x^4}{16}\right) dx$$

$$= \frac{1}{2}\left[2x^2 - \frac{x^5}{80}\right]_0^4 = \frac{1}{2}\left[32 - \frac{1024}{80}\right] = \frac{48}{5}.$$

Example 8. *Evaluate* $\iint (x + y)^2\,dx\,dy$ *over the area bounded by the ellipse*

$$\frac{x^2}{a^2} + \frac{y^2}{b^2} = 1.$$

Sol. For the ellipse

$$\frac{y}{b} = \pm\sqrt{1 - \frac{x^2}{a^2}} \quad \text{or} \quad y = \pm\frac{b}{a}\sqrt{a^2 - x^2}$$

$\therefore$ The region of integration R can be expressed as

$$-a \le x \le a,\ -\frac{b}{a}\sqrt{a^2 - x^2} \le y \le \frac{b}{a}\sqrt{a^2 - x^2}$$

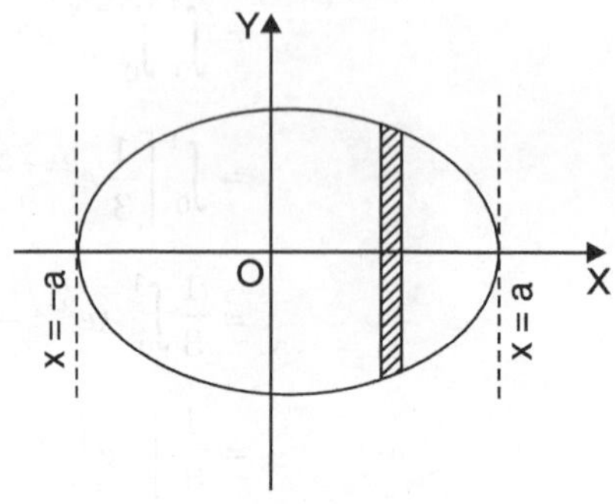

Fig. 3.7

$$\therefore \quad \iint_R (x + y)^2\,dx\,dy = \iint_R (x^2 + y^2 + 2xy)\,dx\,dy$$

$$= \int_{-a}^{a} \int_{-\left(\frac{b}{a}\right)\sqrt{a^2 - x^2}}^{\left(\frac{b}{a}\right)\sqrt{a^2 - x^2}} (x^2 + y^2 + 2xy)\,dy\,dx$$

$$= \int_{-a}^{a} \int_{-\left(\frac{b}{a}\right)\sqrt{a^2 - x^2}}^{\left(\frac{b}{a}\right)\sqrt{a^2 - x^2}} (x^2 + y^2)\,dy\,dx + \int_{-a}^{a} \int_{-\left(\frac{b}{a}\right)\sqrt{a^2 - x^2}}^{\left(\frac{b}{a}\right)\sqrt{a^2 - x^2}} 2xy\,dy\,dx$$

$$= \int_{-a}^{a} \int_{0}^{\left(\frac{b}{a}\right)\sqrt{a^2-x^2}} 2(x^2+y^2)\,dy\,dx + 0$$

[Since $(x^2 + y^2)$ is an even function of y and $2xy$ is an odd function of y]

$$= \int_{-a}^{a} \left[2\left(x^2 y + \frac{y^3}{3} \right) \right]_0^{\left(\frac{b}{a}\right)\sqrt{a^2-x^2}} dx$$

$$= 2\int_{-a}^{a} \left[x^2 \cdot \frac{b}{a}\sqrt{a^2-x^2} + \frac{1}{3}\cdot\frac{b^3}{a^3}(a^2-x^2)^{3/2} \right] dx$$

$$= 4\int_{0}^{a} \left[\frac{b}{a}x^2\sqrt{a^2-x^2} + \frac{b^3}{3a^3}(a^2-x^2)^{3/2} \right] dx$$

Put $x = a \sin\theta$
$\therefore \quad dx = a\cos\theta\, d\theta$

$$= 4\int_{0}^{\pi/2} \left(\frac{b}{a} \times a^2 \sin^2\theta \cdot a\cos\theta + \frac{b^3}{3a^3}\cdot a^3\cos^3\theta \right) a\cos\theta\, d\theta$$

$$= 4\int_{0}^{\pi/2} \left(a^3 b \sin^2\theta\cos^2\theta + \frac{ab^3}{3}\cos^4\theta \right) d\theta$$

$$= 4\left[a^3 b \cdot \frac{1\cdot 1}{4\cdot 2}\cdot\frac{\pi}{2} + \frac{ab^3}{3}\cdot\frac{3\cdot 1}{4\cdot 2}\cdot\frac{\pi}{2} \right] = \frac{\pi}{4}(a^3 b + ab^3) = \frac{\pi}{4}ab(a^2+b^2).$$

EXERCISE 3.1

Evaluate the following integrals (1–10):

1. $\int_0^3\int_0^1 (x^2+3y^2)\,dy\,dx.$

2. $\int_0^3\int_1^2 xy\,(1+x+y)\,dy\,dx.$

3. $\int_1^a\int_1^b \frac{dy\,dx}{xy}.$

4. $\int_1^2\int_0^x \frac{dy\,dx}{x^2+y^2}.$

5. $\int_0^1 dx\int_0^x e^{y/x}\,dy.$

6. $\int_0^1\int_{x^2}^x (x^2+3y+2)\,dy\,dx.$

7. $\int_0^1\int_0^{x^2} e^{y/x}\,dy\,dx.$ (*K.U. Nov. 2009*)

8. $\int_0^1\int_y^{y^2+1} x^2 y\,dx\,dy.$

9. $\int_0^{2a}\int_0^{\sqrt{2ax-x^2}} (x^2+y^2)\,dy\,dx.$

10. $\int_0^1\int_x^{\sqrt{x}} (x^2+y^2)\,dy\,dx.$

11. Evaluate $\iint (x^2+y^2)\,dx\,dy$ over the region in the positive quadrant for which $x + y \le 1$.

12. Evaluate $\iint x^2y^2\,dx\,dy$ over the circle $x^2 + y^2 = 1$.

13. Evaluate $\iint xy\,dx\,dy$ over the positive quadrant of the circle $x^2 + y^2 = a^2$.

14. Compute the value of $\iint_R y\,dx\,dy$, where R is the region in the first quadrant bounded by the ellipse

$$\frac{x^2}{a^2} + \frac{y^2}{b^2} = 1.$$

15. Evaluate $\iint_S \sqrt{xy - y^2}\,dx\,dy$, where S is a triangle with vertices (0, 0), (10, 1) and (1, 1).

16. Evaluate $\iint_R xy\,(x + y)\,dx\,dy$ over the area between $y = x^2$ and $y = x$.

17. Evaluate $\iint_A xy\,dx\,dy$, where A is the domain bounded by x-axis, ordinate $x = 2a$ and the curve $x^2 = 4ay$.

Answers

1. 12	**2.** $30\frac{3}{4}$	**3.** $\log a \log b$	**4.** $\frac{\pi}{4}\log 2$
5. $\frac{1}{2}(e-1)$	**6.** $\frac{7}{12}$	**7.** $\frac{1}{2}$	**8.** $\frac{67}{120}$
9. $\frac{3\pi a^4}{4}$	**10.** $\frac{3}{35}$	**11.** $\frac{1}{6}$	**12.** $\frac{\pi}{24}$
13. $\frac{a^4}{8}$	**14.** $\frac{ab^2}{3}$	**15.** 6	**16.** $\frac{3}{56}$
17. $\frac{a^4}{3}$.			

3.3. EVALUATION OF DOUBLE INTEGRALS IN POLAR CO-ORDINATES

To evaluate $\int_{\theta_1}^{\theta_2}\int_{r_1}^{r_2} f(r,\theta)\,dr\,d\theta$ over the region bounded by the straight lines $\theta = \theta_1$, $\theta = \theta_2$ and the curves $r = r_1$, $r = r_2$, we first integrate w.r.t. r between the limits $r = r_1$ and $r = r_2$ (treating θ as a constant). The resulting expression is then integrated w.r.t. θ between the limits $\theta = \theta_1$ and $\theta = \theta_2$.

Geometrically, AB and CD are the curves $r = f_1(\theta)$ and $r = f_2(\theta)$ bounded by the lines $\theta = \theta_1$ and $\theta = \theta_2$ so that ACDB is the region of integration. PQ is a wedge of angular thickness $\delta\theta$. Then $\int_{r=r_1}^{r=r_2} f(r,\theta)\,dr$ indicates that the integration is performed along PQ (*i.e.*, r varies, θ is constant) and the integration w.r.t. θ means rotation of this strip PQ from AC to BD. (See Fig. 3.8.)

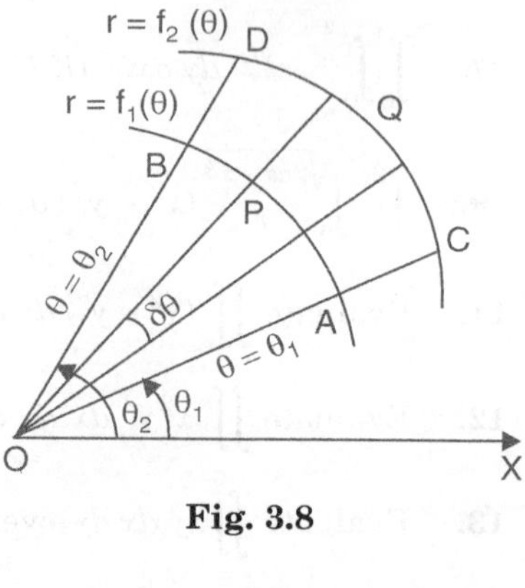

Fig. 3.8

ILLUSTRATIVE EXAMPLES

Example 1. *Evaluate* $\int_0^{\pi/2}\left[\int_0^{a\cos\theta} r\sqrt{a^2-r^2}\,dr\right]d\theta.$

Sol.
$$I = \int_0^{\pi/2}\left[\int_0^{a\cos\theta} -\frac{1}{2}(a^2-r^2)^{1/2}(-2r)\,dr\right]d\theta$$
$$= \int_0^{\pi/2}\left[-\frac{1}{2}\cdot\frac{(a^2-r^2)^{3/2}}{3/2}\right]_0^{a\cos\theta} d\theta$$
$$= -\frac{1}{3}\int_0^{\pi/2}(a^3\sin^3\theta - a^3)\,d\theta = -\frac{a^3}{3}\left[\frac{2}{3}-\frac{\pi}{2}\right] = \frac{a^3}{18}(3\pi-4).$$

Example 2. *Evaluate* $\iint r^3\,dr\,d\theta$, *over the area bounded between the circles* $r = 2\cos\theta$ *and* $r = 4\cos\theta$.

Sol. The region of integration R is shown shaded. Here r varies from $2\cos\theta$ to $4\cos\theta$ while θ varies from $-\frac{\pi}{2}$ to $\frac{\pi}{2}$.

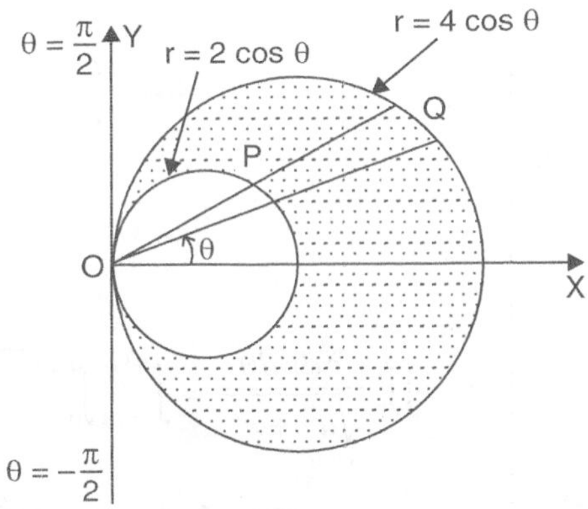

Fig. 3.9

$$\therefore \quad \iint_R r^3\,dr\,d\theta = \int_{-\pi/2}^{\pi/2}\int_{2\cos\theta}^{4\cos\theta} r^3\,dr\,d\theta$$
$$= \int_{-\pi/2}^{\pi/2}\left[\frac{r^4}{4}\right]_{2\cos\theta}^{4\cos\theta} d\theta$$
$$= \int_{-\pi/2}^{\pi/2}\frac{1}{4}(256\cos^4\theta - 16\cos^4\theta)\,d\theta$$
$$= 60\int_{-\pi/2}^{\pi/2}\cos^4\theta\,d\theta$$
$$= 120\int_0^{\pi/2}\cos^4\theta\,d\theta \qquad \text{[since } \cos^4\theta \text{ is an even function of } \theta]$$
$$= 120\times\frac{3\times1}{4\times2}\cdot\frac{\pi}{2} = \frac{45}{2}\pi.$$

Example 3. *Evaluate* $\iint r\sin\theta\,dr\,d\theta$ *over the area of the cardioid* $r = a(1+\cos\theta)$ *above the initial line.*

Sol. The region of integration R is covered by radial strips whose ends are $r = 0$ and $r = a(1+\cos\theta)$, the strips starting from $\theta = 0$ and ending at $\theta = \pi$.

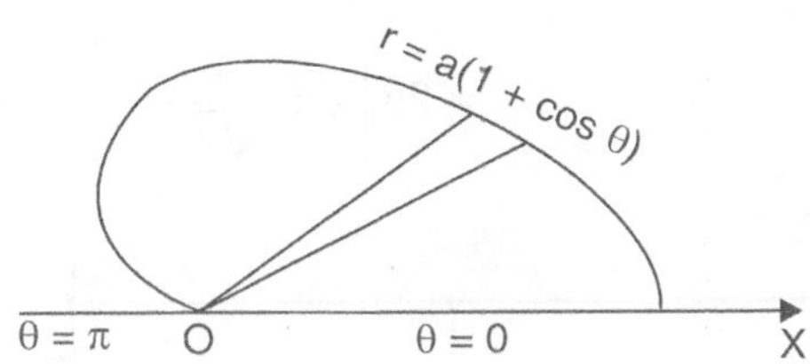

Fig. 3.10

$$\therefore \quad \iint_R r\sin\theta\,dr\,d\theta = \int_0^{\pi}\int_0^{a(1+\cos\theta)} r\sin\theta\,dr\,d\theta\cdot$$
$$= \int_0^{\pi}\sin\theta\left[\frac{r^2}{2}\right]_0^{a(1+\cos\theta)} d\theta = \frac{1}{2}\int_0^{\pi}\sin\theta\cdot a^2(1+\cos\theta)^2\,d\theta$$
$$= \frac{a^2}{2}\int_0^{\pi}2\sin\frac{0}{2}\cos\frac{\theta}{2}\cdot\left(2\cos^2\frac{\theta}{2}\right)^2 d\theta = 4a^2\int_0^{\pi}\sin\frac{\theta}{2}\cos^5\frac{\theta}{2}\,d\theta$$

$$= 4a^2 \int_0^{\pi/2} 2 \sin \phi \cos^5 \phi \, d\phi \qquad \left[\text{Putting } \frac{\theta}{2} = \phi \text{ and } d\theta = 2d\phi\right]$$

$$= -8a^2 \int_0^{\pi/2} \cos^5 \phi \, (-\sin \phi) \, d\phi = -8a^2 \cdot \left[\frac{\cos^6 \phi}{6}\right]_0^{\pi/2} = -\frac{4a^2}{3}(0 - 1) = \frac{4a^2}{3}.$$

Example 4. *Evaluate* $\iint \frac{r \, dr \, d\theta}{\sqrt{a^2 + r^2}}$ *over one loop of the lemniscate* $r^2 = a^2 \cos 2\theta$.

Sol. The region of integration R is covered by radial strips whose ends are $r = 0$ and $r = a\sqrt{\cos 2\theta}$, the strips starting from $\theta = -\frac{\pi}{4}$ and ending at $\theta = \frac{\pi}{4}$.

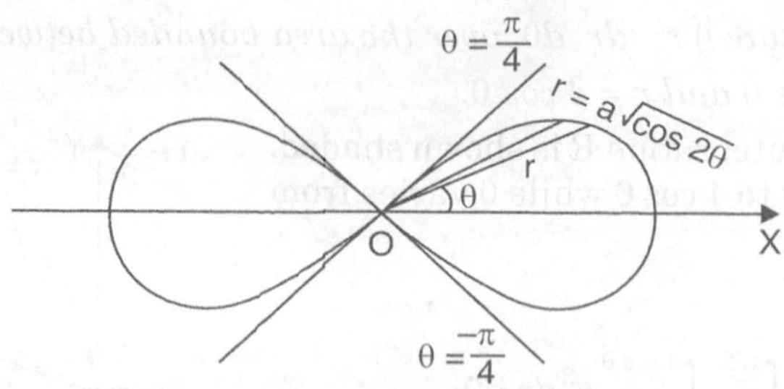

Fig. 3.11

$$\therefore \quad \iint_R \frac{r \, dr \, d\theta}{\sqrt{a^2 + r^2}} = \int_{-\pi/4}^{\pi/4} \int_0^{a\sqrt{\cos 2\theta}} \tfrac{1}{2}(a^2 + r^2)^{-1/2} \cdot 2r \, dr \, d\theta$$

$$= \int_{-\pi/4}^{\pi/4} \left[\frac{1}{2} \cdot \frac{(a^2 + r^2)^{1/2}}{1/2}\right]_0^{a\sqrt{\cos 2\theta}} d\theta = \int_{-\pi/4}^{\pi/4} [(a^2 + a^2 \cos 2\theta)^{1/2} - a] \, d\theta$$

$$= a \int_{-\pi/4}^{\pi/4} [(1 + \cos 2\theta)^{1/2} - 1] \, d\theta = a \int_{-\pi/4}^{\pi/4} [(2 \cos^2 \theta)^{1/2} - 1] \, d\theta$$

$$= a \int_{-\pi/4}^{\pi/4} (\sqrt{2} \cos \theta - 1) \, d\theta = 2a \int_0^{\pi/4} (\sqrt{2} \cos \theta - 1) \, d\theta = 2a \left[\sqrt{2} \sin \theta - \theta\right]_0^{\pi/4}$$

$$= 2a \left[\sqrt{2} \cdot \frac{1}{\sqrt{2}} - \frac{\pi}{4}\right] = 2a \left(1 - \frac{\pi}{4}\right).$$

EXERCISE 3.2

Evaluate the following integrals (1– 4):

1. $\int_0^{\pi} \int_0^{a \sin \theta} r \, dr \, d\theta$
2. $\int_0^{\pi/2} \int_0^{a \cos \theta} r \sin \theta \, dr \, d\theta$
3. $\int_0^{\pi/2} \int_{a(1+\cos \theta)}^{a} r \, dr \, d\theta$
4. $\int_0^{\pi} \int_0^{a(1+\cos \theta)} r^2 \cos \theta \, dr \, d\theta$

5. Show that $\iint_R r^2 \sin\theta \, dr \, d\theta = \frac{2a^3}{3}$, where R is the region bounded by the semi-circle $r = 2a \cos\theta$, above the initial line.

6. Evaluate $\iint r^3 \, dr \, d\theta$ over the area included between the circles $r = 2 \sin\theta$ and $r = 4 \sin\theta$.

Answers

1. $\frac{\pi a^2}{4}$ 2. $\frac{a^2}{6}$ 3. $-a^2\left(1+\frac{\pi}{8}\right)$

4. $\frac{5}{8}\pi a^3$ 6. $\frac{45\pi}{2}$.

3.4. CHANGE OF ORDER OF INTEGRATION

In a double integral, if the limits of integration are constant, then the order of integration is immaterial, provided the limits of integration are changed accordingly. Thus,

$$\int_c^d \int_a^b f(x, y)\, dx\, dy = \int_a^b \int_c^d f(x, y)\, dy\, dx .$$

But if the limits of integration are variable, a change in the order of integration necessitates change in the limits of integration. A rough sketch of the region of integration helps in fixing the new limits of integration.

ILLUSTRATIVE EXAMPLES

Example 1. *Change the order of integration in* $\int_0^a \int_y^a \frac{x\, dx\, dy}{x^2 + y^2}$ *and hence evaluate the same.*

Sol. From the limits of integration, it is clear that the region of integration is bounded by $x = y$, $x = a$, $y = 0$ and $y = a$. Thus the region of integration is the ΔOAB and is divided into horizontal strips. For changing the order of integration, we divide the region of integration into vertical strips. The new limits of integration become: y varies from 0 to x and x varies from 0 to a.

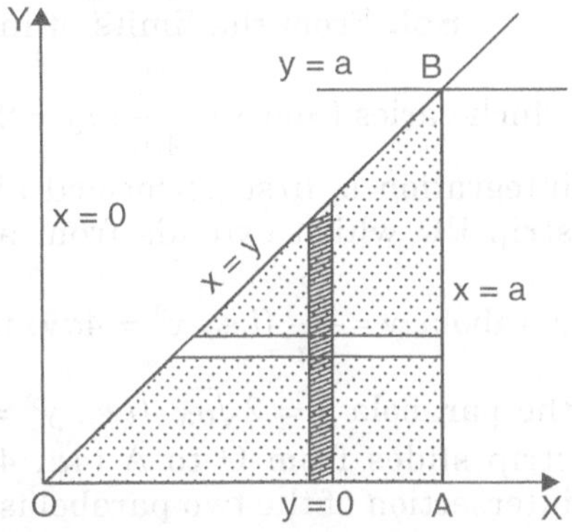

Fig. 3.12

$$\int_0^a \int_y^a \frac{x\, dx\, dy}{x^2 + y^2} = \int_0^a \int_0^x \frac{x\, dy\, dx}{x^2 + y^2}$$

$$= \int_0^a x \cdot \left[\frac{1}{x}\tan^{-1}\frac{y}{x}\right]_0^x dx$$

$$= \int_0^a \frac{\pi}{4}\, dx = \frac{\pi}{4} \cdot \Big[x\Big]_0^a = \frac{\pi a}{4}.$$

Example 2. *Change the order of integration in* $I = \int_0^1 \int_{x^2}^{2-x} xy\, dy\, dx$ *and hence evaluate the same.* (K.U. Nov. 2010, K.U. Dec. 2012)

Sol. From the limits of integration, it is clear that we have to integrate first with respect to y which varies from $y = x^2$ to $y = 2 - x$ and then with respect to x which varies from $x = 0$ to $x = 1$. The region of integration (shown shaded) is divided into vertical strips. For changing the order of integration, we divide the region of integration into horizontal strips.

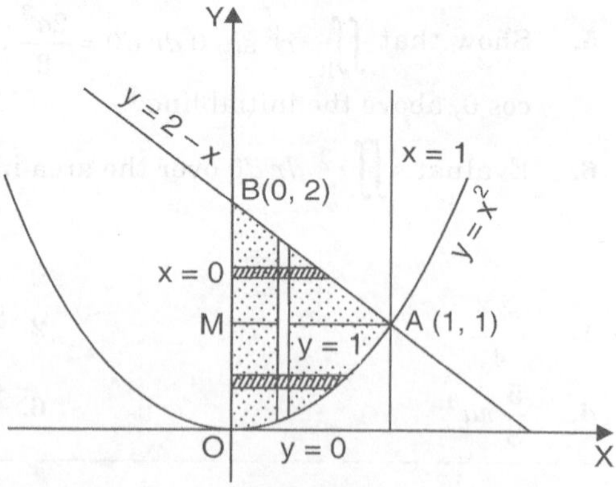

Fig. 3.13

Solving $y = x^2$ and $y = 2 - x$, the co-ordinates of A are (1, 1). Draw AM ⊥ OY. The region of integration is divided into two parts, OAM and MAB.

For the region OAM, x varies from 0 to $\sqrt{y}$ and y varies from 0 to 1. For the region MAB, x varies from 0 to $2 - y$ and y varies from 1 to 2.

$$\therefore \quad \int_0^1 \int_{x^2}^{2-x} xy\, dx\, dy = \int_0^1 \int_0^{\sqrt{y}} xy\, dx\, dy + \int_1^2 \int_0^{2-y} xy\, dx\, dy$$

$$= \int_0^1 y \cdot \left[\frac{x^2}{2}\right]_0^{\sqrt{y}} dy + \int_1^2 y \cdot \left[\frac{x^2}{2}\right]_0^{2-y} dy = \frac{1}{2}\int_0^1 y^2 dy + \frac{1}{2}\int_1^2 y(2-y)^2\, dy$$

$$= \frac{1}{2} \cdot \left[\frac{y^3}{3}\right]_0^1 + \frac{1}{2}\int_1^2 (4y - 4y^2 + y^3)\, dy = \frac{1}{6} + \frac{1}{2}\left[2y^2 - \frac{4y^3}{3} + \frac{y^4}{4}\right]_1^2$$

$$= \frac{1}{6} + \frac{1}{2}\left[\left(8 - \frac{32}{3} + 4\right) - \left(2 - \frac{4}{3} + \frac{1}{4}\right)\right] = \frac{3}{8}.$$

Example 3. *Change the order of integration in the following integral and evaluate:*

$$\int_0^{4a} \int_{x^2/4a}^{2\sqrt{ax}} dy\, dx.$$

Sol. From the limits of integration, it is clear that we have to integrate first w.r.t. y which varies from $y = \dfrac{x^2}{4a}$ to $y = 2\sqrt{ax}$ and then w.r.t. x which varies from $x = 0$ to $x = 4a$. Thus integration is first performed along the vertical strip PQ which extends from a point P on the parabola $y = \dfrac{x^2}{4a}$ (*i.e.*, $x^2 = 4ay$) to the point Q on the parabola $y = 2\sqrt{ax}$ (*i.e.*, $y^2 = 4ax$). Then the strip slides from O to A $(4a, 4a)$, the point of intersection of the two parabolas.

For changing the order of integration, we divide the region of integration OPAQO into horizontal strips P′Q′ which extend from P′ on the parabola $y^2 = 4ax$, *i.e.*, $x = \dfrac{y^2}{4a}$ to Q′ on the parabola $x^2 = 4ay$, *i.e.*, $x = 2\sqrt{ay}$. Then this strip slides from O to A$(4a, 4a)$, *i.e.*, varies from 0 to $4a$.

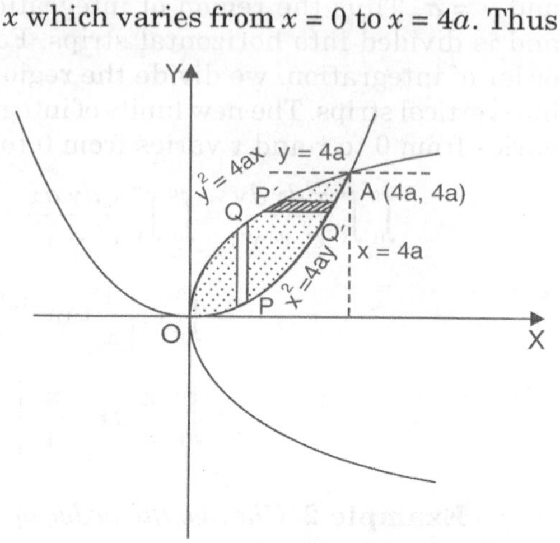

Fig. 3.14

$$\therefore \quad \int_0^{4a}\int_{x^2/4a}^{2\sqrt{ax}} dy\,dx = \int_0^{4a}\int_{y^2/4a}^{2\sqrt{ay}} dx\,dy$$

$$= \int_0^{4a}\Big[x\Big]_{y^2/4a}^{2\sqrt{ay}} dy = \int_0^{4a}\left(2\sqrt{ay} - \frac{y^2}{4a}\right)dy$$

$$= \left[2\sqrt{a}\cdot\frac{y^{3/2}}{3/2} - \frac{y^3}{12a}\right]_0^{4a} = \frac{4}{3}\sqrt{a}\cdot(4a)^{3/2} - \frac{64a^3}{12a}$$

$$= \frac{4}{3}\sqrt{a}\cdot 8a^{3/2} - \frac{16a^2}{3} = \frac{32a^2}{3} - \frac{16a^2}{3} = \frac{16a^2}{3}.$$

Example 4. *Change the order of integration in the integral* $\int_{-a}^{a}\int_0^{\sqrt{a^2-y^2}} f(x, y)\,dx\,dy.$

Sol. From the limits of integration, it is clear that we have to integrate first w.r.t. x which varies from $x = 0$ to $x = \sqrt{a^2 - y^2}$ and then w.r.t. y which varies from $y = -a$ to $y = a$. Thus integration is first performed along the horizontal strip PQ which extends from a point P on $x = 0$ (*i.e.*, y-axis) to the point Q on the circle $x = \sqrt{a^2 - y^2}$ (*i.e.*, $x^2 = a^2 - y^2$ or $x^2 + y^2 = a^2$). Then the strip slides from B′ to B.

For changing the order of integration, we divide the region of integration B′AQBPB′ into vertical strips P′Q′ which extend from P′ on the circle $y = -\sqrt{a^2 - x^2}$ to Q′ on the circle $y = \sqrt{a^2 - x^2}$. [$x^2 + y^2 = a^2 \Rightarrow y = \pm\sqrt{a^2 - x^2}$; for points in the IV quadrant, $y = -\sqrt{a^2 - x^2}$ and for points in the first quadrant, $y = \sqrt{a^2 - x^2}$] . Then this strip slides from y-axis ($x = 0$) to A, where $x = a$.

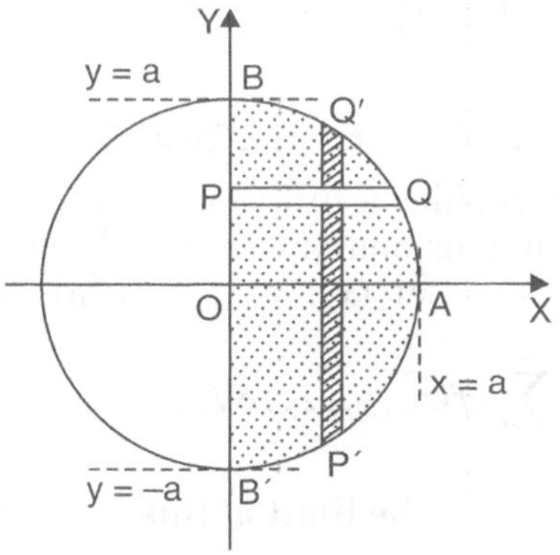

Fig. 3.15

$$\therefore \quad \int_{-a}^{a}\int_0^{\sqrt{a^2-x^2}} f(x,y)\,dx\,dy = \int_0^{a}\int_{-\sqrt{a^2-x^2}}^{\sqrt{a^2-x^2}} f(x,y)\,dy\,dx\,.$$

EXERCISE 3.3

Evaluate the following integrals by changing the order of integration:

1. $\int_0^1\int_0^{\sqrt{1-x^2}} y^2\,dy\,dx$

2. $\int_0^1\int_{4y}^{4} e^{x^2}\,dx\,dy$ (*K.U. April 2012*)

3. $\int_0^4\int_y^4 \frac{x\,dx\,dy}{x^2+y^2}$

4. $\int_0^{a/\sqrt{2}}\int_y^{\sqrt{a^2-y^2}} x\,dx\,dy$

5. $\int_0^1\int_x^{\sqrt{2-x^2}} \frac{x\,dy\,dx}{\sqrt{x^2+y^2}}$

6. $\int_0^a\int_{x^2/a}^{2a-x} xy\,dy\,dx$

7. $\int_0^b\int_0^{a/b\,\sqrt{b^2-y^2}} xy\,dx\,dy$

8. $\int_0^a\int_{\sqrt{ax}}^{a} \frac{y^2\,dy\,dx}{\sqrt{y^4-a^2x^2}}$

9. $\int_0^{\infty}\int_x^{\infty} \frac{e^{-y}}{y}\, dy\, dx$ **10.** $\int_0^{\infty}\int_0^{x} xe^{-\frac{x^2}{y}}\, dy\, dx$

11. $\int_0^{3}\int_1^{\sqrt{4-y}} (x+y)\, dx\, dy$ **12.** $\int_0^{a}\int_{x/a}^{\sqrt{x/a}} (x^2+y^2)\, dy\, dx$

13. $\int_0^{\frac{a}{\sqrt{2}}}\int_y^{\sqrt{a^2-y^2}} \log(x^2+y^2)\, dx\, dy, \ (a > 0)$

Answers

1. $\frac{\pi}{16}$ **2.** $\frac{1}{8}(e^{16}-1)$ **3.** π **4.** $\frac{\sqrt{2}a^3}{6}$

5. $1-\frac{1}{\sqrt{2}}$ **6.** $\frac{3a^4}{8}$ **7.** $\frac{1}{8}a^2b^2$ **8.** $\frac{\pi a^2}{6}$

9. 1 **10.** $\frac{1}{2}$ **11.** $\frac{241}{60}$ **12.** $\frac{a^3}{28}+\frac{a}{20}$

13. $\frac{\pi a^2}{4}\left(\log a - \frac{1}{2}\right)$.

3.5. TRIPLE INTEGRALS

Consider a function $f(x, y, z)$ which is continuous at every point of a finite region V of three dimensional space. Divide the region V into n sub-regions of respective volumes $\delta V_1, \delta V_2, \ldots, \delta V_n$. Let (x_r, y_r, z_r) be an arbitrary point in the rth sub-region. Consider the sum

$$\sum_{r=1}^{n} f(x_r, y_r, z_r)\, \delta V_r .$$

The limit of this sum as $n \to \infty$ and $\delta V_r \to 0$, if it exists, is called the *triple integral* of $f(x, y, z)$ over the region V and is denoted by $\iiint_V f(x, y, z)\, dV$.

For purposes of evaluation, it can be expressed as the repeated integral

$$\int_{z_1}^{z_2}\int_{y_1}^{y_2}\int_{x_1}^{x_2} f(x,y,z)\, dx\, dy\, dz, \qquad \ldots(1)$$

the order of integration depending upon the limits.

Let x_1, x_2 be function of y, z; y_1, y_2 be function of z and z_1, z_2 be constants, *i.e.*,

Let $x_1 = f_1(y, z), x_2 = f_2(y, z), y_1 = \phi_1(z), y_2 = \phi_2(z)$ and $z_1 = a, z_2 = b$.

Then the integral (1) is evaluated as follows:

$$\int_{z_1=a}^{z_2=b}\left[\int_{y_1=\phi_1(z)}^{y_2=\phi_2(z)}\left[\int_{x_1=f_1(y,z)}^{x_2=f_2(y,z)} f(x,y,z)\, dx\right] dy\right] dz$$

First $f(x, y, z)$ is integrated w.r.t. x (keeping y and z constant) between the limits x_1 and x_2. The resulting expression, which is a function of y and z is then integrated w.r.t. y (keeping z constant) between the limits y_1 and y_2. The resulting expression, which is a function of z only is then integrated w.r.t. z between the limits z_1 and z_2. The order of integration is from the innermost rectangle to the outermost rectangle.

Limits involving two variables are kept innermost, then the limits involving one variable and finally the constant limits.

If $x_1, x_2; y_1, y_2$ and z_1, z_2 are all constants, then the order of integration is immaterial, provided the limits are changed accordingly. Thus

$$\int_{z_1}^{z_2}\int_{y_1}^{y_2}\int_{x_1}^{x_2} f(x,y,z)\,dx\,dy\,dz = \int_{x_1}^{x_2}\int_{z_1}^{z_2}\int_{y_1}^{y_2} f(x,y,z)\,dy\,dz\,dx$$

$$= \int_{x_1}^{x_2}\int_{y_1}^{y_2}\int_{z_1}^{z_2} f(x,y,z)\,dz\,dy\,dx\text{, etc.}$$

ILLUSTRATIVE EXAMPLES

Example 1. *Evaluate* $\displaystyle\int_0^1\int_0^{\sqrt{1-x^2}}\int_0^{\sqrt{1-x^2-y^2}} \frac{1}{\sqrt{1-x^2-y^2-z^2}}\,dz\,dy\,dx.$

Sol.
$$I = \int_0^1\int_0^{\sqrt{1-x^2}}\int_0^{\sqrt{1-x^2-y^2}} \frac{1}{\sqrt{(1-x^2-y^2)-z^2}}\,dz\,dy\,dx$$

$$= \int_0^1\int_0^{\sqrt{1-x^2}}\left[\sin^{-1}\frac{z}{\sqrt{1-x^2-y^2}}\right]_0^{\sqrt{1-x^2-y^2}} dy\,dx$$

$$= \int_0^1\int_0^{\sqrt{1-x^2}} (\sin^{-1}1 - \sin^{-1}0)\,dy\,dx$$

$$= \int_0^1\int_0^{\sqrt{1-x^2}} \frac{\pi}{2}\,dy\,dx = \int_0^1 \frac{\pi}{2}\Big[y\Big]_0^{\sqrt{1-x^2}} dx$$

$$= \frac{\pi}{2}\int_0^1 \sqrt{1-x^2}\,dx = \frac{\pi}{2}\left[\frac{x\sqrt{1-x^2}}{2} + \frac{1}{2}\sin^{-1}x\right]_0^1$$

$$= \frac{\pi}{4}[\sin^{-1}1] = \frac{\pi}{4}\cdot\frac{\pi}{2} = \frac{\pi^2}{8}.$$

Example 2. *Evaluate* $\displaystyle\int_1^e\int_1^{\log y}\int_1^{e^x} \log z\,dz\,dx\,dy.$

Sol.
$$I = \int_1^e\int_1^{\log y}\left[\int_1^{e^x} \log z\,dz\right] dx\,dy$$

$$= \int_1^e \int_1^{\log y} \left(\log z \cdot z - \int \frac{1}{z} \cdot z \, dz \right)_1^{e^x} dx\, dy$$

$$= \int_1^e \int_1^{\log y} \Big\{ z (\log z - 1) \Big\}_1^{e^x} dx\, dy$$

$$= \int_1^e \int_1^{\log y} \{(x-1)e^x + 1\}\, dx\, dy$$

Now $\displaystyle\int_1^{\log y} [(x-1)e^x + 1]\, dx = \int_1^{\log y} (x-1)e^x\, dx + \Big[x \Big]_1^{\log y}$

$$= \Big[(x-1)e^x \Big]_1^{\log y} - \int_1^{\log y} 1 \cdot e^x\, dx + \log y - 1$$

$$= (\log y - 1)\, e^{\log y} - \Big[e^x \Big]_1^{\log y} + \log y - 1$$

$$= y(\log y - 1) - (e^{\log y} - e) + \log y - 1 \qquad [\because \; e^{\log y} = y]$$

$$= y(\log y - 1) - y + e + \log y - 1 = (y+1)\log y - 2y + e - 1$$

$\therefore$
$$\text{I} = \int_1^e [\log y \cdot (y+1) - 2y + e - 1]\, dy$$

$$= \left[\log y \cdot \left(\frac{y^2}{2} + y \right) \right]_1^e - \int_1^e \frac{1}{y}\left(\frac{y^2}{2} + y \right) dy - \Big[y^2 \Big]_1^e + (e-1)\Big[y \Big]_1^e$$

$$= \frac{e^2}{2} + e - \int_1^e \left(\frac{y}{2} + 1 \right) dy - (e^2 - 1) + (e-1)^2$$

$$= \frac{e^2}{2} + e - \left[\frac{y^2}{4} + y \right]_1^e - 2e + 2 = \frac{e^2}{2} + e - \left[\left(\frac{e^2}{4} + e \right) - \left(\frac{1}{4} + 1 \right) \right] - 2e + 2$$

$$= \frac{e^2}{4} - 2e + \frac{13}{4} = \frac{1}{4}(e^2 - 8e + 13).$$

EXERCISE 3.4

Evaluate the following integrals (1–11):

1. $\displaystyle\int_0^1 \int_0^1 \int_0^1 e^{x+y+z}\, dx\, dy\, dz$. (*K.U. Dec. 2012*)

2. $\displaystyle\int_0^a \int_0^a \int_0^a (yz + zx + xy)\, dx\, dy\, dz$

3. $\displaystyle\int_{-1}^1 \int_0^z \int_{x-z}^{x+z} (x + y + z)\, dy\, dx\, dz$

4. $\displaystyle\int_1^3 \int_{1/x}^1 \int_0^{\sqrt{xy}} xyz\, dz\, dy\, dx$

5. $\displaystyle\int_0^1 \int_0^{1-x} \int_0^{1-x-y} xyz\, dz\, dy\, dx$

6. $\displaystyle\int_0^1 \int_{y^2}^1 \int_0^{1-x} x\, dz\, dx\, dy$

7. $\int_0^1 \int_0^{\sqrt{1-x^2}} \int_0^{\sqrt{1-x^2-y^2}} xyz\, dz\, dy\, dx$

8. $\int_0^{\log 2} \int_0^{x} \int_0^{x+\log y} e^{x+y+z}\, dz\, dy\, dx$

9. $\int_0^{\pi/2} d\theta \int_0^{a\sin\theta} dr \int_0^{(a^2-r^2)/a} r dz$

10. $\int_0^{\pi/2} \int_0^{a\cos\theta} \int_0^{\sqrt{a^2-r^2}} r dz\, dr\, d\theta$

11. $\int_0^{a} \int_0^{x} \int_0^{x-y} e^{x+y+z}\, dz\, dy\, dx$.

Answers

1. $(e-1)^3$
2. $\frac{3}{4}a^5$
3. 0
4. $\frac{13}{9} - \frac{1}{6}\log 3$
5. $\frac{1}{720}$
6. $\frac{4}{35}$
7. $\frac{1}{48}$
8. $\frac{8}{3}\log 2 - \frac{19}{9}$
9. $\frac{5\pi a^3}{64}$
10. $\frac{a^3}{3}\left(\frac{\pi}{2} - \frac{2}{3}\right)$
11. $\frac{1}{8}e^{4a} - \frac{3}{4}e^{2a} + e^a - \frac{3}{8}$.

3.6. CHANGE OF VARIABLES

Quite often, the evaluation of a double or triple integral is greatly simplified by a suitable change of variables.

Let the variables x, y in the double integral $\iint_R f(x, y)\, dx\, dy$ be changed to u, v by means of the relations $x = \phi(u, v)$, $y = \psi(u, v)$, then the double integral is transformed into

$$\iint_{R'} f\{\phi(u, v), \psi(u, v)\}\, |\, J\, |\, du\, dv, \quad \text{where } J = \frac{\partial(x, y)}{\partial(u, v)} = \begin{vmatrix} \frac{\partial x}{\partial u} & \frac{\partial x}{\partial v} \\ \frac{\partial y}{\partial u} & \frac{\partial y}{\partial v} \end{vmatrix}$$

is the Jacobian of transformation from (x, y) to (u, v) co-ordinates and R′ is the region in the uv-plane which corresponds to the region R in the xy-plane.

(i) To change cartesian co-ordinates (x, y) to polar co-ordinates (r, θ).

Here we have $x = r\cos\theta$, $y = r\sin\theta$ so that $x^2 + y^2 = r^2$

$$J = \begin{vmatrix} \frac{\partial x}{\partial r} & \frac{\partial x}{\partial \theta} \\ \frac{\partial y}{\partial r} & \frac{\partial y}{\partial \theta} \end{vmatrix}$$

$$= \begin{vmatrix} \cos\theta & -r\sin\theta \\ \sin\theta & r\cos\theta \end{vmatrix} = r(\cos^2\theta + \sin^2\theta) = r$$

$$\therefore \iint_R f(x, y)\, dx\, dy = \iint_{R'} f(r\cos\theta, r\sin\theta)\, r\, dr\, d\theta$$

i.e., replace x by $r\cos\theta$, y by $r\sin\theta$ and $dx\, dy$ by $r\, dr\, d\theta$.

(ii) To change cartesian co-ordinates (x, y, z) to spherical polar co-ordinates (r, θ, φ).

Here, we have
$$x = r\sin\theta\cos\phi$$
$$y = r\sin\theta\sin\phi$$
$$z = r\cos\theta$$

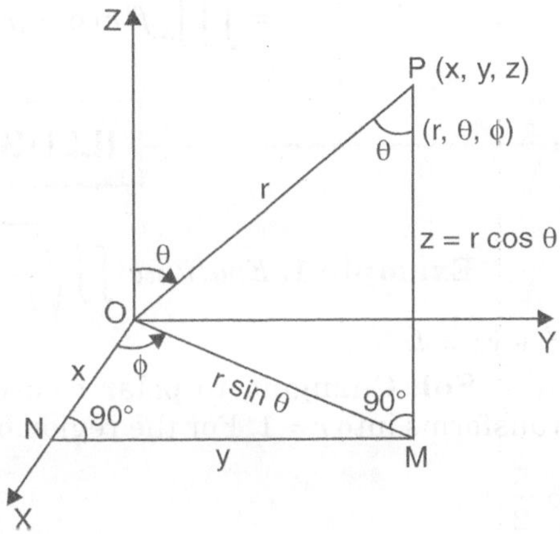

Fig. 3.16

so that $x^2 + y^2 + z^2 = r^2$

$$J = \begin{vmatrix} \frac{\partial x}{\partial r} & \frac{\partial x}{\partial \theta} & \frac{\partial x}{\partial \phi} \\ \frac{\partial y}{\partial r} & \frac{\partial y}{\partial \theta} & \frac{\partial y}{\partial \phi} \\ \frac{\partial z}{\partial r} & \frac{\partial z}{\partial \theta} & \frac{\partial z}{\partial \phi} \end{vmatrix}$$

$$= \begin{vmatrix} \sin\theta\cos\phi & r\cos\theta\cos\phi & -r\sin\theta\sin\phi \\ \sin\theta\sin\phi & r\cos\theta\sin\phi & r\sin\theta\cos\phi \\ \cos\theta & -r\sin\theta & 0 \end{vmatrix} = r^2 \sin\theta$$

$$\therefore \quad \iiint_V f(x, y, z)\, dx\, dy\, dz = \iiint_{V'} f(r\sin\theta\cos\phi, r\sin\theta\sin\phi, r\cos\theta)\, r^2 \sin\theta\, dr\, d\theta\, d\phi.$$

(iii) To change cartesian co-ordinates (x, y , z) to cylindrical co-ordinates (r, φ, z).

Here, we have $x = r\cos\phi$

$y = r\sin\phi$

$z = z$

$$J = \begin{vmatrix} \frac{\partial x}{\partial r} & \frac{\partial x}{\partial \phi} & \frac{\partial x}{\partial z} \\ \frac{\partial y}{\partial r} & \frac{\partial y}{\partial \phi} & \frac{\partial y}{\partial z} \\ \frac{\partial z}{\partial r} & \frac{\partial z}{\partial \phi} & \frac{\partial z}{\partial z} \end{vmatrix}$$

$$= \begin{vmatrix} \cos\phi & -r\sin\phi & 0 \\ \sin\phi & r\cos\phi & 0 \\ 0 & 0 & 1 \end{vmatrix}$$

$$= r(\cos^2\phi + \sin^2\phi) = r$$

$$\therefore \quad \iiint_V f(x, y, z)\, dx\, dy\, dz$$

$$= \iiint_{V'} f(r\cos\phi, r\sin\phi, z)\, r\, dr\, d\phi\, dz.$$

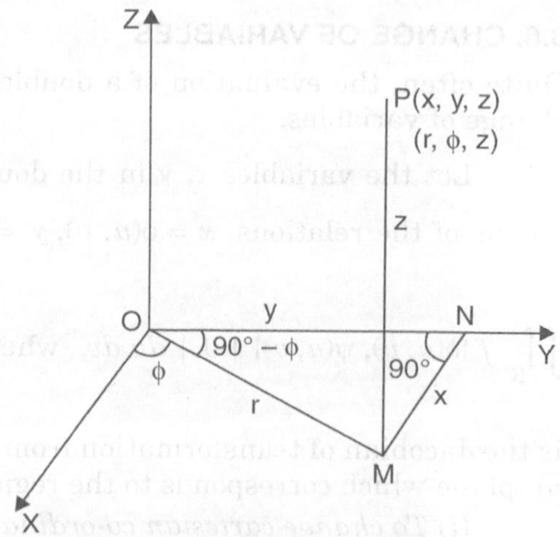

Fig. 3.17

ILLUSTRATIVE EXAMPLES

Example 1. *Evaluate* $\iint \sqrt{\frac{1 - x^2 - y^2}{1 + x^2 + y^2}}\, dx\, dy$ *over the positive quadrant of the circle* $x^2 + y^2 = 1$.

Sol. Changing to polar co-ordinates by putting $x = r\cos\theta$, $y = r\sin\theta$; $x^2 + y^2 = 1$ transforms into $r = 1$. For the region of integration R, r varies from 0 to 1 and θ varies from 0 to $\frac{\pi}{2}$.

$$I = \iint_R \sqrt{\frac{1-x^2-y^2}{1+x^2+y^2}}\, dx\, dy$$

$$= \int_0^{\pi/2}\int_0^1 \sqrt{\frac{1-r^2}{1+r^2}}\; r\, dr\, d\theta$$

| $dx\, dy$ is replaced by $r\, dr\, d\theta$

$$= \int_0^{\pi/2}\int_0^1 \frac{r(1-r^2)}{\sqrt{1-r^4}}\, dr\, d\theta$$

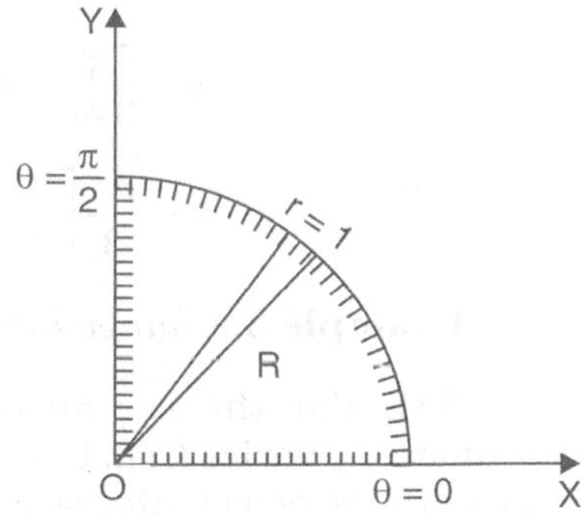

Fig. 3.18

Now $\displaystyle\int_0^1 \frac{r(1-r^2)}{\sqrt{1-r^4}}\, dr = \int_0^1 \left(\frac{r}{\sqrt{1-r^4}} - \frac{r^3}{\sqrt{1-r^4}}\right) dr$

$$= \frac{1}{2}\int_0^1 \frac{2r}{\sqrt{1-r^4}}\, dr + \frac{1}{4}\int_0^1 -4r^3\,(1-r^4)^{-1/2}\, dr$$

$$= \frac{1}{2}\int_0^1 \frac{dt}{\sqrt{1-t^2}} + \frac{1}{4}\cdot\left[\frac{(1-r^4)^{1/2}}{1/2}\right]_0^1, \text{ where } t = r^2$$

$$= \frac{1}{2}\Big[\sin^{-1} t\Big]_0^1 + \frac{1}{2}\,(0-1) = \frac{1}{2}\left(\frac{\pi}{2}\right) - \frac{1}{2} = \frac{\pi}{4} - \frac{1}{2}$$

$$\therefore \qquad I = \int_0^{\pi/2}\left(\frac{\pi}{4} - \frac{1}{2}\right) d\theta = \left(\frac{\pi}{4} - \frac{1}{2}\right)\Big[\theta\Big]_0^{\pi/2} = \left(\frac{\pi}{4} - \frac{1}{2}\right)\frac{\pi}{2} = \frac{\pi^2}{8} - \frac{\pi}{4}.$$

Example 2. *Evaluate* $\iint \sqrt{a^2-x^2-y^2}\; dx\, dy$ *over the semi-circle* $x^2 + y^2 = ax$ *in the positive quadrant.*

Sol. Changing to polar co-ordinates, $x^2 + y^2 = ax$ transforms into $r = a\cos\theta$. For the region of integration R, r varies from 0 to $a\cos\theta$ and θ varies from 0 to $\frac{\pi}{2}$.

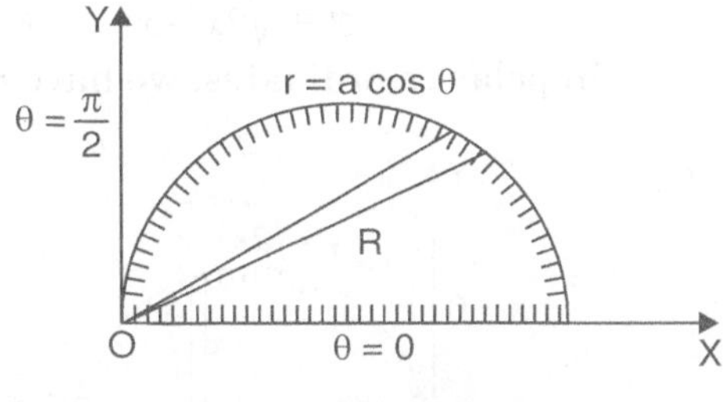

Fig. 3.19

$$\therefore \quad \iint_R \sqrt{a^2-x^2-y^2}\; dx\, dy$$

$$= \int_0^{\pi/2}\int_0^{a\cos\theta} \sqrt{a^2-r^2}\cdot r\, dr\, d\theta$$

$$= \int_0^{\pi/2}\int_0^{a\cos\theta} -\frac{1}{2}(a^2-r^2)^{1/2}\,(-2r)\, dr\, d\theta$$

$$= \int_0^{\pi/2} -\frac{1}{2}\cdot\left[\frac{(a^2-r^2)^{3/2}}{3/2}\right]_0^{a\cos\theta} d\theta$$

$$= -\frac{1}{3}\int_0^{\pi/2} (a^3 \sin^3 \theta - a^3)\, d\theta = -\frac{a^3}{3}\int_0^{\pi/2} (\sin^3 \theta - 1)\, d\theta$$

$$= -\frac{a^3}{3}\left[\frac{2}{3} - \frac{\pi}{2}\right] = \frac{a^3}{3}\left(\frac{\pi}{2} - \frac{2}{3}\right).$$

Example 3. *Change into polar co-ordinates and evaluate* $\int_0^\infty \int_0^\infty e^{-(x^2 + y^2)}\, dy\, dx$.

Sol. For the region of integration in cartesian coordinates, y varies from 0 to ∞ and x also varies from 0 to ∞. Thus the region of integration is the plane XOY. Changing to polar co-ordinates by putting $x = r \cos \theta$, $y = r \sin \theta$ so that $x^2 + y^2 = r^2$; for the region of integration r varies from 0 to ∞ and θ varies from 0 to $\frac{\pi}{2}$.

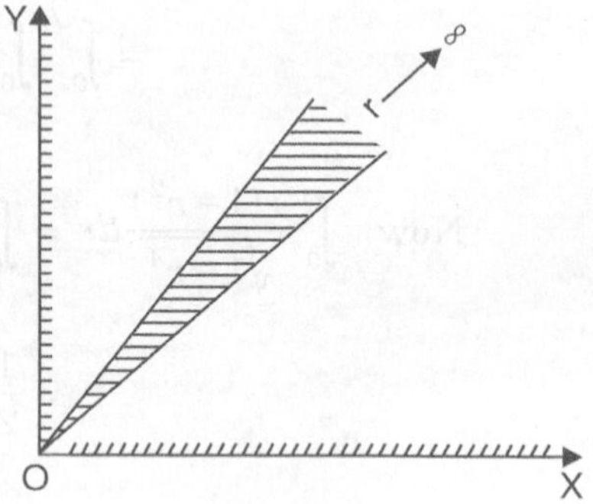

Fig. 3.20

$$\therefore \quad \int_0^\infty \int_0^\infty e^{-(x^2 + y^2)}\, dy\, dx = \int_0^{\pi/2} \int_0^\infty e^{-r^2} \cdot r\, dr\, d\theta$$

$$= \int_0^{\pi/2} \int_0^\infty \frac{1}{2} \cdot e^{-r^2} \cdot 2r\, dr\, d\theta$$

$$= \int_0^{\pi/2} \int_0^\infty \frac{1}{2} e^{-t} dt\, d\theta, \text{ where } t = r^2$$

$$= \int_0^{\pi/2} \left[-\frac{1}{2} e^{-t}\right]_0^\infty d\theta = -\frac{1}{2}\int_0^{\pi/2} (0 - 1)\, d\theta = \frac{1}{2}\Big[\theta\Big]_0^{\pi/2} = \frac{\pi}{4}.$$

Example 4. *Evaluate* $\int_0^2 \int_0^{\sqrt{2x - x^2}} \frac{x\, dy\, dx}{\sqrt{x^2 + y^2}}$ *by changing to polar co-ordinates.*

Sol. In the given integral, y varies from 0 to $\sqrt{2x - x^2}$ and x varies from 0 to 2.

$$y = \sqrt{2x - x^2} \quad \Rightarrow \quad y^2 = 2x - x^2 \quad \Rightarrow \quad x^2 + y^2 = 2x.$$

In polar co-ordinates, we have $r^2 = 2r \cos \theta$ or $r = 2 \cos \theta$.

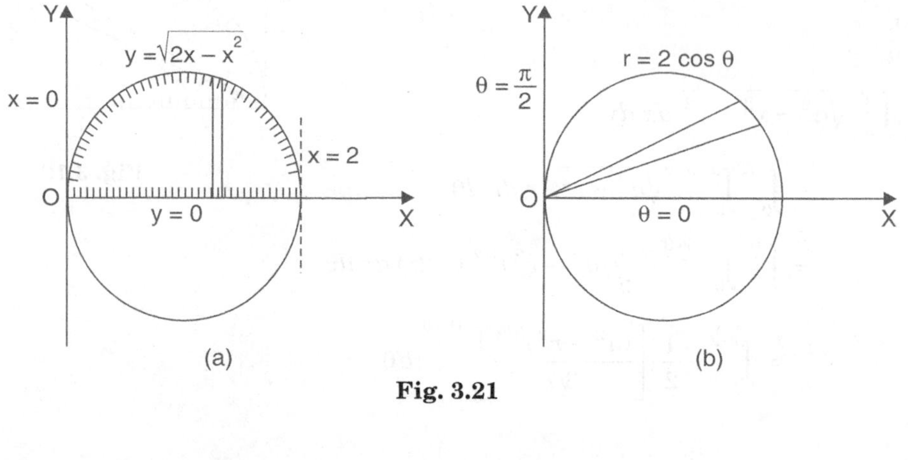

Fig. 3.21

$\therefore$ For the region of integration, r varies from 0 to $2\cos\theta$ and θ varies from 0 to $\frac{\pi}{2}$.

In the given integral, replacing x by $r\cos\theta$, y by $r\sin\theta$, $dy\,dx$ by $r\,dr\,d\theta$, we have

$$I = \int_0^{\pi/2}\int_0^{2\cos\theta} \frac{r\cos\theta \cdot r\,dr\,d\theta}{r} = \int_0^{\pi/2}\int_0^{2\cos\theta} r\cos\theta\,dr\,d\theta$$

$$= \int_0^{\pi/2} \cos\theta \left[\frac{r^2}{2}\right]_0^{2\cos\theta} d\theta = \int_0^{\pi/2} 2\cos^3\theta\,d\theta = 2\cdot\frac{2}{3} = \frac{4}{3}.$$

Example 5. *Evaluate* $\iiint z(x^2 + y^2 + z^2)\,dx\,dy\,dz$ *through the volume of the cylinder* $x^2 + y^2 = a^2$ *intercepted by the planes* $z = 0$ *and* $z = h$.

Sol. Changing to cylindrical co-ordinates by changing x to $r\cos\phi$, y to $r\sin\phi$ and replacing $dx\,dy\,dz$ by $r\,dr\,d\phi\,dz$

$$I = \int_0^h\int_0^{2\pi}\int_0^a z(r^2 + z^2)\,r\,dr\,d\phi\,dz = \int_0^h\int_0^{2\pi}\int_0^a (zr^3 + z^3 r)\,dr\,d\phi\,dz$$

$$= \int_0^h\int_0^{2\pi}\left[z\cdot\frac{r^4}{4} + z^3\cdot\frac{r^2}{2}\right]_0^a d\phi\,dz = \int_0^h\int_0^{2\pi}\left(\frac{a^4}{4}z + \frac{a^2}{2}z^3\right)d\phi\,dz$$

$$= \int_0^h \left(\frac{a^4}{4}z + \frac{a^2}{2}z^3\right)\Big[\phi\Big]_0^{2\pi} dz = \int_0^h 2\pi\left(\frac{a^4}{4}z + \frac{a^2}{2}z^3\right)dz$$

$$= 2\pi\left[\frac{a^4z^2}{8} + \frac{a^2z^4}{8}\right]_0^h = \frac{\pi}{4}(a^4h^2 + a^2h^4) = \frac{\pi}{4}a^2h^2(a^2 + h^2).$$

Example 6. *Evaluate* $\int_0^1\int_0^{\sqrt{1-x^2}}\int_0^{\sqrt{1-x^2-y^2}} \frac{dz\,dy\,dx}{\sqrt{1-x^2-y^2-z^2}}$, *by changing to spherical polar co-ordinates.*

Sol. Here the region of integration is bounded by

$z = 0,\quad z = \sqrt{1-x^2-y^2}\quad$ (*i.e.*, $x^2 + y^2 + z^2 = 1$)

$y = 0,\quad y = \sqrt{1-x^2}\quad$ (*i.e.*, $x^2 + y^2 = 1$)

$x = 0,\quad x = 1$

which is the volume of the sphere $x^2 + y^2 + z^2 = 1$ in the positive octant.

Changing to spherical polar co-ordinates by putting $x = r\sin\theta\cos\phi$, $y = r\sin\theta\sin\phi$, $z = r\cos\theta$ so that $x^2 + y^2 + z^2 = r^2$.

For the volume of sphere $x^2 + y^2 + z^2 = 1$ in the positive octant, r varies from 0 to 1, θ varies from 0 to $\frac{\pi}{2}$ and ϕ varies from 0 to $\frac{\pi}{2}$.

Replacing $dz\,dy\,dx$ by $r^2\sin\theta\,dr\,d\theta\,d\phi$, we have

$$I = \int_0^{\pi/2}\int_0^{\pi/2}\int_0^1 \frac{r^2\sin\theta\,dr\,d\theta\,d\phi}{\sqrt{1-r^2}} = \int_0^{\pi/2}\int_0^{\pi/2}\int_0^1 \frac{1-(1-r^2)}{\sqrt{1-r^2}}\sin\theta\,dr\,d\theta\,d\phi$$

$$= \int_0^{\pi/2}\int_0^{\pi/2}\int_0^1 \left(\frac{1}{\sqrt{1-r^2}} - \sqrt{1-r^2}\right) \sin\theta \, dr \, d\theta \, d\phi$$

$$= \int_0^{\pi/2}\int_0^{\pi/2} \sin\theta \left[\sin^{-1} r - \left(\frac{r\sqrt{1-r^2}}{2} + \frac{1}{2}\sin^{-1} r\right)\right]_0^1 d\theta \, d\phi$$

$$= \int_0^{\pi/2}\int_0^{\pi/2} \sin\theta \left[\frac{\pi}{2} - \frac{1}{2}\cdot\frac{\pi}{2}\right] d\theta \, d\phi = \int_0^{\pi/2} \frac{\pi}{4}\left[-\cos\theta\right]_0^{\pi/2} d\phi = \int_0^{\pi/2} \frac{\pi}{4} \, d\phi$$

$$= \frac{\pi}{4}\cdot\frac{\pi}{2} = \frac{\pi^2}{8}.$$

Note. For the whole volume of the sphere $x^2 + y^2 + z^2 = a^2$.

$0 \le r \le a, 0 \le \theta \le \pi, 0 \le \phi \le 2\pi.$

Example 7. *Evaluate* $\iint_R (x+y)^2 \, dx \, dy$, *where R is the parallelogram in the xy-plane with vertices (1, 0), (3, 1), (2, 2), (0, 1), using the transformation $u = x + y$ and $v = x - 2y$.*

Sol. The vertices A(1, 0), B(3, 1), C(2, 2), D(0, 1) of the parallelogram ABCD in the xy-plane become A′(1, 1), B′(4, 1), C′(4, –2), D′(1, –2) in the uv-plane under the given transformation.

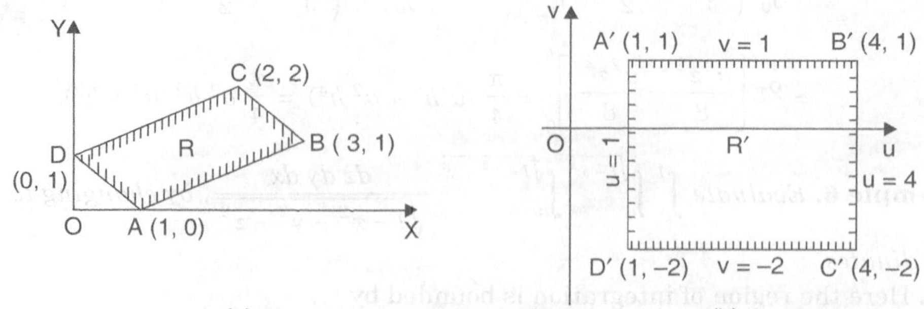

Fig. 3.22

The region R in the xy-plane becomes the region R′ in the uv-plane which is a rectangle bounded by the line $u = 1$, $u = 4$ and $v = -2$, $v = 1$. Solving the given equations for x and y, we have $x = \frac{1}{3}(2u + v)$, $\quad y = \frac{1}{3}(u - v)$

$$\therefore \qquad J = \frac{\partial(x, y)}{\partial(u, v)} = \begin{vmatrix} \dfrac{\partial x}{\partial u} & \dfrac{\partial v}{\partial x} \\ \dfrac{\partial y}{\partial u} & \dfrac{\partial y}{\partial v} \end{vmatrix} = \begin{vmatrix} \dfrac{2}{3} & \dfrac{1}{3} \\ \dfrac{1}{3} & -\dfrac{1}{3} \end{vmatrix} = -\frac{1}{3}$$

$$\therefore \quad \iint_R (x+y)^2 \, dx \, dy = \iint_{R'} u^2 \, |J| \, du \, dv = \int_{-2}^{1}\int_1^4 u^2 \cdot \frac{1}{3} \, du \, dv$$

$$= \int_{-2}^{1} \frac{1}{3}\left[\frac{u^3}{3}\right]_1^4 dv = \int_{-2}^{1} 7 \, dv = 7\left[v\right]_{-2}^{1} = 7 \times 3 = 21.$$

EXERCISE 3.5

1. Evaluate $\iint \sin \pi (x^2 + y^2)\, dx\, dy$ over the region bounded by the circle $x^2 + y^2 = 1$ by changing to polar co-ordinates.

2. Evaluate $\iint (a^2 - x^2 - y^2)\, dx\, dy$ over the semi-circle $x^2 + y^2 = ax$ in the positive quadrant by changing to polar co-ordinates.

3. Evaluate $\iint (x^2 + y^2)^{7/2}\, dx\, dy$ over the circle $x^2 + y^2 = 1$.

4. Evaluate $\iint xy\, (x^2 + y^2)^{3/2}\, dx\, dy$ over the positive quadrant of the circle $x^2 + y^2 = 1$.

5. Evaluate the following by changing into polar co-ordinates:

(i) $\int_0^a \int_y^a \frac{x\, dx\, dy}{x^2 + y^2}$ (ii) $\int_{-a}^a \int_{-\sqrt{a^2 - x^2}}^{\sqrt{a^2 - x^2}} dy\, dx$ (iii) $\int_0^{4a} \int_{y^2/4a}^{y} \frac{x^2 - y^2}{x^2 + y^2}\, dx\, dy$

(iv) $\int_0^a \int_0^{\sqrt{a^2 - y^2}} (x^2 + y^2)\, dx\, dy$ (v) $\int_0^a \int_0^{\sqrt{a^2 - y^2}} y^2 \sqrt{x^2 + y^2}\, dy\, dx$ (vi) $\int_0^a \int_y^a \frac{x^2\, dx\, dy}{\sqrt{x^2 + y^2}}$.

6. Evaluate $\iint_D e^{-(x^2 + y^2)}\, dy\, dx$, where D is the region bounded by $x^2 + y^2 = a^2$.

7. Evaluate $\iint xy\, (x^2 + y^2)^{n/2}\, dx\, dy$ over the positive quadrant of $x^2 + y^2 = 4$, supposing $n + 3 > 0$.

8. Evaluate $\iint \sqrt{\frac{1 - \frac{x^2}{a^2} - \frac{y^2}{b^2}}{1 + \frac{x^2}{a^2} + \frac{y^2}{b^2}}}\, dx\, dy$ over the positive quadrant of the ellipse $\frac{x^2}{a^2} + \frac{y^2}{b^2} = 1$.

9. Transform the following to cartesian form and hence evaluate $\int_0^\pi \int_0^a r^3 \sin\theta \cos\theta\, dr\, d\theta$.

10. Evaluate $\iiint (x + y + z)\, dx\, dy\, dz$ over the tetrahedron bounded by the planes $x = 0, y = 0, z = 0$ and $x + y + z = 1$.

11. Evaluate $\iiint \frac{dx\, dy\, dz}{\sqrt{1 - x^2 - y^2 - z^2}}$, the integral being extended to the positive octant of the sphere $x^2 + y^2 + z^2 = 1$.

12. Evaluate $\iiint \frac{dx\, dy\, dz}{(x + y + z + 1)^3}$ over the tetrahedron bounded by the co-ordinate planes and the plane $x + y + z = 1$.

13. Evaluate $\iiint z(x^2 + y^2)\, dx\, dy\, dz$ over the volume of the cylinder $x^2 + y^2 = 1$ intercepted by the planes $z = 2$ and $z = 3$.

14. Evaluate the following integrals through the volume of the sphere $x^2 + y^2 + z^2 = 1$, by changing into spherical polar co-ordinates :

(i) $\iiint z^2\, dx\, dy\, dz$ (ii) $\iiint (x^2 + y^2 + z^2)^m\, dx\, dy\, dz$. $(m > 0)$

Answers

1. 2 **2.** $\frac{5\pi a^4}{64}$ **3.** $\frac{2\pi}{9}$ **4.** $\frac{1}{14}$

5. (*i*) $\frac{\pi a}{4}$ (*ii*) πa^2 (*iii*) $8\left(\frac{\pi}{2}-\frac{5}{3}\right)a^2$ (*iv*) $\frac{\pi a^4}{8}$

(*v*) $\frac{\pi a^5}{20}$ (*vi*) $\frac{a^3}{3}\log(\sqrt{2}+1)$ **6.** $\pi(1-e^{-a^2})$ **7.** $\frac{2^{n+3}}{n+4}$

8. $\frac{\pi ab}{8}(\pi-2)$ **9.** 0 **10.** $\frac{1}{8}$ **11.** $\frac{\pi^2}{8}$

12. $\frac{1}{2}\log 2-\frac{5}{16}$ **13.** $\frac{5\pi}{4}$ **14.** (*i*) $\frac{4\pi}{15}$ (*ii*) $\frac{4\pi}{2m+3}$.

3.7. AREA BY DOUBLE INTEGRATION

(*a*) *Cartesian Co-ordinates.* The area A of the region bounded by the curves $y = f_1(x)$, $y = f_2(x)$ and the lines $x = a, x = b$ is given by $A = \int_a^b \int_{f_1(x)}^{f_2(x)} dy\, dx$

The area A of the region bounded by the curves $x = f_1(y)$, $x = f_2(y)$ and the lines $y = c$, $y = d$ is given by

$$A = \int_c^d \int_{f_1(y)}^{f_2(y)} dx\, dy .$$

(*b*) *Polar co-ordinates.* The area A of the region bounded by the curves $r = f_1(\theta)$, $r = f_2(\theta)$ and the lines $\theta = \alpha$, $\theta = \beta$ is given by $A = \int_\alpha^\beta \int_{f_1(\theta)}^{f_2(\theta)} r\, dr\, d\theta$.

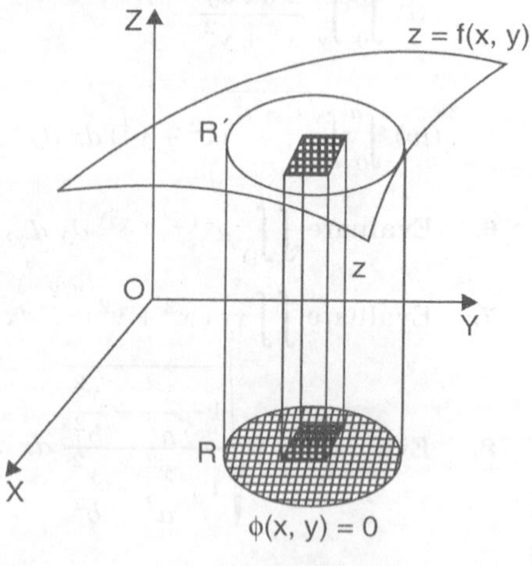

Fig. 3.23

3.8. VOLUME AS A DOUBLE INTEGRAL

(*a*) *Cartesian Co-ordinates.* Consider a surface $z = f(x, y)$ ...(1)

Let the orthogonal projection of its region R′ on the *xy*-plane be the region R given by $\phi(x, y) = 0$. ...(2)

Now (2) represents a cylinder with generators parallel to *z*-axis and the guiding curve given by (2). Let V be the volume of this cylinder between R and R′.

Divide R into elementary rectangles of area $\delta x\, \delta y$ by drawing lines parallel to the axis of *x* and *y*. On each of these rectangles, erect prisms of lengths parallel to *z*-axis. Volume of this prism between R and R′ is $z\, \delta x\, \delta y$. The volume V is composed of such prisms.

$$\therefore \quad V = \lim_{\substack{\delta x \to 0 \\ \delta y \to 0}} \sum z\, \delta x\, \delta y = \iint_R z\, dx\, dy .$$

(*b*) *Cylindrical Co-ordinates.* Let the equation of the surface be $z = f(r, \phi)$. Replacing $dx\, dy$ by $r\, dr\, d\phi$, we get $V = \iint_R zr\, dr\, d\phi$.

ILLUSTRATIVE EXAMPLES

Example 1. *Find the area lying between the parabola $y = 4x - x^2$ and the line $y = x$.*

Sol. The two curves intersect at points whose abscissae are given by $4x - x^2 = x$

or $x^2 - 3x = 0$ *i.e.,* $x = 0, 3$.

Using vertical strips, the required area lies between $x = 0, x = 3$ and $y = x, y = 4x - x^2$.

$$\therefore \quad \text{Required area} = \int_0^3 \int_x^{4x - x^2} dy\, dx$$

$$= \int_0^3 \Big[y \Big]_x^{4x - x^2} dx$$

$$= \int_0^3 (4x - x^2 - x)\, dx = \int_0^3 (3x - x^2)\, dx$$

$$= \left[\frac{3x^2}{2} - \frac{x^3}{3} \right]_0^3 = \frac{27}{2} - 9 = 4.5.$$

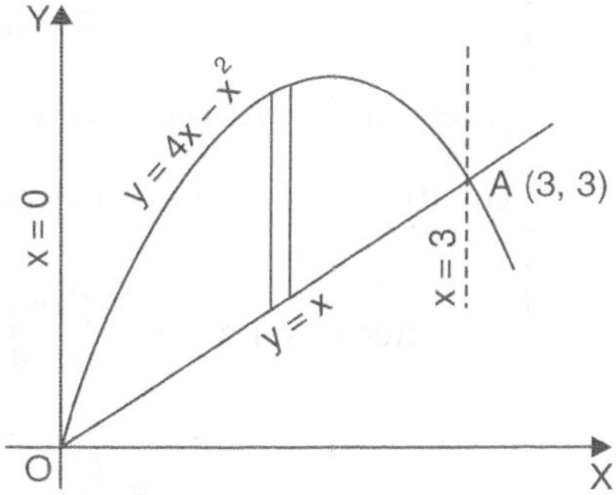

Fig. 3.24

Example 2. *Find the smaller of the areas bounded by the ellipse $4x^2 + 9y^2 = 36$ and the straight line $2x + 3y = 6$.*

Sol. The equation of the ellipse is

$$\frac{x^2}{9} + \frac{y^2}{4} = 1 \qquad \text{...(1)}$$

and the line is

$$\frac{x}{3} + \frac{y}{2} = 1 \qquad \text{...(2)}$$

Both meet x-axis at A(3, 0) and y-axis at B(0, 2).

Using horizontal strips, the required area lies between

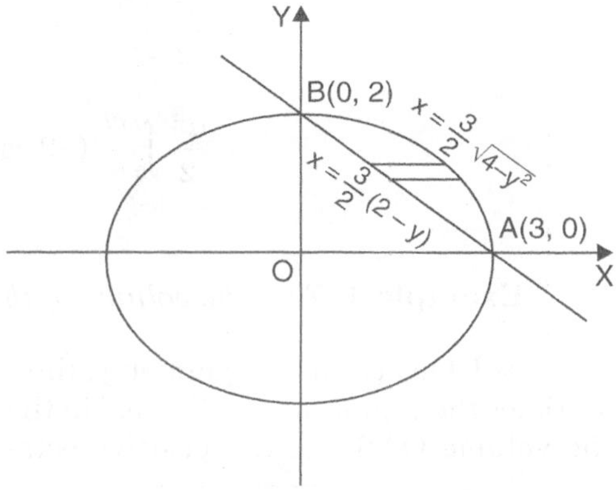

Fig. 3.25

$$x = \frac{3}{2}(2 - y),\ x = \frac{3}{2}\sqrt{4 - y^2} \text{ and } y = 0, y = 2.$$

$$\therefore \quad \text{Required area} = \int_0^2 \int_{(3/2)(2-y)}^{(3/2)\sqrt{4-y^2}} dx\, dy = \int_0^2 \Big[x \Big]_{(3/2)(2-y)}^{(3/2)\sqrt{4-y^2}} dy$$

$$= \int_0^2 \frac{3}{2}\left[\sqrt{4 - y^2} - (2 - y) \right] dy = \frac{3}{2}\left[\frac{y\sqrt{4 - y^2}}{2} + \frac{4}{2}\sin^{-1}\frac{y}{2} - 2y + \frac{y^2}{2} \right]_0^2$$

$$= \frac{3}{2}[2\sin^{-1} 1 - 4 + 2] = \frac{3}{2}\left(2 \cdot \frac{\pi}{2} - 2 \right) = \frac{3}{2}(\pi - 2).$$

Example 3. *Find, by double integration, the area lying inside the circle r = a sin θ and outside the cardioid r = a(1 – cos θ).*

Sol. Eliminating r between the equations of two curves, $\sin\theta = 1 - \cos\theta$ or $\sin\theta + \cos\theta = 1$

Squaring $1 + \sin 2\theta = 1$ or $\sin 2\theta = 0$

$$\therefore \quad 2\theta = 0 \text{ or } \pi$$

i.e., $$\theta = 0 \text{ or } \frac{\pi}{2}$$

For the required area, r varies from $a(1 - \cos\theta)$ to $a \sin\theta$ and θ varies from 0 to $\frac{\pi}{2}$.

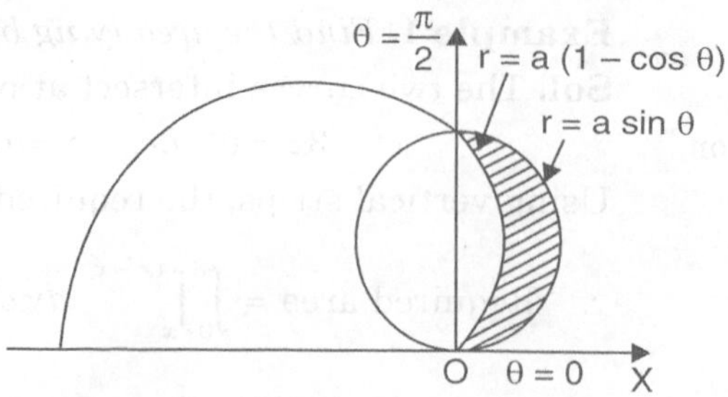

Fig. 3.26

$$\therefore \quad \text{Required area} = \int_0^{\pi/2} \int_{a(1-\cos\theta)}^{a\sin\theta} r\, dr\, d\theta$$

$$= \int_0^{\pi/2} \left[\frac{r^2}{2}\right]_{a(1-\cos\theta)}^{a\sin\theta} d\theta = \frac{1}{2}\int_0^{\pi/2} a^2 [\sin^2\theta - (1-\cos\theta)^2]\, d\theta$$

$$= \frac{a^2}{2}\int_0^{\pi/2} (\sin^2\theta - 1 - \cos^2\theta + 2\cos\theta)\, d\theta$$

$$= \frac{a^2}{2}\int_0^{\pi/2} (-2\cos^2\theta + 2\cos\theta)\, d\theta = a^2\left[-\frac{1}{2}\cdot\frac{\pi}{2} + 1\right] = a^2\left(1 - \frac{\pi}{4}\right).$$

Example 4. *Find the volume of the ellipsoid* $\frac{x^2}{a^2} + \frac{y^2}{b^2} + \frac{z^2}{c^2} = 1.$

Sol. On account of symmetry, the required volume is 8 times the volume of the ellipsoid in the positive octant. The volume OABC in the positive octant lies between the ellipsoid $z = c\sqrt{1 - \frac{x^2}{a^2} - \frac{y^2}{b^2}}$ and the plane XOY, and is bounded on the sides by the planes $x = 0, y = 0$.

The ellipsoid cuts the plane XOY in the ellipse

$$\frac{x^2}{a^2} + \frac{y^2}{b^2} = 1, z = 0$$

∴ The region OAB above which the volume OABC lies is bounded by

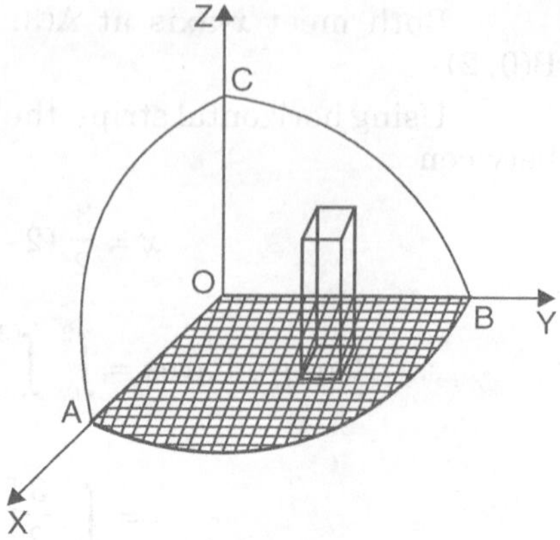

Fig. 3.27

$$x = 0, x = a \text{ and } y = 0, y = b\sqrt{1 - \frac{x^2}{a^2}}$$

Hence the required volume of the ellipsoid

$$= 8\int_0^a \int_0^{b\sqrt{1-x^2/a^2}} z\,dy\,dx = 8\int_0^a \int_0^{b\sqrt{1-x^2/a^2}} c\sqrt{1-\frac{x^2}{a^2}-\frac{y^2}{b^2}}\,dy\,dx$$

$$= 8\int_0^a \int_0^t c\sqrt{\frac{t^2}{b^2}-\frac{y^2}{b^2}}\,dy\,dx, \quad \text{where } b\sqrt{1-\frac{x^2}{a^2}} = t$$

$$= 8\int_0^a \int_0^t \frac{c}{b}\sqrt{t^2-y^2}\,dy\,dx = 8\int_0^a \frac{c}{b}\left[\frac{y\sqrt{t^2-y^2}}{2}+\frac{t^2}{2}\sin^{-1}\frac{y}{t}\right]_0^t dx$$

$$= \frac{4c}{b}\int_0^a t^2 \sin^{-1} 1\,dx = \frac{2\pi c}{b}\int_0^a t^2\,dx$$

$$= \frac{2\pi c}{b}\int_0^a b^2\left(1-\frac{x^2}{a^2}\right)dx = 2\pi bc\left[x-\frac{x^3}{3a^2}\right]_0^a = 2\pi bc\left(a-\frac{a}{3}\right) = \frac{4}{3}\pi abc.$$

Example 5. *Find the volume of the tetrahedron bounded by the co-ordinate planes and the plane* $\frac{x}{a}+\frac{y}{b}+\frac{z}{c} = 1.$ (K.U. Nov. 2009, K.U. April 2012)

Sol. The required volume OABC lies between the plane $z = c\left(1-\frac{x}{a}-\frac{y}{b}\right)$ and the plane XOY, and is bounded on the sides by the planes $x = 0, y = 0$.

The plane $\frac{x}{a}+\frac{y}{b}+\frac{z}{c} = 1$ cuts the plane XOY in the line $\frac{x}{a}+\frac{y}{b} = 1, z = 0$

$\therefore$ The region OAB above which the volume OABC lies is bounded by $x = 0, x = a$ and $y = 0$,

$$y = b\left(1-\frac{x}{a}\right)$$

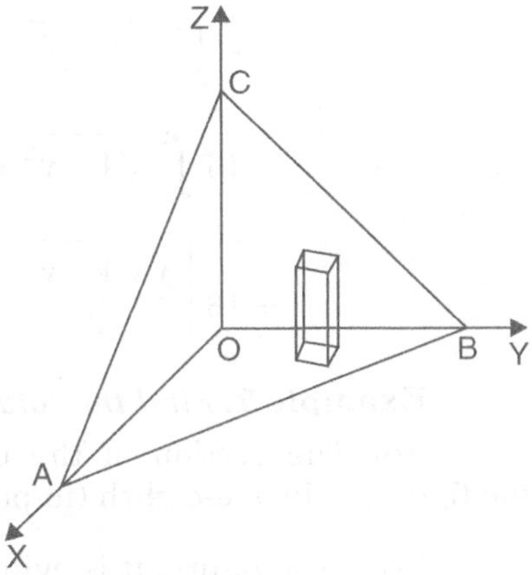

Fig. 3.28

Hence the required volume of the tetrahedron

$$= \int_0^a \int_0^{b(1-x/a)} z\,dy\,dx = \int_0^a \int_0^{b(1-x/a)} c\left(1-\frac{x}{a}-\frac{y}{b}\right)dy\,dx$$

$$= \int_0^a c\left[\left(1-\frac{x}{a}\right)y-\frac{y^2}{2b}\right]_0^{b(1-x/a)} dx = c\int_0^a \left[b\left(1-\frac{x}{a}\right)^2-\frac{b}{2}\left(1-\frac{x}{a}\right)^2\right]dx$$

$$= \frac{bc}{2}\int_0^a \left(1-\frac{x}{a}\right)^2 dx = \frac{bc}{2}\cdot\left[\frac{\left(1-\frac{x}{a}\right)^3}{-\frac{3}{a}}\right]_0^a = -\frac{abc}{6}(0-1) = \frac{1}{6}abc.$$

Example 6. *Find the volume bounded by the cylinder $x^2 + y^2 = 4$ and the planes $y + z = 4$ and $z = 0$.*

Sol. From the figure, it is clear that $z = 4 - y$ is to be integrated over the circle $x^2 + y^2 = 4$ in the xy-plane. To cover the shaded half of this circle, x varies from 0 to $\sqrt{4 - y^2}$ while y varies from -2 to 2.

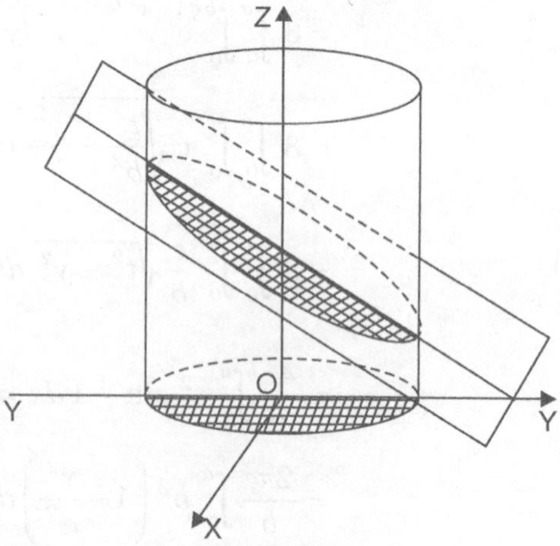

Fig. 3.29

$\therefore$ Required volume

$$= 2\int_{-2}^{2}\int_{0}^{\sqrt{4-y^2}} z\,dx\,dy$$

$$= 2\int_{-2}^{2}\int_{0}^{\sqrt{4-y^2}} (4 - y)\,dx\,dy$$

$$= 2\int_{-2}^{2} (4 - y)\left[x\right]_0^{\sqrt{4-y^2}} dy$$

$$= 2\int_{-2}^{2} (4 - y)\sqrt{4 - y^2}\,dy = 2\int_{-2}^{2} 4\sqrt{4 - y^2}\,dy - 2\int_{-2}^{2} y\sqrt{4 - y^2}\,dy$$

$$= 8\int_{-2}^{2} \sqrt{4 - y^2}\,dy \qquad \left[\begin{array}{l}\text{The second integral is zero since} \\ y\sqrt{4-y^2} \text{ is an odd function of } y\end{array}\right]$$

$$= 16\int_{0}^{2} \sqrt{4 - y^2}\,dy \qquad [\because \sqrt{4 - y^2} \text{ is an even function of } y]$$

$$= 16\left[\frac{y\sqrt{4 - y^2}}{2} + \frac{4}{2}\sin^{-1}\frac{y}{2}\right]_0^2 = 16\,[2\sin^{-1} 1] = 32 \times \frac{\pi}{2} = 16\pi.$$

Example 7. *Find the volume common to the cylinders $x^2 + y^2 = a^2$ and $x^2 + z^2 = a^2$.*

Sol. The section of the cylinder $x^2 + y^2 = a^2$ is the circle $x^2 + y^2 = a^2$ in the xy-plane. In the figure, only one-eighth (in positive octant) of the required volume is shown.

From the figure, it is evident that $z = \sqrt{a^2 - x^2}$ is to be evaluated over the quadrant of the circle $x^2 + y^2 = a^2$ in the first quadrant for which x varies from 0 to a and y varies from 0 to $\sqrt{a^2 - x^2}$.

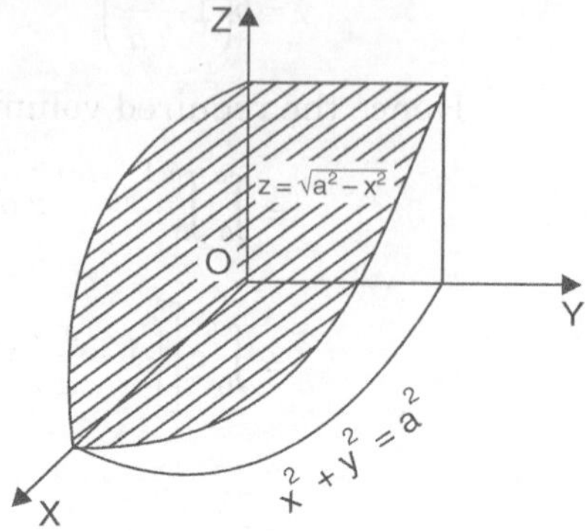

Fig. 3.30

$$\therefore \quad \text{Required volume} = 8\int_0^a\int_0^{\sqrt{a^2-x^2}} z\,dy\,dx$$

$$= 8\int_0^a\int_0^{\sqrt{a^2-x^2}} \sqrt{a^2 - x^2}\,dy\,dx$$

$$= 8\int_0^a \sqrt{a^2 - x^2}\left[y\right]_0^{\sqrt{a^2-x^2}} dx$$

$$= 8\int_0^a (a^2 - x^2)\,dx = 8\left[a^2x - \frac{x^3}{3}\right]_0^a = 8\left(a^3 - \frac{a^3}{3}\right) = \frac{16a^3}{3}.$$

Example 8. *Find the volume common to the sphere $x^2 + y^2 + z^2 = a^2$ and the cylinder $x^2 + y^2 = ay$.*

Sol. The required volume is the part of the sphere $x^2 + y^2 + z^2 = a^2$ lying within the cylinder. On account of the symmetry of the sphere, half of it lies above the plane XOY and half below it.

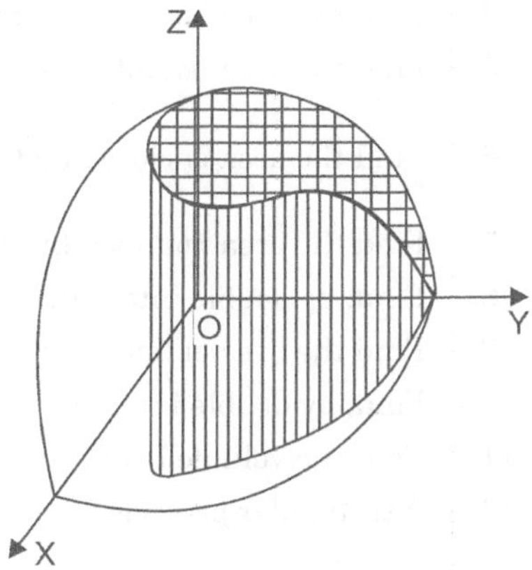

Fig. 3.31

$\therefore$ Required volume $= 2 \iint z \, dy \, dx$

where $z = \sqrt{a^2 - x^2 - y^2}$, and the region of integration is the area inside the circle $x^2 + y^2 = ay$ ...(1)

in the xy-plane.

On account of symmetry, the volumes above the two parts of circle (1) in the first and the second quadrants are equal. (The figure shows only the part in the first quadrant).

$\therefore$ Required volume $= 2 \times 2 \iint_R \sqrt{a^2 - x^2 - y^2} \, dy \, dx$

where R is the half of the circle (1) lying in the first quadrant.

Changing to polar co-ordinates by putting $x = r \cos \theta$, $y = r \sin \theta$ so that $x^2 + y^2 = r^2$, equation (1) becomes

$$r^2 = ar \sin \theta \quad \text{or} \quad r = a \sin \theta.$$

The region of integration is bounded by

$$r = 0, r = a \sin \theta \text{ and } \theta = 0, \theta = \frac{\pi}{2}.$$

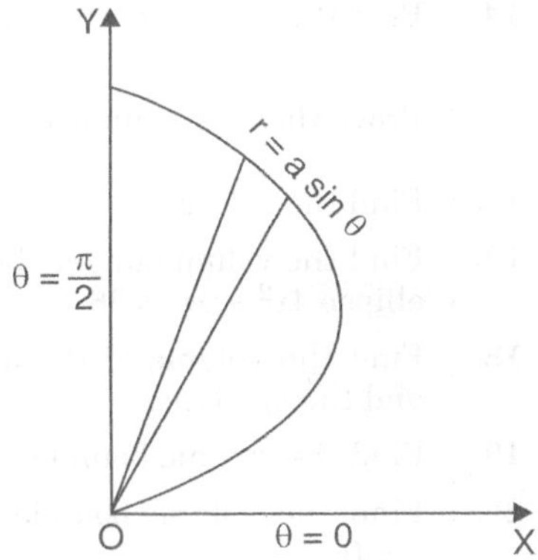

Fig. 3.32

$\therefore$ Required volume

$$= 4 \int_0^{\pi/2} \int_0^{a \sin \theta} \sqrt{a^2 - r^2} \, . \, r \, dr \, d\theta$$

$$= 4 \int_0^{\pi/2} \int_0^{a \sin \theta} -\frac{1}{2} (a^2 - r^2)^{1/2} (-2r) \, dr \, d\theta$$

$$= 4 \int_0^{\pi/2} -\frac{1}{2} . \left[\frac{(a^2 - r^2)^{3/2}}{3/2} \right]_0^{a \sin \theta} d\theta$$

$$= -\frac{4}{3} \int_0^{\pi/2} (a^2 \cos^3 \theta - a^3) = -\frac{4a^3}{3} \left[\frac{2}{3} - \frac{\pi}{2} \right] = \frac{2}{9} a^3 (3\pi - 4).$$

EXERCISE 3.6

1. Find, by double integration, the area enclosed by the ellipse $\frac{x^2}{a^2} + \frac{y^2}{b^2} = 1$. *(K.U. May 2003)*
2. Find, by double integration, the smaller of the areas bounded by the circle $x^2 + y^2 = 9$ and the line $x + y = 3$. *(K.U. Nov. 2009)*

3. Show, by double integration, that the area between the parabolas $y^2 = 4ax$ and $x^2 = 4ay$ is $\frac{16}{3}a^2$.
4. Find the area bounded by the parabola $y = x^2$ and the line $y = 2x + 3$. *(K.U. Nov. 2010)*
5. Find the area bounded by the parabolas $y^2 = 4 - x$ and $y^2 = 4 - 4x$.
6. Find the area enclosed by the curves $y = \frac{3x}{x^2 + 2}$ and $4y = x^2$.
7. Find the area bounded by the circles $r = 2 \sin \theta$ and $r = 4 \sin \theta$.
8. Find, by double integration, the area of one loop of the lemniscate $r^2 = a^2 \cos 2\theta$.
9. Find the area outside the circle $r = a$ and inside the cardioid $r = a(1 + \cos \theta)$.
10. Find, by double integration, the volume of the sphere $x^2 + y^2 + z^2 = 9$.
11. Find the volume bounded by the xy-plane, the cylinder $x^2 + y^2 = 1$ and the plane $x + y + z = 3$.
12. A triangular prism is formed by the planes whose equations are $ay = bx, y = 0$ and $x = a$. Prove that the volume of this prism between the plane $z = 0$ and the surface $z = c + xy$ is $\frac{ab}{8}(4c + ab)$.
13. Find the volume bounded by the xy-plane, the paraboloid $2z = x^2 + y^2$ and the cylinder $x^2 + y^2 = 4$.
14. Find the volume of the region bounded by $z = x^2 + y^2, z = 0, \ x = -a, x = a$ and $y = -a, y = a$.
15. Prove that the volume enclosed between the cylinders $x^2 + y^2 = 2ax$ and $z^2 = 2ax$ is $\frac{128a^3}{15}$.
16. Find the volume bounded by the plane $z = 0$, surface $z = x^2 + y^2 + 2$ and the cylinder $x^2 + y^2 = 4$.
17. Find the volume under the plane $z = x + y$ and above the area cut from the first quadrant by the ellipse $4x^2 + 9y^2 = 36$.
18. Find the volume of the cylinder $x^2 + y^2 - 2ax = 0$ intercepted between the paraboloid $x^2 + y^2 = 2az$ and the xy-plane.
19. Find the volume bounded by the cylinder $x^2 + y^2 = 4$ and the hyperboloid $x^2 + y^2 - z^2 = 1$.
20. Find the volume bounded by the paraboloid $x^2 + y^2 = az$, the cylinder $x^2 + y^2 = 2ay$ and the plane $z = 0$.

Answers

1. πab	**2.** $\frac{9}{4}(\pi - 2)$	**4.** $10\frac{2}{3}$	**5.** 8
6. $\frac{3}{2}\log 3 - \frac{2}{3}$	**7.** 3π	**8.** $\frac{a^2}{2}$	**9.** $\frac{a^2}{4}(8 + \pi)$
10. 36π	**11.** $3\pi - \frac{4}{3}$	**13.** 4π	**14.** $\frac{8a^4}{3}$
15. 16π	**17.** 10	**18.** $\frac{3}{4}\pi a^3$	**19.** $4\sqrt{3}\pi$
20. $\frac{3\pi a^3}{2}$.			

3.9. VOLUME AS A TRIPLE INTEGRAL

The volume V of a three dimensional region is given by $V = \iiint_V dx\, dy\, dz$.

If the region is bounded by $x = f_1(y, z)$, $x = f_2(y, z)$; $y = \phi_1(z)$, $y = \phi_2(z)$ and $z = a$, $z = b$, then

$$V = \int_a^b \int_{\phi_1(z)}^{\phi_2(z)} \int_{f_1(y,z)}^{f_2(y,z)} dx\, dy\, dz$$

The order of integration may be changed with a suitable change in the limits of integration.

In cylindrical co-ordinates, we have $\quad V = \iiint_V r\, dr\, d\phi\, dz$

In spherical polar co-ordinates, we have $\quad V = \iiint_V r^2 \sin\theta\, dr\, d\theta\, d\phi$.

ILLUSTRATIVE EXAMPLES

Example 1. *Find by triple integration, the volume of the paraboloid of revolution $x^2 + y^2 = 4z$ cut off by the plane $z = 4$.*

Sol. By symmetry, the required volume is 4 times the volume in the positive octant. The volume in the positive octant is bounded on the sides by the *zx* and *yz*-planes; from above by the plane $z = 4$ and below by the curved surface $x^2 + y^2 = 4z$.

The section of the paraboloid by the plane $z = 4$ is the circle $x^2 + y^2 = 16$, $z = 4$ and its projection on the *xy*-plane is the circle $x^2 + y^2 = 16$, $z = 0$.

The volume in the positive octant is bounded by $z = \dfrac{x^2 + y^2}{4}$, $z = 4$, $y = 0$, $y = \sqrt{16 - x^2}$ and $x = 0$, $x = 4$.

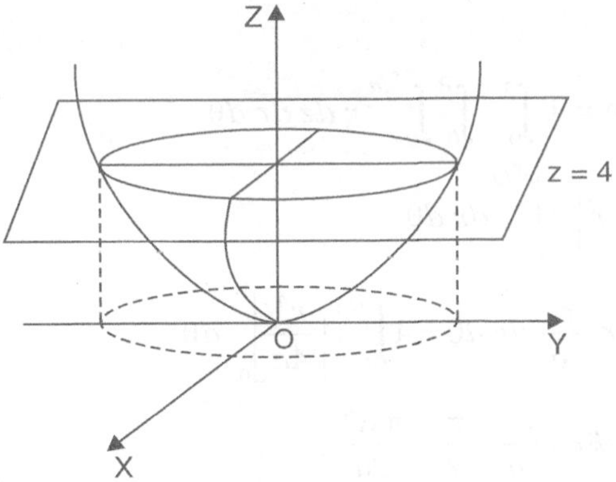

Fig. 3.33

$\therefore$ Required volume

$$= 4 \int_0^4 \int_0^{\sqrt{16-x^2}} \int_{(x^2+y^2)/4}^{4} dz\, dy\, dx = 4 \int_0^4 \int_0^{\sqrt{16-x^2}} \Big[z \Big]_{(x^2+y^2)/4}^{4} dy\, dx$$

$$= 4 \int_0^4 \int_0^{\sqrt{16-x^2}} \left(4 - \frac{x^2 + y^2}{4} \right) dy\, dx = 4 \int_0^4 \left[\left(4 - \frac{x^2}{4} \right) y - \frac{1}{4} \cdot \frac{y^3}{3} \right]_0^{\sqrt{16-x^2}} dx$$

$$= 4\int_0^4 \left[\left(4 - \frac{x^2}{4}\right)\sqrt{16 - x^2} - \frac{1}{12}(16 - x^2)^{3/2}\right] dx$$

$$= 4\int_0^4 \left[\frac{1}{4}(16 - x^2)\sqrt{16 - x^2} - \frac{1}{12}(16 - x^2)^{3/2}\right] dx$$

$$= 4\int_0^4 \frac{1}{6}(16 - x^2)^{3/2}\, dx = \frac{2}{3}\int_0^4 (16 - x^2)^{3/2}\, dx$$

$$= \frac{2}{3}\int_0^{\pi/2} (16)^{3/2} \cdot \cos^3\theta \cdot 4\cos\theta\, d\theta, \text{ where } x = 4\sin\theta$$

$$= \frac{512}{3}\int_0^{\pi/2} \cos^4\theta\, d\theta = \frac{512}{3} \cdot \frac{3 \cdot 1}{4 \cdot 2} \cdot \frac{\pi}{2} = 32\pi.$$

Example 2. *Find, by triple integration, the volume of the region bounded by the paraboloid $az = x^2 + y^2$ and the cylinder $x^2 + y^2 = R^2$.*

Sol. Changing to cylindrical co-ordinates, by putting $x = r\cos\theta$, $y = r\sin\theta$ the equation of the paraboloid becomes $az = r^2$ or $z = \frac{r^2}{a}$ and the equation of the cylinder becomes $r^2 = R^2$ or $r = R$. On account of symmetry, the required volume is four times the volume in the positive octant. Thus, in the common region, z varies from 0 to $\frac{r^2}{a}$, r varies from 0 to R and θ varies from 0 to $\frac{\pi}{2}$.

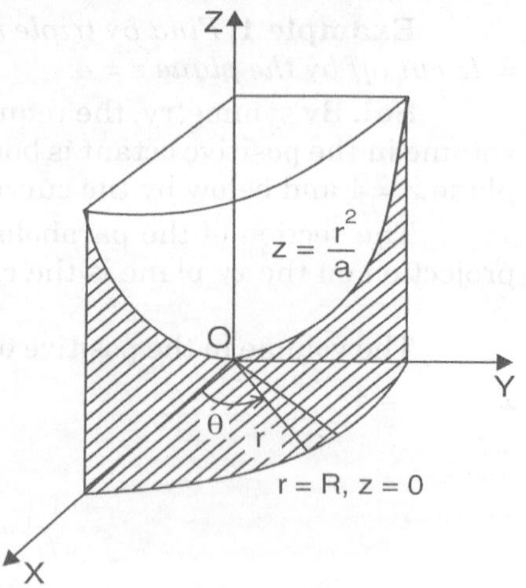

Fig. 3.34

$$\therefore \quad \text{Required volume} = 4\int_0^{\pi/2}\int_0^R\int_0^{r^2/a} r\, dz\, dr\, d\theta$$

$$= 4\int_0^{\pi/2}\int_0^R r\left[z\right]_0^{r^2/a} dr\, d\theta$$

$$= 4\int_0^{\pi/2}\int_0^R r \cdot \frac{r^2}{a}\, dr\, d\theta = 4\int_0^{\pi/2}\left[\frac{r^4}{4a}\right]_0^R d\theta$$

$$= \frac{1}{a}\int_0^{\pi/2} R^4\, d\theta = \frac{R^4}{a} \cdot \frac{\pi}{2} = \frac{\pi R^4}{2a}.$$

Example 3. *Find, by triple integration, the volume of a sphere of radius a.*

Sol. Changing to spherical polar co-ordinates by putting

$x = r\sin\theta\cos\phi$, $y = r\sin\theta\sin\phi$, $z = r\cos\theta$ so that $x^2 + y^2 + z^2 = r^2$.

The equation of a sphere of radius a in cartesian co-ordinates is $x^2 + y^2 + z^2 = a^2$

The same equation in spherical polar co-ordinates is $r^2 = a^2$ or $r = a$.

On account of symmetry, the required volume is 8 times the volume of the sphere in the positive octant for which r varies from 0 to a, θ varies from 0 to $\frac{\pi}{2}$ and ϕ varies from 0 to $\frac{\pi}{2}$.

$$\therefore \quad \text{Required volume} = 8\int_0^{\pi/2}\int_0^{\pi/2}\int_0^a r^2\sin\theta\, dr\, d\theta\, d\phi$$

$$= 8\int_0^{\pi/2}\int_0^{\pi/2} \sin\theta \cdot \left[\frac{r^3}{3}\right]_0^a d\theta\, d\phi = 8\int_0^{\pi/2}\int_0^{\pi/2} \frac{a^3}{3}\sin\theta\, d\theta\, d\phi$$

$$= 8\int_0^{\pi/2} -\frac{a^3}{3}\left[\cos\theta\right]_0^{\pi/2} d\phi = \frac{8}{3}a^3\int_0^{\pi/2} d\phi = \frac{8}{3}a^3 \cdot \frac{\pi}{2} = \frac{4}{3}\pi a^3.$$

EXERCISE 3.7

1. Find the volume bounded by the cylinder $x^2 + y^2 = 4$ and the planes $y + z = 3$ and $z = 0$.
2. Find by triple integration the volume of a solid bounded by the sphere $x^2 + y^2 + z^2 = 4$ and the paraboloid $x^2 + y^2 = 3z$.
3. Find the volume bounded by the cylinder $x^2 + y^2 = 4$ and the planes $y + z = 4$ and $z = 0$.
4. Find the volume common to the cylinders $x^2 + y^2 = a^2$ and $x^2 + z^2 = a^2$.
5. Find the volume of a solid bounded by the spherical surface $x^2 + y^2 + z^2 = 4a^2$ and the cylinder $x^2 + y^2 - 2ay = 0$.
6. Find by triple integration, the volume in the positive octant bounded by the co-ordinate planes and the plane $x + 2y + 3z = 4$.
7. Find the volume of the portion of the sphere $x^2 + y^2 + z^2 = a^2$ lying inside the cylinder $x^2 + y^2 = ax$.
8. Find the volume bounded above by the sphere $x^2 + y^2 + z^2 = 2a^2$ and below by the paraboloid $az = x^2 + y^2$.
9. Find by double integration, the volume of the solid generated by revolving the ellipse $\dfrac{x^2}{a^2} + \dfrac{y^2}{b^2} = 1$ about the x-axis.

Answers

1. 12π
2. $\dfrac{19\pi}{6}$
3. 16π
4. $\dfrac{16a^2}{3}$
5. $\dfrac{32a^2}{3}\left[\dfrac{\pi}{2} - \dfrac{2}{3}\right]$
6. $\dfrac{16}{9}$
7. $\dfrac{2a^3}{9}(3\pi - 4)$
8. $\left(\dfrac{4}{3}\sqrt{2} - \dfrac{7}{6}\right)\pi a^3$
9. $\dfrac{4}{3}\pi ab^2$.

4

Laplace Transforms

4.1. THE TRANSFORM CONCEPT

A transformation is a mathematical device which converts one function into another. For example, when the differential operator $D\left(=\dfrac{d}{dx}\right)$ operates on $f(x) = \sin x$, it gives a new function

$$g(x) = Df(x) = \cos x.$$

Laplace transforms or Laplace transformation is widely used by scientists and engineers. It is particularly effective in solving linear differential equations – ordinary as well as partial. It reduces an ordinary differential equation into an algebraic equation.

Laplace transform directly gives the solution of differential equations with given initial conditions without the necessity of first finding the general solution and then evaluating the arbitrary constants.

4.2. DEFINITION

Let $f(t)$ be a function of t defined for all $t \geq 0$. Then the Laplace transform of $f(t)$, denoted by $L\{f(t)\}$, is defined as $L\{f(t)\} = \int_0^\infty e^{-st} f(t)\,dt$ provided that the integral exists, 's' is a parameter which may be real or complex.

$L\{f(t)\}$ is clearly a function of s and is briefly written as $\bar{f}(s)$ *i.e.*, $L\{f(t)\} = \bar{f}(s)$.

4.3. LINEARITY PROPERTY

If c_1, c_2 are constants and f, g are functions of t, then

$$L\{c_1 f(t) + c_2 g(t)\} = c_1 L\{f(t)\} + c_2 L\{g(t)\}$$

By definition,
$$L\{c_1 f(t) + c_2 g(t)\} = \int_0^\infty e^{-st} \{c_1 f(t) + c_2 g(t)\}\,dt$$
$$= c_1 \int_0^\infty e^{-st} f(t)\,dt + c_2 \int_0^\infty e^{-st} g(t)\,dt$$
$$= c_1 L\{f(t)\} + c_2 L\{g(t)\}$$

The result can easily be generalised.

4.4. LAPLACE TRANSFORMS OF SOME ELEMENTARY FUNCTIONS

(1) $$L\{1\} = \frac{1}{s},\ s > 0$$

$$L\{1\} = \int_0^\infty e^{-st} \cdot 1\, dt = \left[-\frac{e^{-st}}{s}\right]_0^\infty = \frac{1}{s},\ \text{if } s > 0$$

(2) $$L\{t^n\} = \frac{n!}{s^{n+1}},\ \text{where } n \text{ is a positive integer.}$$

$$L\{t^n\} = \int_0^\infty e^{-st} \cdot t^n\, dt = \int_0^\infty e^{-x} \left(\frac{x}{s}\right)^n \frac{dx}{s},\ \text{on putting } st = x$$

$$= \frac{1}{s^{n+1}} \int_0^\infty x^n e^{-x}\, dx = \frac{\Gamma(n+1)}{s^{n+1}}$$

provided that $s > 0$ and $n + 1 > 0$ *i.e.*, $n > -1$.

If n is a positive integer, $\Gamma(n + 1) = n$

$$\therefore \quad L(t^n) = \frac{n!}{s^{n+1}}$$

(3) $$L\{e^{at}\} = \frac{1}{s-a},\ s > a$$

$$L\ \{e^{at}\} = \int_0^\infty e^{-st} \cdot e^{at}\, dt = \int_0^\infty e^{-(s-a)t}\, dt = \left[-\frac{e^{-(s-a)t}}{s-a}\right]_0^\infty$$

$$= \frac{1}{s-a},\ \text{if } s > a$$

(4) $$L\ \{\sin at\} = \frac{a}{s^2 + a^2},\ s > 0.$$

$$L\ \{\sin at\} = \int_0^\infty e^{-st} \cdot \sin at\, dt = \left[\frac{e^{-st}}{s^2 + a^2}(-s \sin at + a \cos at)\right]_0^\infty$$

$$= \frac{a}{s^2 + a^2}.$$

(5) $$L\{\cos at\} = \frac{s}{s^2 + a^2},\ s > 0.$$

$$L\ \{\cos at\} = \int_0^\infty e^{-st} \cdot \cos at\, dt = \left[\frac{e^{-st}}{s^2 + a^2}(-s \cos at - a \sin at)\right]_0^\infty$$

$$= \frac{s}{s^2 + a^2}.$$

(6) $$L\{\sinh at\} = \frac{a}{s^2 - a^2},\ s > |\, a \,|.$$

$$L\{\sinh at\} = \int_0^\infty e^{-st} \cdot \sinh at\, dt = \int_0^\infty e^{-st} \cdot \left[\frac{e^{at} - e^{-at}}{2}\right] dt$$

$$= \frac{1}{2}\left[\int_0^\infty e^{-(s-a)t}\, dt - \int_0^\infty e^{-(s+a)t}\, dt\right]$$

$$= \frac{1}{2}\left[\frac{1}{s-a} - \frac{1}{s+a}\right] = \frac{a}{s^2 - a^2}, \text{ for } s > |a|.$$

Note. We can also prove it by using linearity property.

Thus, $$L\{\sinh at\} = L\left\{\frac{1}{2}(e^{at} - e^{-at})\right\} = \frac{1}{2} L\{e^{at}\} - \frac{1}{2} L\{e^{-at}\}$$

$$= \frac{1}{2}\left(\frac{1}{s-a}\right) - \frac{1}{2}\left(\frac{1}{s+a}\right) = \frac{a}{s^2 - a^2}.$$

(7) $$L\{\cosh at\} = \frac{s}{s^2 - a^2}, s > |a|.$$

$$L\{\cosh at\} = L\left\{\frac{1}{2}(e^{at} + e^{-at})\right\} = \frac{1}{2} L\{e^{at}\} + \frac{1}{2} L\{e^{-at}\}$$

$$= \frac{1}{2}\left(\frac{1}{s-a}\right) + \frac{1}{2}\left(\frac{1}{s+a}\right) = \frac{s}{s^2 - a^2}.$$

4.5. (A) FIRST SHIFTING THEOREM (Multiplication by e^{at})

If $L\{f(t)\} = \bar{f}(s)$, then $L\{e^{at} f(t)\} = \bar{f}(s-a)$. By definition, $L\{e^{at} f(t)\} = \int_0^\infty e^{-st} \cdot e^{at} f(t)\, dt$

$$= \int_0^\infty e^{-(s-a)t} f(t)\, dt = \int_0^\infty e^{-pt} f(t)\, dt, \text{ where } p = s - a$$

$$= \bar{f}(p) = \bar{f}(s-a)$$

Similarly, $L\{e^{-at} f(t)\} = \bar{f}(s+a)$

Applying this theorem to the elementary functions of Art. 4.4, we get the following useful results :

(1) $$L\{e^{at} t^n\} = \frac{n!}{(s-a)^{n+1}}$$

(2) $$L\{e^{at} \sin bt\} = \frac{b}{(s-a)^2 + b^2}$$

(3) $$L\{e^{at} \cos bt\} = \frac{s-a}{(s-a)^2 + b^2}$$

(4) $$L\{e^{at} \sinh bt\} = \frac{b}{(s-a)^2 - b^2}$$

(5) $$L\{e^{at} \cosh bt\} = \frac{s-a}{(s-a)^2 - b^2}.$$

4.5. (B) CHANGE OF SCALE PROPERTY

If $L\{f(t)\} = \bar{f}(s)$, then $L\{f(at)\} = \dfrac{1}{a}\bar{f}\left(\dfrac{s}{a}\right)$.

Proof.
$$L\{f(at)\} = \int_0^\infty e^{-st} \cdot f(at)\, dt \qquad \left|\begin{array}{l} \text{Put } at = u \\ \text{so that } dt = \dfrac{du}{a}\end{array}\right.$$

$$= \int_0^\infty e^{-\frac{su}{a}} f(u) \cdot \frac{du}{a}$$

$$= \frac{1}{a}\int_0^\infty e^{-\lambda u} f(u)\, du, \text{ where } \lambda = \frac{s}{a}$$

$$= \frac{1}{a} L\{f(u)\} = \frac{1}{a}\bar{f}(\lambda) = \frac{1}{a}\bar{f}\left(\frac{s}{a}\right).$$

ILLUSTRATIVE EXAMPLES

Example 1. *Find the Laplace transforms of*

(i) sin 2t cos 3t *(ii) sin³ (2t)* (K.U. May 2003)

(iii) cosh³ (2t) *(iv) sin (at + b).*

Sol. (*i*) Since $\sin 2t \cos 3t = \dfrac{1}{2}(2\cos 3t \sin 2t) = \dfrac{1}{2}(\sin 5t - \sin t)$

$$\therefore \quad L\{\sin 2t \cos 3t\} = L\left\{\frac{1}{2}(\sin 5t - \sin t)\right\} = \frac{1}{2}[L\{\sin 5t\} - L\{\sin t\}]$$

$$= \frac{1}{2}\left[\frac{5}{s^2+5^2} - \frac{1}{s^2+1^2}\right] = \frac{2(s^2-5)}{(s^2+25)(s^2+1)}.$$

(*ii*) Since $\sin 6t = 3\sin 2t - 4\sin^3 2t$

$$\Rightarrow \quad \sin^3 2t = \frac{3}{4}\sin 2t - \frac{1}{4}\sin 6t$$

$$\therefore \quad L\{\sin^3 2t\} = L\left\{\frac{3}{4}\sin 2t - \frac{1}{4}\sin 6t\right\}$$

$$= \frac{3}{4} L\{\sin 2t\} - \frac{1}{4} L\{\sin 6t\}$$

$$= \frac{3}{4}\cdot\frac{2}{s^2+2^2} - \frac{1}{4}\cdot\frac{6}{s^2+6^2} = \frac{48}{(s^2+4)(s^2+36)}.$$

(*iii*) Since $\cosh 6t = 4\cosh^3 2t - 3\cosh 2t$

$$\Rightarrow \quad \cosh^3 2t = \frac{3}{4}\cosh 2t + \frac{1}{4}\cosh 6t$$

$$\therefore \quad L\{\cosh^3 2t\} = L\left\{\frac{3}{4}\cosh 2t + \frac{1}{4}\cosh 6t\right\} = \frac{3}{4} L\{\cosh 2t\} + \frac{1}{4} L\{\cosh 6t\}$$

$$= \frac{3}{4}\cdot\frac{s}{s^2-2^2} + \frac{1}{4}\cdot\frac{s}{s^2-6^2} = \frac{s(s^2-28)}{(s^2-4)(s^2-36)}.$$

(*iv*) Since $\sin(at + b) = \sin at \cos b + \cos at \sin b$

$\therefore \quad L[\sin(at + b)] = L[\sin at \cos b + \cos at \sin b]$

$= \cos b\, L(\sin at) + \sin b\, L(\cos at)$

$= \cos b \cdot \dfrac{a}{s^2 + a^2} + \sin b \cdot \dfrac{s}{s^2 + a^2} = \dfrac{a \cos b + s \sin b}{s^2 + a^2}.$

Example 2. *Find the Laplace transforms of*

(*i*) $\left(\sqrt{t} + \dfrac{1}{\sqrt{t}}\right)^3$ (*ii*) $(1 + te^{-t})^3$

(*iii*) $e^{-3t} \cos^2 t$ (K.U. Nov. 2001) (*iv*) $\sinh 3t \cos^2 t$.

Sol. (*i*) Since, $\left(\sqrt{t} + \dfrac{1}{\sqrt{t}}\right)^3 = (t^{1/2} + t^{-1/2})^3$

$= t^{3/2} + t^{-3/2} + 3(t^{1/2} + t^{-1/2})$

$= t^{3/2} + 3t^{1/2} + 3t^{-1/2} + t^{-3/2}$

$$\therefore \quad L\left\{\left(\sqrt{t} + \frac{1}{\sqrt{t}}\right)^3\right\} = L(t^{3/2}) + 3L(t^{1/2}) + 3L(t^{-1/2}) + L(t^{-3/2})$$

$$= \frac{\Gamma\left(\frac{3}{2}+1\right)}{s^{3/2+1}} + 3\frac{\Gamma\left(\frac{1}{2}+1\right)}{s^{1/2+1}} + 3\frac{\Gamma\left(-\frac{1}{2}+1\right)}{s^{-1/2+1}} + \frac{\Gamma\left(-\frac{3}{2}+1\right)}{s^{-3/2+1}}$$

$$= \frac{\Gamma\left(\frac{5}{2}\right)}{s^{5/2}} + 3\frac{\Gamma\left(\frac{3}{2}\right)}{s^{3/2}} + 3\frac{\Gamma\left(\frac{1}{2}\right)}{s^{1/2}} + \frac{\Gamma\left(-\frac{1}{2}\right)}{s^{-1/2}}$$

$$= \frac{\frac{3}{2}\cdot\frac{1}{2}\cdot\Gamma\left(\frac{1}{2}\right)}{s^{5/2}} + 3\frac{\frac{1}{2}\Gamma\left(\frac{1}{2}\right)}{s^{3/2}} + 3\frac{\Gamma\left(\frac{1}{2}\right)}{s^{1/2}} - \frac{2\Gamma\left(\frac{1}{2}\right)}{s^{-1/2}}$$

$$\left[\text{Since } \Gamma(n+1) = n\Gamma n \quad \therefore \quad \Gamma n = \frac{\Gamma n + 1}{n} \Rightarrow \Gamma(-1/2) = \frac{\Gamma(1/2)}{-1/2} = -2\,\Gamma(1/2)\right]$$

$$= \Gamma\left(\frac{1}{2}\right)\left[\frac{3}{4s^{5/2}} + \frac{3}{2s^{3/2}} + \frac{3}{s^{1/2}} - \frac{2}{s^{-1/2}}\right]$$

$$= \frac{\sqrt{\pi}}{4}\left[\frac{3}{s^{5/2}} + \frac{6}{s^{3/2}} + \frac{12}{\sqrt{s}} - 8\sqrt{s}\right].$$

(*ii*) Since, $(1 + te^{-t})^3 = 1 + 3t\,e^{-t} + 3t^2 e^{-2t} + t^3 e^{-3t}$

$\therefore \quad L\{(1 + te^{-t})^3\} = L\{1\} + 3L\{te^{-t}\} + 3L\{t^2 e^{-2t}\} + L\{t^3 e^{-3t}\}$

$$= \frac{1}{s} + 3\frac{1!}{(s+1)^2} + 3\frac{2!}{(s+2)^3} + \frac{3!}{(s+3)^4} \qquad \left[\because L(e^{at}t^n) = \frac{n!}{(s-a)^{n+1}}\right]$$

$$= \frac{1}{s} + \frac{3}{(s+1)^2} + \frac{6}{(s+2)^3} + \frac{6}{(s+3)^4}$$

(*iii*) Since, $\quad L(\cos^2 t) = \frac{1}{2} L(1 + \cos 2t) = \frac{1}{2}\left(\frac{1}{s} + \frac{s}{s^2 + 2^2}\right)$

$\therefore$ By first shifting property

$$L(e^{-3t}\cos^2 t) = \frac{1}{2}\left[\frac{1}{s+3} + \frac{s+3}{(s+3)^2 + 2^2}\right]$$

(*iv*) Since, $\quad L(\cos^2 t) = \frac{1}{2} L(1 + \cos 2t) = \frac{1}{2}\left(\frac{1}{s} + \frac{s}{s^2 + 2^2}\right)$

$$\therefore \quad L(\sinh 3t \cos^2 t) = L\left(\frac{e^{+3t} - e^{-3t}}{2}\cos^2 t\right)$$

$$= \frac{1}{2} L(e^{3t}\cos^2 t) - \frac{1}{2} L(e^{-3t}\cos^2 t)$$

$$= \frac{1}{4}\left[\frac{1}{s-3} + \frac{s-3}{(s-3)^2 + 4}\right] - \frac{1}{4}\left[\frac{1}{s+3} + \frac{s+3}{(s+3)^2 + 4}\right]$$

$$= \frac{1}{4}\left[\frac{1}{s-3} - \frac{1}{s+3}\right] + \frac{1}{4}\left[\frac{s-3}{s^2 - 6s + 13} - \frac{s+3}{s^2 + 6s + 13}\right]$$

$$= \frac{3}{2(s^2 - 9)} + \frac{1}{4}\left[\frac{(s-3)(s^2 - 6s + 13) - (s+3)(s^2 - 6s + 13)}{(s^2 + 13)^2 - (6s)^2}\right]$$

$$= \frac{3}{2(s^2 - 9)} + \frac{1}{4}\left[\frac{6s^2 - 78}{s^4 - 10s^2 + 169}\right]$$

$$= \frac{3}{2}\left[\frac{1}{s^2 - 9} + \frac{s^2 - 13}{s^4 - 10s^2 + 169}\right].$$

Example 3. *Show that*

(*i*) $L(t \sin at) = \dfrac{2as}{(s^2 + a^2)^2}$ $\qquad$ (*ii*) $L(t \cos at) = \dfrac{s^2 - a^2}{(s^2 + a^2)^2}.$

Sol. Since, $\quad L(t) = \frac{1}{s^2}$

$$\therefore \quad L\{te^{iat}\} = \frac{1}{(s - ia)^2}, \text{ by first shifting property}$$

$$= \frac{(s + ia)^2}{(s - ia)^2 (s + ia)^2} = \frac{(s^2 - a^2) + 2ias}{[(s - ia)(s + ia)]^2}$$

$$\Rightarrow \quad L[t(\cos at + i \sin at)] = \frac{(s^2 - a^2) + i(2as)}{(s^2 + a^2)^2}$$

Equating real parts $L(t \cos at) = \dfrac{s^2 - a^2}{(s^2 + a^2)^2}$

Equating imaginary parts $L(t \sin at) = \dfrac{2as}{(s^2 + a^2)^2}.$

Example 4. *Find the Laplace transform of e^{-3t} (cos 4t + 3 sin 4t).*

Sol. $L[e^{-3t}(\cos 4t + 3\sin 4t)] = L(e^{-3t}\cos 4t) + 3L(e^{-3t}\sin 4t)$

$$= \frac{s+3}{(s+3)^2+4^2} + 3\cdot\frac{4}{(s+3)^2+4^2} = \frac{s+5}{s^2+6s+25}.$$

Example 5. *Find the Laplace transform of t^2 sin at.*

Sol. Since, $L(t^2) = \frac{2!}{s^3} = \frac{2}{s^3}$

∴ By first shifting theorem, we have

$$L\{t^2 e^{iat}\} = \frac{2}{(s-ia)^3} = \frac{2(s+ia)^3}{[(s-ia)(s+ia)]^3}$$

or $$L\{t^2(\cos at + i\sin at)\} = \frac{2[(s^3-3a^2s)+i(3as^2-a^3)]}{(s^2+a^2)^3}$$

Equating the imaginary parts on both sides, we get $L\{t^2 \sin at\} = \frac{2a(3s^2-a^2)}{(s^2+a^2)^3}$.

Example 6. *Find the Laplace transform of t e^{-4t} sin 3t.*

Sol. Since, $L(t) = \frac{1}{s^2}$

∴ By first shifting theorem, we have

$$L\{t\, e^{3it}\} = \frac{1}{(s-3i)^2} = \frac{(s+3i)^2}{[(s-3i)(s+3i)]^2}$$

or $$L\{t(\cos 3t + i\sin 2t)\} = \frac{(s^2-9)+6is}{(s^2+9)^2}$$

Equating the imaginary parts on both sides, we get

$$L\{t\sin 3t\} = \frac{6s}{(s^2+9)^2}$$

Again applying the first shifting theorem, we have

$$L\{e^{-4t}\cdot t\sin 3t\} = \frac{6(s+4)}{((s+4)^2+9)^2} = \frac{6(s+4)}{(s^2+8s+25)^2}.$$

(For another method, see examples after Art. 4.10.)

Example 7. *Find the Laplace transform of f(t) defined as*

$$f(t) = \begin{cases} \frac{t}{T} & when\quad 0<t<T \\ 1 & when\quad t>T \end{cases}.$$

Sol. $$L\{f(t)\} = \int_0^\infty e^{-st} f(t)\,dt = \int_0^T e^{-st}\cdot\frac{t}{T}\,dt + \int_T^\infty e^{-st}\cdot 1\,dt$$

$$= \frac{1}{T}\left[\left| t\cdot\frac{e^{-st}}{-s}\right|_0^T - \int_0^T 1\cdot\frac{e^{-st}}{-s}\,dt + \left(\frac{e^{-st}}{-s}\right)_T^\infty\right]$$

$$= \frac{1}{T}\left[-\frac{T}{s}e^{-sT} + \frac{1}{s}\left\{\frac{e^{-st}}{-s}\right\}_0^T\right] + \frac{e^{-sT}}{s}$$

$$= -\frac{e^{-sT}}{s} - \frac{e^{-sT}-1}{Ts^2} + \frac{e^{-sT}}{s} = \frac{1-e^{-sT}}{Ts^2}.$$

Example 8. *Find the Laplace transform of the function f(t) defined as*

$$f(t) = |\ t-1\ | + |\ t+1\ | + |\ t+2\ | + |\ t-2\ |,\ t \geq 0.$$

Sol. $f(t)$ is defined differently in the *nbd* of $t = -2, -1, 1, 2$

(Obtained by equating to zero each of the four modulus functions)

Since $t \geq 0$, we have

For $0 \leq t \leq 1$, $f(t) = -(t-1) + (t+1) + (t+2) - (t-2) = 6$

For $1 \leq t \leq 2$, $f(t) = (t-1) + (t+1) + (t+2) - (t-2) = 2t + 4$

For $t > 2$, $f(t) = (t-1) + (t+1) + (t+2) + (t-2) = 4t$

$$\therefore\ L\{f(t)\} = \int_0^\infty e^{-st} f(t)\, dt$$

$$= \int_0^1 e^{-st} \cdot 6\, dt + \int_1^2 e^{-st} \cdot (2t+4)\, dt + \int_2^\infty e^{-st} \cdot 4t\, dt$$

$$= 6\int_0^1 e^{-st}\, dt + 2\int_1^2 (t+2)\, e^{-st} dt + 4\int_2^\infty t\, e^{-st}\, dt$$

$$= 6\left[\frac{e^{-st}}{-s}\right]_0^1 + 2\left[(t+2)\cdot\frac{e^{-st}}{-s} - 1\cdot\frac{e^{-st}}{(-s)^2}\right]_1^2 + 4\left[t\cdot\frac{e^{-st}}{-s} - 1\cdot\frac{e^{-st}}{(-s)^2}\right]_2^\infty$$

$$= \frac{-6}{5}(e^{-s}-1) + 2\left[\frac{-4}{s}e^{-2s} - \frac{1}{s^2}e^{-2s} + \frac{3}{s}e^{-s} + \frac{1}{s^2}e^{-s}\right] + 4\left[0 + \frac{2}{s}e^{-2s} + \frac{1}{s^2}e^{-2s}\right]$$

$$= \frac{1}{s^2}\left[6s + (-6s + 6s + 2)e^{-s} + (-8s - 2 + 8s + 4)\, e^{-2s}\right]$$

$$= \frac{1}{s^2}(6s + 2e^{-s} + 2e^{-2s}) = \frac{2}{s^2}(3s + e^{-s} + e^{-2s}).$$

Example 9. *If* $L\left(\frac{\sin t}{t}\right) = \tan^{-1}\left(\frac{1}{s}\right)$, *find* $L\left(\frac{\sin at}{t}\right)$.

Sol. Given $L\left(\frac{\sin t}{t}\right) = \tan^{-1}\left(\frac{1}{s}\right)$

By change of scale property

$$L\left(\frac{\sin at}{at}\right) = \frac{1}{a}\tan^{-1}\frac{1}{(s/a)} \Rightarrow \frac{1}{a}L\left(\frac{\sin at}{t}\right) = \frac{1}{a}\tan^{-1}\left(\frac{a}{s}\right)$$

$$\therefore\ L\left(\frac{\sin at}{t}\right) = \tan^{-1}\left(\frac{a}{s}\right).$$

EXERCISE 4.1

Find the Laplace transforms of (1 to 14):

1. $\cosh at - \cos at$ **2.** $\cos(at + b)$ **3.** $\sin at \sin bt$

4. $\sin^2 3t$ **5.** $\cos^3 2t$ **6.** $\sinh^3 2t$

7. $e^{-2t} \sin 4t$ **8.** $t^3 e^{-3t}$ *(K.U. Nov. 2002)* **9.** $\cosh at \cos at$

10. $e^{-3t}(2\cos 5t - 3\sin 4t)$ **11.** $\sin at - at\cos at$ **12.** $e^{-t}(\sin 2t - 2t\cos 2t)$

13. $\sinh t/2 \sin \dfrac{\sqrt{3}t}{2}$ **14.** $t\, e^{at} \sin at$

15. If $L\{f(t)\} = \bar{f}(s)$, show that $L\left\{f\left(\dfrac{t}{a}\right)\right\} = a\bar{f}(as)$.

16. Show that $L\{\sin kt \sinh kt\} = \dfrac{2k^2 s}{s^4 + 4k^4}$.

17. If $L[f(t)] = \bar{f}(s)$, show that

(*i*) $L[(\sinh at) f(t)] = \dfrac{1}{2}[\bar{f}(s-a) - \bar{f}(s+a)]$

(*ii*) $L[(\cosh at) f(t)] = \dfrac{1}{2}[\bar{f}(s-a) + \bar{f}(s+a)]$.

18. Find the Laplace transforms of

(*i*) $e^{3t}\sin^2 t$ (*ii*) $e^{4t}\sin 2t \cos t$

(*iii*) $(\sin t - \cos t)^2$ (*iv*) $\cosh at \sin at$.

19. Find the Laplace transforms of $f(t)$ defined as

(*i*) $f(t) = \begin{cases} t, & 0 < t < 4 \\ 5, & t > 4 \end{cases}$ (*ii*) $f(t) = \begin{cases} \sin t, & 0 < t < \pi \\ 0, & t > \pi \end{cases}$

(*iii*) $f(t) = \begin{cases} \cos t, & 0 < t < 2\pi \\ 0, & t > 2\pi \end{cases}$ (*iv*) $f(t) = \begin{cases} t^2, & 0 < t < 2 \\ t-1, & 2 < t < 3 \\ 7, & t > 3 \end{cases}$

20. Find the Laplace transforms of $f(t)$ defined as $f(t) = |t - 1| + |t + 1|, t \geq 0$.

$\left[\textbf{Hint.}\ f(t) = \begin{cases} 2, & 0 \le t \le 1 \\ 2t, & t > 1 \end{cases}\right]$

21. Find the Laplace transform of

(*i*) $f(t) = \begin{cases} e^t, & 0 < t < 1 \\ 0, & t > 1 \end{cases}$ (*ii*) $f(t) = \begin{cases} \cos\left(t - \dfrac{2\pi}{3}\right), & t > \dfrac{2\pi}{3} \\ 0, & t < \dfrac{2\pi}{3} \end{cases}$.

Answers

1. $\dfrac{2a^4 s}{s^4 - a^4}$ **2.** $\dfrac{s\cos b - a\sin b}{s^2 + a^2}$ **3.** $\dfrac{2abs}{[s^2 + (a+b)^2][s^2 + (a-b)^2]}$

4. $\dfrac{18}{s(s^2 + 36)}$ **5.** $\dfrac{s(s^2 + 28)}{(s^2 + 4)(s^2 + 36)}$ **6.** $\dfrac{48}{(s^2 - 4)(s^2 - 36)}$

7. $\dfrac{4}{s^2 + 4s + 20}$ **8.** $\dfrac{6}{(s+3)^4}$ **9.** $\dfrac{a(s^2 + 2a^2)}{s^4 + 4a^4}$

10. $\dfrac{2s-9}{s^2+6s+34}$ **11.** $\dfrac{2a^3}{(s^2+a^2)^2}$ **12.** $\dfrac{16}{(s^2+2s+5)^2}$

13. $\dfrac{\sqrt{3}s}{2(s^4+s^2+1)}$ **14.** $\dfrac{2a(s-a)}{(s^2-2as+2a^2)^2}$ **18.** (*i*) $\dfrac{1}{2}\left[\dfrac{1}{s-3}-\dfrac{s-3}{s^2-6s+13}\right]$

(*ii*) $\dfrac{1}{2}\left[\dfrac{3}{s^2-8s+25}+\dfrac{1}{s^2-8s+17}\right]$ (*iii*) $\dfrac{s^2-2s+4}{s(s^2+4)}$ (*iv*) $\dfrac{a(s^2+2a^2)}{s^4+4a^4}$

19. (*i*) $\dfrac{1}{s^2}+e^{-4s}\left(\dfrac{1}{s}-\dfrac{1}{s^2}\right)$ (*ii*) $\dfrac{1+e^{-\pi s}}{s^2+1}$ (*iii*) $\dfrac{s(1-e^{-2\pi s})}{s^2+1}$

(*iv*) $\dfrac{2}{s^3}-\dfrac{e^{-2s}}{s^3}(2+3s+3s^2)+\dfrac{e^{-3s}}{s^2}(5s-1)$.

20. $\dfrac{2}{s}\left(1+\dfrac{e^{-s}}{s}\right)$ **21.** (*i*) $\dfrac{e^{1-s}-1}{1-s}$ (*ii*) $e^{-2\pi s/3}\cdot\dfrac{s}{s^2+1}$.

4.6. INVERSE LAPLACE TRANSFORM

Definition. If $L\{f(t)\} = \bar{f}(s)$, then $f(t)$ is called the inverse Laplace transform of $\bar{f}(s)$ and is denoted by

$$L^{-1}\{\bar{f}(s)\} = f(t)$$

Here L^{-1} denotes the inverse Laplace transform.

For example, since $$L\{e^{5t}\} = \frac{1}{s-5}$$

$$\therefore \quad L^{-1}\left\{\frac{1}{s-5}\right\} = e^{5t}.$$

The inverse Laplace transforms given below follow at once from the results of Laplace transforms given earlier:

(1) $L^{-1}\left\{\dfrac{1}{s}\right\} = 1$ (2) $L^{-1}\left\{\dfrac{1}{s-a}\right\} = e^{at}$

(3) $L^{-1}\left\{\dfrac{1}{s^n}\right\} = \dfrac{t^{n-1}}{(n-1)!}$, if n is a positive integer. $\left[\text{Otherwise} = \dfrac{t^{n-1}}{\Gamma(n)}\right]$.

(4) $L^{-1}\left\{\dfrac{1}{(s-a)^n}\right\} = e^{at}\cdot\dfrac{t^{n-1}}{(n-1)!}$ (5) $L^{-1}\left\{\dfrac{1}{s^2+a^2}\right\} = \dfrac{1}{a}\sin at$

(6) $L^{-1}\left\{\dfrac{s}{s^2+a^2}\right\} = \cos at$ (7) $L^{-1}\left\{\dfrac{1}{s^2-a^2}\right\} = \dfrac{1}{a}\sinh at$

(8) $L^{-1}\left\{\dfrac{s}{s^2-a^2}\right\} = \cosh at$ (9) $L^{-1}\left\{\dfrac{1}{(s-a)^2+b^2}\right\} = \dfrac{1}{b}e^{at}\sin bt$

(10) $L^{-1}\left\{\dfrac{s-a}{(s-a)^2+b^2}\right\} = e^{at}\cos bt$ (11) $L^{-1}\left\{\dfrac{s}{(s^2+a^2)^2}\right\} = \dfrac{1}{2a}t\sin at$

(12) $L^{-1}\left\{\frac{1}{(s^2+a^2)}\right\} = \frac{1}{2a^3}$ (sin $at - at$ cos at).

All the above results must be remembered.

ILLUSTRATIVE EXAMPLES

Example 1. *Find the inverse Laplace transforms of*

(i) $\frac{3(s^2-1)^2}{2s^5}$ (ii) $\frac{3s+5\sqrt{2}}{s^2+8}$

(iii) $\frac{4s+15}{16s^2-25}$ (iv) $\frac{s+1}{s^2+s+1}$.

Sol. (*i*) Since, $\frac{3(s^2-1)^2}{2s^5} = \frac{3s^4-6s^2+3}{2s^5} = \frac{3}{2}\cdot\frac{1}{s} - 3\cdot\frac{1}{s^3} + \frac{3}{2}\cdot\frac{1}{s^5}$

$$\therefore \quad L^{-1}\left\{\frac{3(s^2-1)^2}{2s^5}\right\} = \frac{3}{2}L^{-1}\left\{\frac{1}{s}\right\} - 3L^{-1}\left\{\frac{1}{s^3}\right\} + \frac{3}{2}L^{-1}\left\{\frac{1}{s^5}\right\}$$

$$= \frac{3}{2}(1) - 3\left\{\frac{t^2}{2!}\right\} + \frac{3}{2}\left\{\frac{t^4}{4!}\right\} \quad \text{since } L^{-1}\left\{\frac{1}{s^n}\right\} = \frac{t^{n-1}}{(n-1)!}$$

$$= \frac{3}{2} - \frac{3}{2}t^2 + \frac{1}{16}t^4.$$

(*ii*) Since, $\frac{3s+5\sqrt{2}}{s^2+8} = 3\frac{s}{s^2+(2\sqrt{2})^2} + 5\sqrt{2}\frac{1}{s^2+(2\sqrt{2})^2}$

$$\therefore \quad L^{-1}\left\{\frac{3s+5\sqrt{2}}{s^2+8}\right\} = 3L^{-1}\left\{\frac{s}{s^2+(2\sqrt{2})^2}\right\} + 5\sqrt{2}\,L^{-1}\left\{\frac{1}{s^2+(2\sqrt{2})^2}\right\}$$

$$= 3\cos(2\sqrt{2}\,t) + 5\sqrt{2}\cdot\frac{1}{2\sqrt{2}}\sin(2\sqrt{2}\,t)$$

$$= 3\cos(2\sqrt{2}\,t) + \frac{5}{2}\sin(2\sqrt{2}\,t).$$

(*iii*) Since, $\frac{4s+15}{16s^2-25} = \frac{4s+15}{16\left(s^2-\frac{25}{16}\right)} = \frac{1}{4}\cdot\frac{s}{s^2-\left(\frac{5}{4}\right)^2} + \frac{15}{16}\cdot\frac{1}{s^2-\left(\frac{5}{4}\right)^2}$

$$\therefore \quad L^{-1}\left\{\frac{4s+15}{16s^2-25}\right\} = \frac{1}{4}L^{-1}\left\{\frac{s}{s^2-\left(\frac{5}{4}\right)^2}\right\} + \frac{15}{16}\cdot L^{-1}\left\{\frac{1}{s^2-\left(\frac{5}{4}\right)^2}\right\}$$

$$= \frac{1}{4}\cosh\left(\frac{5}{4}t\right) + \frac{15}{16}\cdot\frac{1}{\frac{5}{4}}\sinh\left(\frac{5}{4}t\right)$$

$$= \frac{1}{4} \cosh\left(\frac{5}{4}t\right) + \frac{3}{4} \sinh\left(\frac{5}{4}t\right).$$

(*iv*) Since,
$$\frac{s+1}{s^2+s+1} = \frac{\left(s+\frac{1}{2}\right)+\frac{1}{2}}{\left(s+\frac{1}{2}\right)^2+\frac{3}{4}} = \frac{s+\frac{1}{2}}{\left(s+\frac{1}{2}\right)^2+\left(\frac{\sqrt{3}}{2}\right)^2} + \frac{1}{2}\cdot\frac{1}{\left(s+\frac{1}{2}\right)^2+\left(\frac{\sqrt{3}}{2}\right)^2}$$

$$\therefore \quad L^{-1}\left\{\frac{s+1}{s^2+s+1}\right\} = L^{-1}\left[\frac{s+\frac{1}{2}}{\left(s+\frac{1}{2}\right)^2+\left(\frac{\sqrt{3}}{2}\right)^2}\right] + \frac{1}{2}L^{-1}\left[\frac{1}{\left(s+\frac{1}{2}\right)^2+\left(\frac{\sqrt{3}}{2}\right)^2}\right]$$

$$= e^{-\frac{1}{2}t}\cos\left(\frac{\sqrt{3}}{2}t\right) + \frac{1}{2}\cdot\frac{1}{\frac{\sqrt{3}}{2}}e^{-\frac{1}{2}t}\sin\left(\frac{\sqrt{3}}{2}t\right)$$

$$= \frac{1}{\sqrt{3}}e^{-\frac{1}{2}t}\left[\sqrt{3}\cos\left(\frac{\sqrt{3}}{2}t\right) + \sin\left(\frac{\sqrt{3}}{2}t\right)\right].$$

Example 2. *Find the inverse Laplace transforms of*

(*i*) $\dfrac{2s+5}{s^2+4s+13}$ (K.U. Nov. 2000) (*ii*) $\dfrac{2s^2-4}{(s+1)(s-2)(s-3)}$

(*iii*) $\dfrac{2s^2-1}{(s^2+1)(s^2+4)}$ (*iv*) $\dfrac{21s-33}{(s+1)(s-2)^3}$.

Sol. (*i*) Here the denominator $s^2 + 4s + 13$ is not factorizable.

Since,
$$\frac{2s+5}{s^2+4s+13} = \frac{2s+5}{(s+2)^2-4+13} = \frac{2(s+2)+1}{(s+2)^2+9}$$

$$= 2\cdot\frac{s+2}{(s+2)^2+3^2} + \frac{1}{(s+2)^2+3^2}$$

$$\therefore \quad L^{-1}\left\{\frac{2s+5}{s^2+4s+13}\right\} = 2\cdot L^{-1}\left\{\frac{s+2}{(s+2)^2+3^2}\right\} + L^{-1}\left\{\frac{1}{(s+2)^2+3^2}\right\}$$

$$= 2\,e^{-2t}\cos 3t + e^{-2t}\,\frac{\sin 3t}{3}$$

$$= \frac{e^{-2t}}{3}[6\cos 3t + \sin 3t].$$

(*ii*) Here the denominator has non-repeated linear factors.

Let
$$\frac{2s^2-4}{(s+1)(s-2)(s-3)} = \frac{A}{s+1} + \frac{B}{s-2} + \frac{C}{s-3}$$

Multiplying both sides by $(s + 1)(s - 2)(s - 3)$, we get

$$2s^2 - 4 = A(s-2)(s-3) + B(s+1)(s-3) + C(s+1)(s-2)$$

Putting $s = -1$, we get $-2 = A(-3)(-4) \Rightarrow A = \frac{-1}{6}$

Putting $s = 2$, we get $4 = B(3)(-1) \Rightarrow B = \frac{-4}{3}$

Putting $s = 3$, we get $14 = C(4)(1) \Rightarrow C = \frac{7}{2}$

Thus
$$\frac{2s^2 - 4}{(s+1)(s-2)(s-3)} = \frac{-1}{6(s+1)} - \frac{4}{3(s-2)} + \frac{7}{2(s-3)}$$

$$\therefore \quad L^{-1}\left\{\frac{2s^2 - 4}{(s+1)(s-2)(s-3)}\right\} = \frac{-1}{6} L^{-1}\left(\frac{1}{s+1}\right) - \frac{4}{3} L^{-1}\left(\frac{1}{s-2}\right) + \frac{7}{2} L^{-1}\left(\frac{1}{s-3}\right)$$

$$= \frac{-1}{6} e^{-t} - \frac{4}{3} e^{2t} + \frac{7}{2} e^{3t}.$$

(*iii*) Since the fraction involves only even powers of s, we put $s^2 = p$ so that the given fraction becomes $\frac{2p-1}{(p+1)(p+4)}$ which has only non-repeated linear factors in this denominator.

Let
$$\frac{2p-1}{(p+1)(p+4)} = \frac{A}{p+1} + \frac{B}{p+4} \quad \ldots(1)$$

Multiplying both sides by $(p + 1)(p + 4)$, we get

$$2p - 1 = A(p + 4) + B(p + 1)$$

Putting $p = -1$, we get $-3 = A(3) \Rightarrow A = -1$

Putting $p = -4$, we get $-9 = B(-3) \Rightarrow B = 3$

Putting these values of A and B in (1) and replacing p by s^2, we have

$$\frac{2s^2 - 1}{(s^2+1)(s^2+4)} = \frac{-1}{s^2+1} + \frac{3}{s^2+4}$$

$$\therefore \quad L^{-1}\left\{\frac{2s^2 - 1}{(s^2+1)(s^2+4)}\right\} = -L^{-1}\left\{\frac{1}{s^2+1}\right\} + 3 \cdot L^{-1}\left\{\frac{1}{s^2+4}\right\}$$

$$= -\sin t + \frac{3}{2}\sin 2t.$$

(*iv*) Here the denominator has a linear factor repeated thrice.

Let
$$\frac{21s - 33}{(s+1)(s-2)^3} = \frac{A}{s+1} + \frac{B}{s-2} + \frac{C}{(s-2)^3} + \frac{D}{(s-2)^4}$$

Multiplying both sides $(s + 1)(s - 2)^3$, we get

$$21s - 33 = A(s-2)^3 + B(s+1)(s-2)^2 + C(s+1)(s-2) + D(s+1)$$

Putting $s = -1$, we get $-54 = A(-27) \Rightarrow A = 2$

Putting $s = 2$, we get $9 = D(3) \Rightarrow D = 3$

Equating co-efficient of s^3, $0 = A + B \Rightarrow B = -A = -2$

Putting $s = 0$, we get $-33 = -8A + 4B - 2C + D$

$$-33 = 16 - 8 - 2C + 3 \quad \text{or} \quad C = 6$$

Thus $$L^{-1}\left\{\frac{21s-33}{(s+1)(s-2)^3}\right\} = 2\,L^{-1}\left\{\frac{1}{s+1}\right\} - 2\,L^{-1}\left\{\frac{1}{s-2}\right\} + 6\,L^{-1}\left\{\frac{1}{(s-2)^2}\right\} + 3\,L^{-1}\left\{\frac{1}{(s-2)^3}\right\}$$

$$= 2e^{-t} - 2e^{2t} + 6\cdot e^{2t}\cdot\frac{t}{1!} + 3\cdot e^{2t}\cdot\frac{t^2}{2!} \quad \text{since } L^{-1}\left\{\frac{1}{(s-a)^n}\right\} = e^{at}\,\frac{t^{n-1}}{(n-1)!}$$

$$= 2e^{-t} - 2e^{2t} + 6t\,e^{2t} + \frac{3}{2}\,t^2\,e^{2t}.$$

Example 3. *Find the inverse Laplace transforms of*

(*i*) $\dfrac{5s+3}{(s-1)(s^2+2s+5)}$ (*ii*) $\dfrac{s}{s^4+s^2+1}$

(*iii*) $\dfrac{s}{s^4+4a^4}$ (*iv*) $\dfrac{s^3}{s^4-a^4}$

Sol. (*i*) Here the denominator has a non-factorisable quadratic factor $s^2 + 2s + 5$.

Let $$\frac{5s+3}{(s-1)(s^2+2s+5)} = \frac{A}{s-1} + \frac{Bs+C}{s^2+2s+5}$$

Multiplying both sides by $(s-1)(s^2+2s+5)$, we get

$$5s + 3 = A(s^2 + 2s + 5) + (Bs + C)(s - 1)$$

Putting $s = 1$, we get $8 = 8A$ $\Rightarrow$ $A = 1$

Equating the co-efficients of s^2, $0 = A + B$ $\Rightarrow$ $B = -A = -1$

Putting $s = 0$, we get $3 = 5A - C$ $\Rightarrow$ $C = 5A - 3 = 5 \times 1 - 3 = 2$

Thus $$\frac{5s+3}{(s-1)(s^2+2s+5)} = \frac{1}{s-1} + \frac{-s+2}{s^2+2s+5} = \frac{1}{s-1} - \frac{(s+1)-3}{(s+1)^2+4}$$

$$= \frac{1}{s-1} - \frac{s+1}{(s+1)^2+2^2} + 3\cdot\frac{1}{(s+1)^2+2^2}$$

$$\therefore \quad L^{-1}\left\{\frac{5s+3}{(s-1)(s^2+2s+5)}\right\} = L^{-1}\left\{\frac{1}{s-1}\right\} - L^{-1}\left\{\frac{s+1}{(s+1)^2+2^2}\right\} + 3\,L^{-1}\left\{\frac{1}{(s+1)^2+2^2}\right\}$$

$$= e^t - e^{-t}\cos 2t + \frac{3}{2}\,e^{-t}\sin 2t.$$

(*ii*) Since $$s^4 + s^2 + 1 = (s^4 + 2s^2 + 1) - s^2 = (s^2 + 1)^2 - s^2 = (s^2 + 1 + s)(s^2 + 1 - s)$$

$$\therefore \quad \frac{s}{s^4+s^2+1} = \frac{s}{(s^2+s+1)(s^2-s+1)}$$

Let $$\frac{s}{(s^2+s+1)(s^2-s+1)} = \frac{As+B}{s^2+s+1} + \frac{Cs+D}{s^2-s+1}$$

Multiplying both sides by $(s^2 + s + 1)(s^2 - s + 1)$, we get

$$s = (As + B)(s^2 - s + 1) + (Cs + D)(s^2 + s + 1)$$

Equating co-efficients of s^3, $0 = A + C$ or $C = -A$ ...(1)

Equating co-efficients of s^2, $0 = -A + B + C + D$ ...(2)

Equating co-efficients of s, $1 = A - B + C + D$ ...(3)

putting $s = 0$, $0 = B + D$ or $D = -B$ ...(4)

From (2), $0 = -A + B - A - B$

$\Rightarrow$ $A = 0$ $\therefore$ From (1), $C = 0$

From (3), $1 = A - B - A - B$

$\Rightarrow$ $B = -\frac{1}{2}$ $\therefore$ From (4), $D = \frac{1}{2}$

Thus
$$\frac{s}{s^4 + s^2 + 1} = \frac{-\frac{1}{2}}{s^2 + s + 1} + \frac{\frac{1}{2}}{s^2 - s + 1} = \frac{1}{2}\left[\frac{1}{\left(s - \frac{1}{2}\right)^2 + \frac{3}{4}} - \frac{1}{\left(s + \frac{1}{2}\right)^2 + \frac{3}{4}}\right]$$

$$\therefore \quad L^{-1}\left[\frac{s}{s^4 + s^2 + 1}\right] = \frac{1}{2}\left[L^{-1}\left\{\frac{1}{\left(s - \frac{1}{2}\right)^2 + \left(\frac{\sqrt{3}}{2}\right)^2}\right\} - L^{-1}\left\{\frac{1}{\left(s + \frac{1}{2}\right)^2 + \left(\frac{\sqrt{3}}{2}\right)^2}\right\}\right]$$

$$= \frac{1}{2}\left[\frac{1}{\sqrt{3}/2} e^{\frac{1}{2}t} \sin\frac{\sqrt{3}}{2}t - \frac{1}{\sqrt{3}/2} e^{-\frac{1}{2}t} \sin\frac{\sqrt{3}}{2}t\right]$$

$$= \frac{2}{\sqrt{3}} \sin\frac{\sqrt{3}}{2}t \left(\frac{e^{\frac{1}{2}t} - e^{-\frac{1}{2}t}}{2}\right) = \frac{2}{\sqrt{3}} \sin\frac{\sqrt{3}}{2}t \sinh\frac{t}{2}.$$

Note. In certain cases, with a little insight, the partial fractions can be written down directly.

Thus,
$$\frac{s}{(s^2 + s + 1)(s^2 - s + 1)} = \frac{1}{2} \cdot \frac{(s^2 + s + 1) - (s^2 - s + 1)}{(s^2 + s + 1)(s^2 - s + 1)} = \frac{1}{2}\left[\frac{1}{s^2 - s + 1} - \frac{1}{s^2 + s + 1}\right].$$

(*iii*) Since
$$s^4 + 4a^4 = (s^2 + 2a^2)^2 - (2as)^2$$
$$= (s^2 + 2as + 2a^2)(s^2 - 2as + 2a^2)$$

$$\therefore \quad \frac{s}{s^4 + 4a^4} = \frac{s}{(s^2 + 2as + 2a^2)(s^2 - 2as + 2a^2)}$$

$$= \frac{1}{4a}\left(\frac{1}{s^2 - 2as + 2a^2} - \frac{1}{s^2 + 2as + 2a^2}\right)$$

$$= \frac{1}{4a}\left(\frac{1}{(s - a)^2 + a^2} - \frac{1}{(s + a)^2 + a^2}\right)$$

$$\therefore \quad L^{-1}\left(\frac{s}{s^4 + 4a^4}\right) = \frac{1}{4a}\left[L^{-1}\left(\frac{1}{(s - a)^2 + a^2}\right) - L^{-1}\left(\frac{1}{(s + a)^2 + a^2}\right)\right]$$

$$= \frac{1}{4a}\left[\frac{1}{a} e^{at} \sin at - \frac{1}{a} e^{-at} \sin at\right]$$

$$= \frac{1}{2a^2} \sin at \left[\frac{e^{at} - e^{-at}}{2}\right]$$

$$= \frac{1}{2a^2} \sin at \sinh at.$$

(*iv*) $$\frac{s^3}{s^4 - a^4} = s\left[\frac{s^2}{(s^2 - a^2)(s^2 + a^2)}\right] = \frac{1}{2}s\left[\frac{1}{s^2 - a^2} + \frac{1}{s^2 + a^2}\right]$$

$$= \frac{1}{2}\left[\frac{s}{s^2 - a^2} + \frac{s}{s^2 + a^2}\right]$$

$$\therefore \quad L^{-1}\left\{\frac{s^3}{s^4 - a^4}\right\} = \frac{1}{2}\left[L^{-1}\left\{\frac{s}{s^2 - a^2}\right\} + L^{-1}\left\{\frac{s}{s^2 + a^2}\right\}\right]$$

$$= \frac{1}{2}(\cosh at + \cos at).$$

Example 4. *Find the inverse Laplace transforms of*

(*i*) $\dfrac{s^2 + 6}{(s^2 + 1)(s^2 + 4)}$ (*ii*) $\dfrac{s^2 + s}{(s^2 + 1)(s^2 + 2s + 2)}$.

Sol. (*i*) $$\frac{s^2 + 6}{(s^2 + 1)(s^2 + 4)} = \frac{A}{s^2 + 1} + \frac{B}{s^2 + 4}$$

Multiplying by $(s^2 + 1)(s^2 + 4)$, we get

$$(s^2 + 6) = A(s^2 + 4) + B(s^2 + 1)$$

Equating co-efficients of s^2, $1 = A + B$

Equating constant terms, $6 = 4A + B$

Solving, $A = \dfrac{5}{3}, \quad B = \dfrac{-2}{3}$

$$\Rightarrow \quad \frac{s^2 + 6}{(s^2 + 1)(s^2 + 4)} = \frac{5}{3(s^2 + 1)} - \frac{2}{3(s^2 + 4)}$$

$$\therefore \quad L^{-1}\left[\frac{s^2 + 6}{(s^2 + 1)(s^2 + 4)}\right] = \frac{5}{3}L^{-1}\left(\frac{1}{s^2 + 1^2}\right) - \frac{2}{3}L^{-1}\left(\frac{1}{s^2 + 2^2}\right)$$

$$= \frac{5}{3}\sin t - \frac{1}{3}\sin 2t = \frac{1}{3}(5 \sin t - \sin 2t).$$

Note. For resolution into partial fractions, we can also put $s^2 = x$.

(*ii*) $$\frac{s^2 + s}{(s^2 + 1)(s^2 + 2s + 2)} = \frac{As + B}{s^2 + 1} + \frac{Cs + D}{s^2 + 2s + 2}$$

Multiplying by $(s^2 + 1)(s^2 + 2s + 2)$, we get

$$s^2 + s = (As + B)(s^2 + 2s + 2) + (Cs + D)(s^2 + 1)$$

Equating co-efficients of s^3, $A + C = 0$ ...(1)

Equating co-efficients of s^2, $2A + B + D = 1$ ...(2)

Equating co-efficients of s, $2A + 2B + C = 1$ ...(3)

Equating constant terms, $2B + D = 0$ ...(4)

From (1) $C = -A$ From (4) $D = -2B$

$\therefore$ Equations (2) and (3) reduce to

$$2A - B = 1 \quad \text{and} \quad A + 2B = 1$$

Solving, $A = \frac{3}{5}, \quad B = \frac{1}{5}$

$\Rightarrow$ $C = -\frac{3}{5}, \quad D = -\frac{2}{5}$

$$\therefore \quad \frac{s^2 + s}{(s^2+1)(s^2+2s+2)} = \frac{3s+1}{5(s^2+1)} - \frac{3s+2}{5(s^2+2s+2)}$$

$$L^{-1}\left[\frac{s^2+s}{(s^2+1)(s^2+2s+2)}\right] = \frac{3}{5} L^{-1}\left(\frac{s}{s^2+1}\right) + \frac{1}{5} L^{-1}\left(\frac{1}{s^2+1}\right) - \frac{1}{5} L^{-1}\left(\frac{3s+2}{(s+1)^2+1}\right)$$

$$= \frac{3}{5}\cos t + \frac{1}{5}\sin t - \frac{1}{5} L^{-1}\left[\frac{3(s+1)-1}{(s+1)^2+1}\right]$$

$$= \frac{3}{5}\cos t + \frac{1}{5}\sin t - \frac{3}{5} L^{-1}\left[\frac{s+1}{(s+1)^2+1^2}\right] + \frac{1}{5} L^{-1}\left[\frac{1}{(s+1)^2+1^2}\right]$$

$$= \frac{3}{5}\cos t + \frac{1}{5}\sin t - \frac{3}{5} e^{-t}\cos t + \frac{1}{5} e^{-t}\sin t$$

$$= \frac{3}{5}(1 - e^{-t})\cos t + \frac{1}{5}(1 + e^{-t})\sin t.$$

EXERCISE 4.2

Find the inverse Laplace transforms of:

1. $\left(\frac{1-\sqrt{s}}{s^2}\right)^2$
2. $\frac{2s+6}{s^2+4}$
3. $\frac{s+2}{s^2-4s+13}$
4. $\frac{4s+12}{s^2+8s+16}$
5. $\frac{s-a}{(s-a)^2+b^2}$
6. $\frac{s+8}{s^2+4s+5}$
7. $\frac{s}{(s+a)^2}$
8. $\frac{s+1}{s^2+2s}$
9. $\frac{s^2+1}{s^3+3s^2+2s}$
10. $\frac{2s^2-6s+5}{s^3-6s^2+11s-6}$
11. $\frac{s^2+2s-3}{s(s-3)(s+2)}$
12. $\frac{4s+5}{(s-1)^2(s+2)}$
13. $\frac{s}{(s^2-1)^2}$
14. $\frac{1+2s}{(s+2)^2(s-1)^2}$
15. $\frac{1}{s(s^2+4)}$
16. $\frac{1}{(s+1)(s^2+2s+2)}$
17. $\frac{s}{(s+1)^2(s^2+1)}$
18. $\frac{1}{s^2(s^2+a^2)}$

19. (*i*) $\dfrac{s+2}{(s^2+4s+5)^2}$ (*ii*) $\dfrac{s+3}{(s^2+6s+13)^2}$ (*iii*) $\dfrac{1}{(s^2+2s+5)^2}$

20. $\dfrac{1}{s^3-a^3}$

21. $\dfrac{s}{(s^2+1)(s^2+4)}$

22. $\dfrac{3s+1}{(s+1)^4}$

23. $\dfrac{a(s^2-2a^2)}{s^4+4a^4}$.

Answers

1. $\dfrac{t^3}{6}+\dfrac{t^2}{2}-\dfrac{16}{15\sqrt{\pi}}t^{5/2}$

2. $2\cos 2t+3\sin 2t$

3. $e^{2t}\cos 3t+\dfrac{4}{3}e^{2t}\sin 3t$

4. $4e^{-4t}(1-t)$

5. $e^{at}\cos bt$

6. $e^{-2t}(\cos t+6\sin t)$

7. $(1-at)e^{-at}$

8. $1+\dfrac{1}{2}e^{-2t}$

9. $\dfrac{1}{2}-2e^{-t}+\dfrac{5}{2}e^{-2t}$

10. $\dfrac{1}{2}e^{t}-e^{2t}+\dfrac{5}{2}e^{3t}$

11. $\dfrac{1}{2}t+\dfrac{4}{5}e^{3t}-\dfrac{3}{10}e^{-2t}$

12. $\dfrac{1}{3}e^{t}+3te^{t}-\dfrac{1}{3}e^{-2t}$

13. $\dfrac{1}{2}t\cosh t$

14. $\dfrac{1}{3}t(e^{t}-e^{-2t})$

15. $\dfrac{1}{4}(1-\cos 2t)$

16. $e^{-t}(1-\cos t)$

17. $\dfrac{1}{2}(\sin t-t\,e^{-t})$

18. $\dfrac{1}{a^3}(at-\sin at)$

19. (*i*) $\dfrac{1}{2}t\,e^{-2t}\sin t$ (*ii*) $\dfrac{1}{4}te^{-3t}\sin 2t$ (*iii*) $\dfrac{1}{16}e^{-t}(\sin 2t-2t\cos 2t)$

20. $\dfrac{1}{3a^2}\left[e^{at}-e^{-\frac{1}{2}at}\left\{\cos\dfrac{\sqrt{3}}{2}at+\sqrt{3}\sin\dfrac{\sqrt{3}}{2}at\right\}\right]$

21. $\dfrac{1}{3}(\cos t-\cos 2t)$

22. $e^{-t}\left(\dfrac{3}{2}t^2-\dfrac{1}{3}t^3\right)$

23. $\cos at\sinh at$.

4.7. LAPLACE TRANSFORMS OF DERIVATIVES

Theorem. If $f'(t)$ be continuous and $L\{f(t)\}=\bar{f}(s)$, then $L\{f'(t)\}=s\bar{f}(s)-f(0)$, provided, $\underset{t\to\infty}{\text{Lt}}[e^{-st}f(t)]=0$.

Proof. $L\{f'(t)\}=\int_0^\infty e^{-st}f'(t)\,dt$ Integrating by parts

$$=\Big[e^{-st}f(t)\Big]_0^\infty-\int_0^\infty -se^{-st}\cdot f(t)\,dt$$

$$=\underset{t\to\infty}{\text{Lt}}e^{-st}f(t)-f(0)+s\int_0^\infty e^{-st}f(t)\,dt$$

$$=0-f(0)+sL\{f(t)\}=s\bar{f}(s)-f(0).$$

Note 1. If $f(t)$ be such that $\underset{t\to\infty}{\text{Lt}}[e^{-st}f(t)]=0$, then $f(t)$ is said to be of exponential order s.

Note 2. The above theorem can be generalised.

Generalisation. If $f(t)$ and its first $(n-1)$ derivatives be continuous, then
$L\{f^n(t)\} = s^n \bar{f}(s) - s^{n-1} f(0) - s^{n-2} f'(0) - s^{n-3} f''(0) - \ldots - f^{n-1}(0)$.

Using the general rule of integration by parts, $L\{f^n(t)\} = \int_0^\infty e^{-st} f^n(t)\, dt$

$$= [e^{-st} f^{n-1}(t) - (-s) e^{-st} f^{n-2}(t) + (-s)^2 e^{-st} f^{n-3}(t) - \ldots$$
$$+ (-1)^{n-1} (-s)^{n-1} e^{-st} f(t)]_0^\infty + (-1)^n (-s)^n \int_0^\infty e^{-st} f(t)\, dt$$
$$= -f^{n-1}(0) - sf^{n-2}(0) - s^2 f^{n-3}(0) - \ldots - s^{n-1} f(0) + s^n L\{f(t)\}$$
$$= s^n \bar{f}(s) - s^{n-1} f(0) - s^{n-2} f'(0) - s^{n-3} f''(0) - \ldots - f^{n-1}(0)$$

assuming that $\underset{t \to \infty}{\text{Lt}} [e^{-st} f^m(t)] = 0$, where $m = 0, 1, 2, \ldots, n-1$.

Note. In particular, when $n = 2, 3, 4, \ldots$, we have

$$L\{f''(t)\} = s^2 \bar{f}(s) - s f(0) - f'(0)$$
$$L\{f'''(t)\} = s^3 \bar{f}(s) - s^2 f(0) - sf'(0) - f''(0)$$
$$L\{f^{iv}(t)\} = s^4 \bar{f}(s) - s^3 f(0) - s^2 f'(0) - sf''(0) - f'''(0) \text{ and so on.}$$

These results will be used in the solution of differential equations.

4.8. LAPLACE TRANSFORMS OF INTEGRALS

Theorem. If $L\{f(t)\} = \bar{f}(s)$, then

$$L\left\{\int_0^t f(u)\, du\right\} = \frac{1}{s} \bar{f}(s) \quad \text{or} \quad L^{-1}\left\{\frac{1}{s} \bar{f}(s)\right\} = \int_0^t f(u)\, du.$$

Proof. Let $\phi(t) = \int_0^t f(u)\, du$, then $\phi'(t) = f(t)$ and $\phi(0) = 0$ [by Art. 4.7]

$\therefore \quad L\{\phi'(t)\} = s\bar{\phi}(s) - \phi(0)$

$\Rightarrow \quad L\{f(t)\} = s\bar{\phi}(s) \Rightarrow \bar{f}(s) = sL\{\phi(t)\}$

$\Rightarrow \quad L\{\phi(t)\} = \frac{1}{s} \bar{f}(s)$

$$\therefore \quad L\left[\int_0^t f(u)\, du\right] = \frac{1}{s} \bar{f}(s) \quad \text{or} \quad L^{-1}\left[\frac{1}{s} \bar{f}(s)\right] = \int_0^t f(u)\, du$$

Hence, if $L^{-1}\{\bar{f}(s)\} = f(t)$ then $L^{-1}\left[\frac{1}{s} \bar{f}(s)\right] = \int_0^t f(u)\, du.$

ILLUSTRATIVE EXAMPLES

Example 1. *If* $L\{t \sin wt\} = \dfrac{2ws}{(s^2 + w^2)^2}$, *evaluate*

(i) $L\{wt \cos wt + \sin wt\}$ *(ii)* $L\{2 \cos wt - wt \sin wt\}$.

Sol. Let $f(t) = t \sin wt$, then $f'(t) = wt \cos wt + \sin wt$

and $\quad f''(t) = 2w \cos wt - w^2 t \sin wt$

Also $$f(0) = 0, f'(0) = 0, f''(0) = \frac{2ws}{(s^2 + w^2)^2}$$

Since $$L\{f^n(t)\} = s^n \bar{f}(s) - s^{n-1} f(0) - s^{n-2} f'(0) - s^{n-3} f''(0) - \ldots\ldots$$

(*i*) $$L\{f'(t)\} = s\bar{f}(s) - f(0)$$

$$\Rightarrow \quad L\{wt \cos wt + \sin wt\} = s.\frac{2ws}{(s^2 + w^2)^2} - 0 = \frac{2ws^2}{(s^2 + w^2)^2}.$$

(*ii*) $$L\{f''(t)\} = s^2\bar{f}(s) - sf(0) - f'(0)$$

$$\Rightarrow \quad L\{2w \cos wt - w^2 t \sin wt\} = s^2 . \frac{2ws}{(s^2 + w^2)^2} - s . 0 - 0 = \frac{2ws^3}{(s^2 + w^2)^2}$$

$$\therefore \quad L\{2 \cos wt - wt \sin wt\} = \frac{2ws^3}{(s^2 + w^2)^2}.$$

Example 2. *Find the inverse Laplace transforms of*

(*i*) $\dfrac{1}{s^3(s^2 + a^2)}$ (*ii*) $\dfrac{1}{s(s+1)^3}$.

Sol. (*i*) Since $L^{-1}\left[\dfrac{1}{s^2 + a^2}\right] = \dfrac{1}{a}\sin at$

therefore, by Art. 4.8, we have

$$L^{-1}\left[\frac{1}{s(s^2 + a^2)}\right] = \int_0^t \frac{1}{a}\sin au \, du = \frac{-1}{a^2}\Big[\cos au\Big]_0^t = \frac{1}{a^2}(1 - \cos at)$$

$$L^{-1}\left[\frac{1}{s^2(s^2 + a^2)}\right] = \int_0^t \frac{1}{a^2}(1 - \cos au)\, du = \frac{1}{a^2}\left[u - \frac{\sin au}{a}\right]_0^t = \frac{1}{a^2}\left[t - \frac{\sin at}{a}\right]$$

$$L^{-1}\left[\frac{1}{s^2(s^2 + a^2)}\right] = \int_0^t \frac{1}{a^2}\left(u - \frac{\sin au}{a}\right) du = \frac{1}{a^2}\left[\frac{u^2}{2} + \frac{\cos au}{a^2}\right]_0^t$$

$$= \frac{1}{a^2}\left[\frac{t^2}{2} + \frac{\cos at}{a^2} - \frac{1}{a^2}\right].$$

(*ii*) Since $$L^{-1}\left(\frac{1}{(s+1)^3}\right) = \frac{e^{-t} . t^2}{2!} = \frac{1}{2}t^2 e^{-t}$$

$$L^{-1}\left[\frac{1}{s(s+1)^3}\right] = \frac{1}{2}\int_0^t t^2 e^{-t}\, dt = \frac{1}{2}\left[t^2(-e^{-t}) - 2t(e^{-t}) + 2(-e^{-t})\right]_0^t$$

$$= \frac{1}{2}\left[(-t^2 - 2t - 2)e^{-t} + 2\right] = 1 - e^{-t}\left(1 + t + \frac{t^2}{2}\right).$$

EXERCISE 4.3

1. Given $\left\{2\sqrt{\frac{t}{\pi}}\right\} = \frac{1}{s^{3/2}}$, show that $L\left\{\frac{1}{\sqrt{\pi t}}\right\} = \frac{1}{\sqrt{s}}$.
2. If $L\{f(t)\} = \bar{f}(s)$, show that $L\{f'(t)\} = s\bar{f}(s) - f(0)$. Use it to evaluate $L\{f'(t)\}$, when $f(t) = e^{-5t}\sin t$.
3. If $L\{f(t)\} = \bar{f}(s)$, show that $L\{f''(t)\} = s^2\bar{f}(s) - sf(0) - f'(0)$. If $f(t) = e^{-2t}\sin 4t$, find $L\{f'(t)\}$.
4. If $L^{-1}\{\bar{f}(s)\} = f(t)$, show that $L^{-1}\left\{\frac{1}{s}\bar{f}(s)\right\} = \int_0^t f(x)\,dx$. Use this result to obtain $L^{-1}\left\{\frac{1}{s(s+a)}\right\}$ and $L^{-1}\left\{\frac{1}{s^2(s+a)}\right\}$.
5. Find the inverse transform of $\frac{1}{s(s^2+a^2)}$.

Answers

2. $\frac{s}{s^2+10s+26}$ 3. $\frac{-16(s+5)}{s^2+4s+20}$
4. $\frac{1}{a}(1-e^{-at})$; $\frac{1}{a^2}(at+e^{-at}-1)$ 5. $\frac{1}{a^2}(1-\cos at)$.

4.9. MULTIPLICATION BY t^n

If $L\{f(t)\} = \bar{f}(s)$, then $L\{t^n f(t)\} = (-1)^n \frac{d^n(\bar{f}(s))}{ds^n}$, where $n = 1, 2, 3, ...$

Proof. We prove the theorem by induction

$$L\{f(t)\} = \bar{f}(s) \Rightarrow \int_0^\infty e^{-st} f(t)\,dt = \bar{f}(s)$$

Differentiating both sides w.r.t. s (Using Leibnitz's rule for differentiation under the integral sign), we have

$$\frac{d}{ds}\int_0^\infty e^{-st} f(t)\,dt = \frac{d}{ds}(\bar{f}(s)) \text{ or } \int_0^\infty \frac{\partial}{\partial s}[e^{-st} f(t)]\,dt = \frac{d}{ds}[f'(s)]$$

or
$$\int_0^\infty -t\,e^{-st} f(t)\,dt = \frac{d}{ds}[\bar{f}(s)] \text{ or } \int_0^\infty e^{-st}[t\,f(t)]\,dt = -\frac{d}{ds}[\bar{f}(s)]$$

or
$$L\{t\,f(t)\} = -\frac{d}{ds}[\bar{f}(s)]$$ which confirms the truth of the theorem for $n = 1$.

Now assume the theorem to be true for $n = m$, so that $L\{t^m f(t)\} = (-1)^m \frac{d^m}{ds^m}[\bar{f}(s)]$ or

$$\int_0^\infty e^{-st} t^m f(t)\,dt = (-1)^m \frac{d^m}{ds^m}[\bar{f}(s)]$$

Differentiating both sides w.r.t. s, we have

$$\frac{d}{ds}\int_0^\infty e^{-st} t^m f(t)\,dt = (-1)^m \frac{d^{m+1}[\bar{f}(s)]}{ds^{m+1}}$$

or $$\int_0^\infty \frac{\partial}{\partial s} (e^{-st}\, t^m\, f(t))\, dt = (-1)^m \frac{d^{m+1}}{ds^{m+1}} [\bar{f}(s)]$$

or $$\int_0^\infty -t\, e^{-st}\, t^m f(t)\, dt = (-1)^m \frac{d^{m+1} [\bar{f}(s)]}{ds^{m+1}}$$

or $$\int_0^\infty e^{-st}\, t^{m+1} f(t)\, dt = (-1)^{m+1} \frac{d^{m+1} [\bar{f}(s)]}{ds^{m+1}}$$

or $$\mathrm{L}\,\{t^{m+1} f(t)\} = (-1)^{m+1} \frac{d^{m+1} [\bar{f}(s)]}{ds^{m+1}}$$

Which shows that the theorem is true for $n = m + 1$. Hence by mathematical induction, the theorem is true for all positive integers n.

Cor. If $\mathrm{L}^{-1}\,\{\bar{f}(s)\} = f(t)$, then $\mathrm{L}^{-1}\left[\frac{d^n}{ds^n}\{\bar{f}(s)\}\right] = (-1)^n\, t^n\, f(t)$.

In particular $\mathrm{L}^{-1}\left[\frac{d}{ds}\{\bar{f}(s)\}\right] = (-t)\, f(t)$.

ILLUSTRATIVE EXAMPLES

Example 1. *Find the Laplace transforms of*

(i) $t \sin^2 3t$ *(ii)* $t^3 e^{-3t}$

Sol. (*i*) Since $\mathrm{L}\{\sin^2 3t\} = \mathrm{L}\left\{\frac{1-\cos 6t}{2}\right\}$

$$= \frac{1}{2}\left\{\frac{1}{s} - \frac{s}{s^2+36}\right\} = \frac{18}{s(s^2+36)}$$

$$\therefore \quad \mathrm{L}\,\{t \sin^2 3t\} = (-1)\frac{d}{ds}\left(\frac{18}{s(s^2+36)}\right)$$

$$= -18 \frac{d}{ds} [(s^3 + 36s)^{-1}]$$

$$= -18(-1)(s^3 + 36s)^{-2}(3s^2 + 36) = \frac{54(s^2+12)}{s^2(s^2+36)^2}.$$

(*ii*) Since $\mathrm{L}\{e^{-3t}\} = \frac{1}{s+3}$

$$\therefore \quad \mathrm{L}\{t^3 e^{-3t}\} = (-1)^3 \frac{d^3\left(\frac{1}{s+3}\right)}{ds^3} = \frac{(-1)^3.(3\,!)}{(s+3)^4} = \frac{6}{(s+3)^4}$$

Example 2. *Find the inverse Laplace transform of* $\log \dfrac{s+1}{s-1}$ *(K.U. May 2009)*

Sol. Let $L^{-1}\left\{\log \dfrac{s+1}{s-1}\right\} = f(t)$, then $L\{f(t)\} = \bar{f}(s) = \log \dfrac{s+1}{s-1}$

$$\therefore \quad L\{t\, f(t)\} = (-1)\frac{d}{ds}\left[\log \frac{s+1}{s-1}\right] = -\frac{d}{ds}[\log (s+1) - \log (s-1)]$$

$$= -\left[\frac{1}{s+1} - \frac{1}{s-1}\right] = \frac{1}{s-1} - \frac{1}{s+1}$$

$$\Rightarrow \quad t\, f(t) = L^{-1}\left(\frac{1}{s-1}\right) - L^{-1}\left(\frac{1}{s+1}\right) = e^{t} - e^{-t} = 2 \sinh t$$

or $$f(t) = \frac{2 \sinh t}{t}$$

Hence $$L^{-1}\left\{\log \frac{s+1}{s-1}\right\} = \frac{2 \sinh t}{t}$$

4.10. DIVISION BY *t*

If $L\{f(t)\} = \bar{f}(s)$, *then* $L\left\{\dfrac{1}{t} f(t)\right\} = \displaystyle\int_s^{\infty} \bar{f}(s)ds$ *provided the integral exists.*

Proof. We have $\bar{f}(s) = \displaystyle\int_0^{\infty} e^{-st} f(t)\, dt$

Integrating both sides w.r.t. s from s to ∞, we have

$$\int_0^{\infty} \bar{f}(s)\, ds = \int_0^{\infty}\left[\int_0^{\infty} e^{-st} f(t)\, dt\right] ds$$

Since s and t are independent, changing the order of integration on the right-hand side, we have

$$\int_0^{\infty} \bar{f}(s) ds = \int_0^{\infty}\left[\int_s^{\infty} e^{-st}\, ds\right] f(t)\, dt$$

$$= \int_0^{\infty}\left[\frac{e^{-st}}{-t}\right]_s^{\infty} f(t)\, dt = \int_0^{\infty} e^{-st} \frac{f(t)}{t}\, dt = L\left\{\frac{1}{t} f(t)\right\}$$

ILLUSTRATIVE EXAMPLES

Example 1. *Find the Laplace transforms of*

(i) $\dfrac{e^{-at} - e^{-bt}}{t}$ *(K.U. April 2013)* *(ii)* $\dfrac{\cos at - \cos bt}{t}$

Sol. *(i)* Since $L\{e^{-at} - e^{-bt}\} = \dfrac{1}{s+a} - \dfrac{1}{s+b}$

$$\therefore \qquad L\left\{\frac{e^{-at}-e^{-bt}}{t}\right\}=\int_s^\infty\left(\frac{1}{s+a}-\frac{1}{s+b}\right)ds$$

$$=\left[\log(s+a)-\log(s+b)\right]_s^\infty=\left[\log\frac{s+a}{s+b}\right]_s^\infty=\left[\log\frac{1+a/s}{1+b/s}\right]_s^\infty$$

$$=\log 1-\log\frac{1+a/s}{1+b/s}=0-\log\frac{s+a}{s+b}=\log\frac{s+b}{s+a}.$$

(*ii*) Since $L\{\cos at-\cos bt\}=\frac{s}{s^2+a^2}-\frac{s}{s^2+b^2}$

$$\therefore \qquad L\left\{\frac{\cos at-\cos bt}{t}\right\}=\int_s^\infty\left(\frac{s}{s^2+a^2}-\frac{s}{s^2+b^2}\right)ds$$

$$=\left[\frac{1}{2}\log(s^2+a^2)-\frac{1}{2}\log(s^2+b^2)\right]_s^\infty$$

$$=\frac{1}{2}\left[\log\frac{s^2+a^2}{s^2+b^2}\right]_s^\infty=\frac{1}{2}\left[\log\frac{1+\frac{a^2}{s^2}}{1+\frac{b^2}{s^2}}\right]_s^\infty$$

$$=\frac{1}{2}\log 1-\log\frac{s^2+a^2}{s^2+b^2}=\frac{1}{2}\log\frac{s^2+b^2}{s^2+a^2}.$$

Example 2. *Find the Laplace transforms of the following functions:*

(*i*) $t\sin 3t\cos 2t$ (*ii*) $t\,e^{-t}\cosh t$

(*iii*) $t^2e^{-2t}\cos t$.

Sol. (*i*) $L(\sin 3t\cos 2t)=\frac{1}{2}L(\sin 5t+\sin t)$

$$=\frac{1}{2}\left(\frac{5}{s^2+5^2}+\frac{1}{s^2+1^2}\right)$$

$$\Rightarrow \qquad L\{t\sin 3t\cos 2t\}=-\frac{d}{ds}\left[\frac{1}{2}\left(\frac{5}{s^2+5^2}+\frac{1}{s^2+1}\right)\right]$$

$$=\frac{-1}{2}\left[\frac{-5}{(s^2+25)^2}(2s)-\frac{1}{(s^2+1)^2}(2s)\right]$$

$$=\frac{5s}{(s^2+25)^2}+\frac{s}{(s^2+1)^2}.$$

(*ii*) $$L(e^{-t}\cosh t)=L\left[e^{-t}\left(\frac{e^t+e^{-t}}{2}\right)\right]=\frac{1}{2}L(1+e^{-2t})$$

$$= \frac{1}{2}\left(\frac{1}{s} + \frac{1}{s+2}\right)$$

$$\Rightarrow \quad L(t\, e^{-t} \cosh t) = -\frac{d}{ds}\left[\frac{1}{2}\left(\frac{1}{s} + \frac{1}{s+2}\right)\right]$$

$$= -\frac{1}{2}\left[\frac{-1}{s^2} - \frac{1}{(s+2)^2}\right] = \frac{1}{2}\,\frac{(s+2)^2 + s^2}{s^2\,(s+2)^2} = \frac{s^2 + 2s + 2}{s^2\,(s+2)^2}.$$

(iii) $L(e^{-2t} \cos t) = \dfrac{s+2}{(s+2)^2 + 1} = \dfrac{s+2}{s^2 + 4s + 5}$

$$\Rightarrow L(t^2\, e^{-2t} \cos t) = (-1)^2 \frac{d^2}{ds^2}\left[\frac{s+2}{s^2 + 4s + 5}\right]$$

$$= \frac{d}{ds}\left[\frac{(s^2 + 4s + 5) \times 1 - (s+2)\,(2s+4)}{(s^2 + 4s + 5)^2}\right] = \frac{d}{ds}\left[-\frac{s^2 + 4s + 3}{(s^2 + 4s + 5)^2}\right]$$

$$= -\frac{(s^2 + 4s + 5)^2\,(2s+4) - (s^2 + 4s + 3).\,2\,(s^2 + 4s + 5)\,(2s+4)}{(s^2 + 4s + 5)^4}$$

$$= -\frac{(s^2 + 4s + 5)\,(2s+4) - 2\,(2s+4)\,(s^2 + 4s + 3)}{(s^2 + 4s + 5)^3}$$

$$= -\frac{2(s^3 + 6s^2 + 9s + 2)}{(s^2 + 4s + 5)^3}.$$

Example 3. *Find the Laplace transforms of the following functions:*

(i) $\dfrac{e^{-t} \sin t}{t}$ *(ii)* $\dfrac{1 - \cos 2t}{t}$.

Sol. *(i)* $L(e^{-t} \sin t) = \dfrac{1}{(s+1)^2 + 1}$

$$\Rightarrow \quad L\left(\frac{e^{-t} \sin t}{t}\right) = \int_s^\infty \frac{1}{(s+1)^2 + 1}\, ds$$

$$= \left[\tan^{-1}(s+1)\right]_s^\infty = \frac{\pi}{2} - \tan^{-1}(s+1) = \cot^{-1}(s+1).$$

(ii) $L(1 - \cos 2t) = \dfrac{1}{s} - \dfrac{s}{s^2 + 2^2}$

$$\Rightarrow \quad L\left(\frac{1 - \cos 2t}{t}\right) = \int_s^\infty \left(\frac{1}{s} - \frac{s}{s^2 + 2^2}\right) ds$$

$$= \left(\log s - \frac{1}{2}\log(s^2 + 4)\right)_s^\infty = \frac{1}{2}\left[\log\left(\frac{s^2}{s^2 + 4}\right)\right]_s^\infty$$

$$= \frac{1}{2}\left[\lim_{n\to\infty} \log \frac{1}{1+\frac{4}{s^2}} - \log \frac{s^2}{s^2+4}\right]$$

$$= \frac{1}{2}\left[\log 1 + \log \frac{s^2+4}{s^2}\right] = \frac{1}{2}\log\left(\frac{s^2+4}{s^2}\right)$$

Evaluation of integrals by Laplace Transform

Example 4. *Evaluate*

(*i*) $\int_0^\infty t^3\, e^{-t} \sin t\, dt$ (*ii*) $\int_0^\infty \frac{e^{-2t} \sin^2 t}{t}\, dt.$

Sol. (*i*) $\int_0^\infty t^3\, e^{-t} \sin t\, dt = \int_0^\infty e^{-st}\, (t^3 \sin t)\, dt$, where $s = 1$

$= \text{L}\,(t^3 \sin t)$ by definition

$$= (-1)^3 \frac{d^3}{ds^3}\left(\frac{1}{s^2+1}\right) = \frac{d^2}{ds^2}\left[\frac{2s}{(s^2+1)^2}\right]$$

$$= \frac{d}{ds}\left[\frac{(s^2+1)^2 \times 2 - 2s \times 2\,(s^2+1) \times 2s}{(s^2+1)^4}\right]$$

$$= \frac{d}{ds}\left[\frac{2\,(s^2+1) - 8s^2}{(s^2+1)^3}\right] = \frac{d}{ds}\left[\frac{2\,(1-3s^2)}{(s^2+1)^3}\right]$$

$$= \frac{2 \times (s^2+1)^3\,(-6s) - (1-3s^2) \times 3\,(s^2+1)^2 \times 2s}{(s^2+1)^6}$$

$$= 2 \times \frac{-6s\,(s^2+1) - 6s\,(1-3s^2)}{(s^2+1)^4} = \frac{24s\,(s^2-1)}{(s^2+1)^4} = 0 \text{ as } s \to 1$$

(*ii*) $$\int_0^\infty \frac{e^{-2t} \sin^2 t}{t}\, dt = \int_0^\infty e^{-st}\left(\frac{\sin^2 t}{t}\right) dt, \text{ where } s = 2$$

$= \text{L}\left(\frac{\sin^2 t}{t}\right)$ by definition

$= \text{L}\left(\frac{1-\cos 2t}{t}\right) = \frac{1}{2}\log\left(\frac{s^2+4}{s^2}\right)$ as in example 3(*ii*)

$= \frac{1}{2}\log 2$ as $s \to 2$

Example 5. *Evaluate* $L\left[\int_0^t e^t \frac{\sin t}{t}\, dt\right]$.

Sol. Since $\text{L}(\sin t) = \frac{1}{s^2+1}$

$$\therefore \quad L\left(\frac{\sin t}{t}\right) = \int_s^\infty \frac{ds}{s^2+1} = \left[\tan^{-1} s\right]_s^\infty = \frac{\pi}{2} - \tan^{-1} s = \cot^{-1} s$$

$$\Rightarrow \quad L\left(e^t \frac{\sin t}{t}\right) = \cot^{-1}(s-1) = \bar{f}(s)$$

$$\therefore \quad L\left[\int_0^t e^t \frac{\sin t}{t}\, dt\right] = \frac{1}{s}\bar{f}(s) = \frac{1}{s}\cot^{-1}(s-1).$$ **[Using Theorem 4.8]**

Example 6. *Find the inverse Laplace transforms of the following:*

(*i*) $\dfrac{s^2 - s + 1}{(s-2)^3}$ (*ii*) $\dfrac{s+2}{s^2(s+1)(s-2)}$

(*iii*) $\dfrac{s+2}{(s^2+4s+5)^2}$.

Sol. (*i*) Put $s - 2 = u$ so that $s = u + 2$,

$$s^2 - s + 1 = (u+2)^2 - (u+2) + 1 = u^2 + 3u + 3$$

$$\therefore \quad \frac{s^2 - s + 1}{(s-2)^3} = \frac{u^2 + 3u + 3}{u^3} = \frac{1}{u} + \frac{3}{u^2} + \frac{3}{u^3}$$

$$= \frac{1}{s-2} + \frac{3}{(s-2)^2} + \frac{3}{(s-2)^3}$$

$$L^{-1}\left(\frac{s^2 - s + 1}{(s-2)^3}\right) = L^{-1}\left(\frac{1}{s-2}\right) + 3\,L^{-1}\left(\frac{1}{(s-2)^2}\right) + 3\,L^{-1}\left(\frac{1}{(s-2)^3}\right)$$

$$= e^{2t} + 3e^{2t}.\frac{t}{1!} + 3e^{2t}.\frac{t^2}{2!} \quad \left[\because \; L^{-1}\left\{\frac{1}{(s-a)^n}\right\} = e^{at}\cdot\frac{t^{n-1}}{(n-1)!}\right]$$

$$= e^{2t}\left(1 + 3t + \frac{3}{2}t^2\right).$$

Note. The above method is useful when the denominator has a single repeated linear factor.

(*ii*) With usual procedure

$$\frac{s+2}{(s+1)(s-2)} = -\frac{1}{(s+1)} + \frac{4}{3(s-1)}$$

$$\therefore \quad L^{-1}\left[\frac{s+2}{(s+1)(s-2)}\right] = \frac{-1}{3}L^{-1}\left(\frac{1}{s+1}\right) + \frac{4}{3}L^{-1}\left(\frac{1}{s-2}\right)$$

$$= -\frac{1}{3}e^{-t} + \frac{4}{3}e^{2t}$$

Using $\quad L^{-1}\{\bar{f}(s)\} = f(t) \Rightarrow L^{-1}\left[\dfrac{\bar{f}(s)}{s}\right] = \displaystyle\int_0^t f(t)\,dt$

$$L^{-1}\left[\frac{s+2}{s(s+1)(s-2)}\right] = \int_0^t \left(-\frac{1}{3}e^{-t} + \frac{4}{3}e^{2t}\right) dt$$

$$= \left(\frac{1}{3}e^{-t} + \frac{2}{3}e^{2t}\right)_0^t = \frac{1}{3}e^{-t} + \frac{2}{3}e^{2t} - 1$$

$$L^{-1}\left[\frac{s+2}{s^2(s+1)(s-2)}\right] = \int_0^t \left(\frac{1}{3}e^{-t} + \frac{2}{3}e^{2t} - 1\right)dt$$

$$= \left(-\frac{1}{3}e^{-t} + \frac{1}{3}e^{2t} - t\right)_0^t = -\frac{1}{3}e^{-t} + \frac{1}{3}e^{2t} - t$$

$$= \frac{1}{3}(e^{2t} - e^{t}) - t$$

(*iii*) We know that

$$L^{-1}\left(\frac{1}{s^2+4s+5}\right) = L^{-1}\left(\frac{1}{(s+2)^2+1}\right) = e^{-2t}\sin t$$

Using $L^{-1}\{\bar{f}(s)\} = f(t) \Rightarrow L^{-1}\left[\frac{d}{ds}\{\bar{f}(s)\}\right] = (-t)f(t)$

$$L^{-1}\left[\frac{d}{ds}\left(\frac{1}{s^2+4s+5}\right)\right] = (-t)e^{-2t}\sin t$$

or
$$L^{-1}\left[\frac{-(2s+4)}{(s^2+4s+5)^2}\right] = -te^{-2t}\sin t$$

or
$$L^{-1}\left[\frac{s+2}{(s^2+4s+5)^2}\right] = -\frac{1}{2}(-te^{-2t}\sin t) = \frac{1}{2}te^{-2t}\sin t.$$

Example 7. *Find the inverse Laplace transforms of the following:*

(*i*) $\log \frac{s^2+1}{s(s+1)}$ (*ii*) $\tan^{-1}(2/s^2)$.

Sol. (*i*) Let $f(t) = L^{-1}\left[\log \frac{s^2+1}{s(s+1)}\right]$

Using $f(t) = L^{-1}\{\bar{f}(s)\} \Rightarrow t\,f(t) = L^{-1}\left[-\frac{d}{ds}\{\bar{f}(s)\}\right]$

$$tf(t) = L^{-1}\left[-\frac{d}{ds}\left(\log\frac{s^2+1}{s(s+1)}\right)\right]$$

$$= -L^{-1}\left[\frac{d}{ds}\{\log(s^2+1) - \log s - \log(s+1)\}\right]$$

$$= -L^{-1}\left[\frac{2s}{s^2+1} - \frac{1}{s} - \frac{1}{s+1}\right]$$

$$= -2L^{-1}\left(\frac{s}{s^2+1}\right) + L^{-1}\left(\frac{1}{s}\right) + L^{-1}\left(\frac{1}{s+1}\right)$$

$$= -2\cos t + 1 + e^{-t}$$

$\Rightarrow$
$$f(t) = \frac{1}{t}(-2\cos t + 1 + e^{-t})$$

Hence $L^{-1}\left[\log \frac{s^2+1}{s(s+1)}\right] = \frac{1}{t}(-2\cos t + 1 + e^{-t})$.

(*ii*) Let $f(t) = L^{-1}\left(\tan^{-1}\left(\frac{2}{s^2}\right)\right)$

Then $$t\,f(t) = L^{-1}\left[-\frac{d}{ds}\left(\tan^{-1}\frac{2}{s^2}\right)\right] = L^{-1}\left[-\frac{1}{1+\frac{4}{s^4}}\cdot\left(\frac{-4}{s^3}\right)\right]$$

$$= L^{-1}\left(\frac{4s}{s^4+4}\right) = L^{-1}\left(\frac{4s}{(s^2+2)^2-(2s)^2}\right)$$

$$= L^{-1}\left[\frac{4s}{(s^2+2s+2)(s^2-2s+2)}\right] = L^{-1}\left[\frac{(s^2+2s+2)-(s^2-2s+2)}{(s^2+2s+2)(s^2-2s+2)}\right]$$

$$= L^{-1}\left[\frac{1}{s^2-2s+2}+\frac{1}{s^2+2s+2}\right] = L^{-1}\left(\frac{1}{(s-1)^2+1}+\frac{1}{(s+1)^2+1}\right)$$

$$= e^t \sin t - e^{-t}\sin t = 2\left(\frac{e^t-e^{-t}}{2}\right)\sin t = 2\sinh t \sin t.$$

$\Rightarrow$ $$f(t) = \frac{2\sinh t\sin t}{t}$$

Hence $L^{-1}\left(\tan^{-1}\frac{2}{s^2}\right) = \frac{2\sinh t \sin t}{t}$.

4.11. CONVOLUTION THEOREM

If $L^{-1}\{\bar{f}(s)\} = f(t)$ and $L^{-1}\{\bar{g}(s)\} = g(t)$, then

$$L^{-1}\{\bar{f}(s)\,\bar{g}(s)\} = \int_0^t f(u)\,g(t-u)\,du = f * g.$$

[$f * g$ is called the convolution of f and g.]

Example. *Apply convolution theorem to evaluate:*

(*i*) $L^{-1}\left\{\frac{s}{(s^2+a^2)^2}\right\}$ *(K.U. May 2009)*

(*ii*) $L^{-1}\left\{\frac{s^2}{(s^2+a^2)(s^2+b^2)}\right\}$

Sol. (*i*) Since $L^{-1}\left\{\frac{s}{s^2+a^2}\right\} = \cos at$ and $L^{-1}\left\{\frac{1}{s^2+a^2}\right\} = \frac{1}{a}\sin at$

$\therefore$ By convolution theorem, we have

$$L^{-1}\left\{\frac{1}{(s^2+a^2)^2}\right\} = L^{-1}\left\{\frac{s}{s^2+a^2}\cdot\frac{1}{s^2+a^2}\right\} = \int_0^t \cos a(t-u)\frac{\sin au}{a}\,du$$

$$= \frac{1}{2a}\int_0^t 2 \sin au \cos a\,(t-u)\,du$$

$$= \frac{1}{2a}\int_0^t [\sin at + \sin(2au - at)]\,du$$

$$= \frac{1}{2a}\left[u \sin at - \frac{1}{2a}\cos(2au - at)\right]_0^t = \frac{1}{2a}\,t \sin at.$$

(*ii*) Since $\quad L^{-1}\left(\frac{s}{s^2+a^2}\right) = \cos at$ and $L^{-1}\left(\frac{s}{s^2+b^2}\right) = \cos bt$

$\therefore$ By convolution theorem, we have

$$L^{-1}\left[\frac{s^2}{(s^2+a^2)(s^2+b^2)}\right] = L^{-1}\left[\frac{s^2}{s^2+a^2}\cdot\frac{s}{s^2+b^2}\right]$$

$$= \int_0^t \cos a\,(t-u)\cos bu\,du$$

$$= \frac{1}{2}\int_0^t [\cos\{at - (a-b)u\} + \cos\{at - (a+b)u\}]\,du$$

$$= \frac{1}{2}\left[-\frac{1}{a-b}\sin\{at-(a+b)u\} - \frac{1}{a+b}\sin\{at-(a+b)u\}\right]_0^t$$

$$= -\frac{1}{2}\left[\frac{\sin bt - \sin at}{a-b} - \frac{\sin at + \sin bt}{a+b}\right] = \frac{a \sin at - b \sin bt}{a^2 - b^2}.$$

EXERCISE 4.4

1. If $L\{f(t)\} = \bar{f}(s)$, show that $L\{t\,f(t)\} = -\frac{d}{ds}[\bar{f}(s)]$.

2. If $L\{f(t)\} = \bar{f}(s)$, show that $L\{t^2 f(t)\} = \frac{d^2}{ds^2}[\bar{f}(s)]$. Use this result to obtain $L\{t^2 \cos at\}$.

3. Find the Laplace transforms of

(*i*) $t \sinh at$ (*ii*) $t^2 \sin at$ (*iii*) $t^2 e^{-2t}$

(*iv*) $t e^{at} \sin at$ (*v*) $\sin 2t - 2t \cos 2t$ (*vi*) $t e^{-t} \cos t$

(*K.U. May 2011*)

4. If $L\left\{\frac{1-\cos at}{a^2}\right\} = \frac{1}{s(s^2+a^2)}$, show that $L\left\{\frac{t(1-\cos at)}{a^2}\right\} = \frac{3s^2+a^2}{s^2(s^2+a^2)^2}$

5. Find the inverse Laplace transform of

(*i*) $\log\left(1+\frac{1}{s^2}\right)$ (*ii*) $\log\left(\frac{1+s}{s}\right)$

6. Find Laplace transforms of

(*i*) $\frac{\sin 2t}{t}$ (*ii*) $\frac{1-e^t}{t}$ (*iii*) $\frac{\cos 2t - \cos 3t}{t}$

(iv) $\dfrac{\sin^2 t}{t}$ *(v)* $\dfrac{\sinh t}{t}$ *(vi)* $\dfrac{1-\cos t}{t}$

(K.U. May 2010)

(vii) $\dfrac{e^{at} - \cos ft}{t}$ *(viii)* $\dfrac{e^{-4t} \sin 3t}{t}$ *(K.U. May 2012)*

7. Evaluate:

(i) $\displaystyle\int_0^\infty t\, e^{-3t} \sin t\, dt$ *(ii)* $\displaystyle\int_0^\infty \frac{\sin t}{t}\, dt$ *(iii)* $\displaystyle\int_0^\infty \frac{e^{-t} - e^{-3t}}{t}\, dt$

(iv) $\displaystyle\int_0^\infty t\, e^{-2t} \sin 3t\, dt$ *(v)* $\displaystyle\int_0^\infty \frac{e^{-at} - e^{-bt}}{t}\, dt$

8. Evaluate:

(i) $L\left[\displaystyle\int_0^t e^{-t} \cos t\, dt\right]$ *(ii)* $L\left[\displaystyle\int_0^t \frac{\sin t}{t} dt\right]$ *(iii)* $L\left[\displaystyle\int_0^t \frac{\cos at - \cos bt}{t} dt\right]$.

9. Apply convolution theorem to evaluate:

(i) $L^{-1}\left[\dfrac{1}{(s+a)(s+b)}\right]$ *(ii)* $L^{-1}\left(\dfrac{s^2}{(s^2+4)^2}\right)$

(K.U. April 2000, K.U. June 2003)

(iii) $L^{-1}\left[\dfrac{1}{s(s^2-a^2)}\right]$ *(K.U. Nov. 2002)* *(iv)* $L^{-1}\left(\dfrac{1}{s(s^2+4)}\right)$

(v) $L^{-1}\left(\dfrac{s^2}{s^4-a^4}\right)$ *(vi)* $L^{-1}\left(\dfrac{s}{(s^2+1)(s^2+4)}\right)$

(vii) $L^{-1}\left(\dfrac{1}{(s^2+a^2)^2}\right)$ *(viii)* $L^{-1}\left(\dfrac{1}{s^2(s+1)^2}\right)$

(ix) $L^{-1}\left(\dfrac{1}{s(s^2-1)}\right)$ *(K.U. May 2010)*

10. Find the inverse transforms of

(i) $\cot^{-1}\left(\dfrac{s}{a}\right)$ *(ii)* $\tan^{-1}\left(\dfrac{1}{s}\right)$ *(K.U. May 2011)* *(iii)* $\cot^{-1}(s+1)$

(iv) $\log \dfrac{s(s+1)}{s^2+4}$ *(v)* $\dfrac{s^2}{(s+1)^3}$ *(vi)* $\log\left(\dfrac{s+a}{s+b}\right)$

(vii) $\log\left(\dfrac{s+5}{s-3}\right)$ *(K.U. May 2010)*

11. Show that:

(i) $L^{-1}(1/s \sin 1/s) = t - \dfrac{t^3}{(3!)^2} + \dfrac{t^5}{(5!)^2} - \dfrac{t^7}{(7!)^2} + \ldots$

(ii) $L^{-1}\left(\dfrac{1}{s}\cos\dfrac{1}{s}\right) = 1 - \dfrac{t^2}{(2!)^2} + \dfrac{t^4}{(4!)^2} - \dfrac{t^6}{(6!)^2} + \ldots$

12. Find the Laplace transform of $t \sin^2(2t)$.

Answers

2. $\dfrac{2s(s^2-3a^2)}{(s^2+a^2)^3}$

3. (*i*) $\dfrac{2as}{(s^2-a^2)^2}$ (*ii*) $\dfrac{2a(3s^2-a^2)}{(s^2+a^2)^3}$ (*iii*) $\dfrac{2}{(s+2)^4}$

(*iv*) $\dfrac{2a(s-a)}{(s^2-2as+2a^2)^2}$ (*v*) $\dfrac{16}{(s^2+4)^2}$ (*vi*) $\dfrac{s^2+2s}{s^2+2s+2}$

5. (*i*) $\dfrac{2(1-\cos t)}{t}$ (*ii*) $\dfrac{1}{t}(1-e^{-t})$

6. (*i*) $\cos^{-1}\left(\dfrac{s}{2}\right)$ (*ii*) $\log\left(\dfrac{s-1}{s}\right)$ (*iii*) $\dfrac{1}{2}\log\left(\dfrac{s^2+9}{s^2+4}\right)$

(*iv*) $\dfrac{1}{4}\log\dfrac{s^2+4}{s^2}$ (*v*) $\dfrac{1}{2}\log\dfrac{s-1}{s+1}$ (*vi*) $\dfrac{1}{2}\log\left(\dfrac{s^2+1}{s^2}\right)$

(*vii*) $\log\dfrac{\sqrt{s^2+b^2}}{s-a}$ (*viii*) $\cot^{-1}\left(\dfrac{s+4}{3}\right)$

7. (*i*) $\dfrac{3}{50}$ (*ii*) $\dfrac{\pi}{2}$ (*iii*) $\log 3$ (*iv*) $\dfrac{12}{169}$ (*v*) $\log\dfrac{b}{a}$

8. (*i*) $\dfrac{s+1}{s(s^2+2s+2)}$ (*ii*) $\dfrac{1}{s}\cot^{-1}s$ (*iii*) $\dfrac{1}{2s}\log\left(\dfrac{s^2+b^2}{s^2+a^2}\right)$

9. (*i*) $\dfrac{e^{-bt}-e^{-at}}{a-b}$ (*ii*) $\dfrac{1}{4}[\sin 2t+2t\cos 2t]$ (*iii*) $\dfrac{1}{a^2}(\cosh at-1)$

(*iv*) $\dfrac{1}{4}(1-\cos 2t)$ (*v*) $\dfrac{1}{2a}(\sinh at+\sin at)$ (*vi*) $\dfrac{1}{3}(\cos t-\cos 2t)$

(*vii*) $\dfrac{1}{2a^3}(\sin at-at\cos at)$ (*viii*) $t(e^{-t}+1)+2(e^{-t}-1)$ (*ix*) $\cosh t-1$

10. (*i*) $\dfrac{\sin at}{t}$ (*ii*) $\dfrac{\sin t}{t}$ (*iii*) $e^{-t}\dfrac{\sin t}{t}$

(*iv*) $\dfrac{2\cos 2t-e^{-t}-1}{t}$ (*v*) $\dfrac{1}{2}(t^2-4t+2)e^{-t}$

(*vi*) $\dfrac{e^{-bt}-e^{-at}}{t}$ (*vii*) $\dfrac{e^{3t}-e^{-5t}}{t}$

12. $\dfrac{1}{2s^2}+\dfrac{1}{2}\dfrac{16-s^2}{(s^2+16)^2}$.

4.12. APPLICATION TO DIFFERENTIAL EQUATIONS

Laplace transforms can be used to solve ordinary as well as partial differential equations. We shall apply this method to solve only ordinary linear differential equations with constant coefficients. The advantage of this method is that it yields the particular solution directly without the necessity of first finding the general solution and then evaluating the orbitrary constants.

Method 1. Take Laplace transforms of both sides of the given differential equation; using initial conditions. This gives an algebraic equation.

2. Solve the algebraic equation to get $\bar{y}$ in terms of s.

3. Take inverse Laplace transform of both sides. This gives y as a function of t which is the desired solution.

Remember. $L\{f^n(t)\} = s^n \bar{f}(s) - s^{n-1} f(0) - s^{n-2} f'(0) \dots - f^{n-1}(0)$.

ILLUSTRATIVE EXAMPLES

Example 1. *Solve the equation:*

$$\frac{d^3y}{dt^3} + 2\frac{d^2y}{dt^2} - \frac{dy}{dt} - 2y = 0, \text{ where } y = 1,$$

$$\frac{dy}{dt} = 2, \frac{d^2y}{dt^2} = 2 \text{ at } t = 0$$

Sol. The given eqn is $y''' + 2y'' - y' - 2y = 0$

Taking the Laplace transform of both sides, we get

$$[s^3\bar{y} - s^2y(0) - s\,y'(0) - y'(0)] + 2\,[s^2\bar{y} - sy(0) - y'(0)]$$

$$- [s\bar{y} - y(0)] - 2\bar{y} = 0 \qquad \dots(1)$$

Using the given conditions $y(0) = 1, y'(0) = 2, y''(0) = 2$, equation (1) reduces to

$$(s^3 + 2s^2 - s - 2)\bar{y} - s^2 - 2s - 2 - 2s - 4 + 1 = 0$$

or $$(s^3 - 2s^2 - s - 2)\bar{y} = s^2 + 4s + 5$$

$$\therefore \quad \bar{y} = \frac{s^2 + 4s + 5}{s^3 + 2s^2 - s - 2} = \frac{s^2 + 4s + 5}{(s-1)(s+1)(s+2)}$$

$$= \frac{5}{3(s-1)} - \frac{1}{s+1} + \frac{1}{3(s+2)} \qquad \text{(Partial Fractions)}$$

Taking the inverse Laplace transforms of both sides,

We get

$$y = \frac{5}{3}L^{-1}\left(\frac{1}{s-1}\right) - L^{-1}\left(\frac{1}{s+1}\right) + \frac{1}{3}L^{-1}\left(\frac{1}{s+2}\right)$$

$$= \frac{5}{3}e^t - e^{-t} + \frac{1}{3}e^{-2t}$$

$$\therefore \quad y = \frac{1}{3}(5e^t + e^{-2t}) - e^{-t} \text{ which is the required solution.}$$

Example 2. *Solve the equation*

$$y'' - 3y' + 2y = 4t + e^{3t}, \text{ when } y(0) = 1 \text{ and } y'(0) = -1.$$

Sol. Taking the Laplace transform of both sides, we get

$$[s^2\bar{y} - sy(0) - y'(0)] - 3[s\bar{y} - y(0)] + 2\bar{y} = \frac{4}{s^2} + \frac{1}{s-3}$$

Using the given conditions, it reduces to

$$(s^2 - 3s + 2)\bar{y} - s + 1 + 3 = \frac{4}{s^2} + \frac{1}{s-3}$$

or $$(s^2 - 3s + 2)\bar{y} = \frac{4}{s^2} + \frac{1}{s-3} + s - 4$$

$\therefore$ $$\bar{y} = \frac{s^4 - 7s^3 + 13s^2 + 4s - 12}{s^2 (s-1)(s-2)(s-3)}$$

$$= \frac{3}{s} + \frac{2}{s^2} - \frac{1}{2}\frac{1}{(s-1)} - \frac{2}{s-2} + 2\frac{1}{2(s-3)} \quad \text{(Partial Fractions)}$$

Taking the inverse Laplace transforms of both sides, we get

$$y = 3\,L^{-1}\left(\frac{1}{s}\right) + 2\,L^{-1}\left(\frac{1}{s^2}\right) - \frac{1}{2}L^{-1}\left(\frac{1}{s-1}\right) - 2\,L^{-1}\left(\frac{1}{s-2}\right) + \frac{1}{2}L^{-1}\left(\frac{1}{s-3}\right)$$

$$= 3 + 2t - \frac{1}{2}e^t - 2\,e^{2t} + \frac{1}{2}e^{3t}$$

or $$y = 3 + 2t + \frac{1}{2}(e^{3t} - e^t) - 2e^{2t}$$ which is the required solution.

Example 3. *Solve the equation* $\dfrac{d^2x}{dt^2} + \dfrac{2dx}{dt} + 5x = e^{-t} \sin t,\ x(0) = 0,\ x'(0) = 1.$

(K.U. May 2008)

Sol. The given equation is $x'' + 2x' + 5x = e^{-t} \sin t$

Taking the Laplace transform of both sides, we get

$$[s^2\bar{x} - sx(0) - x'(0)] + 2\,[s\bar{x} - x(0)] + 5\bar{x} = \frac{1}{(x+1)^2 + 1}$$

Using the given conditions, it reduces to

$$(s^2 + 2s + 5)\bar{x} - 1 = \frac{1}{s^2 + 2s + 2}$$

or $$\bar{x} = \frac{1}{(s^2 + 2s + 2)(s^2 + 2s + 5)} + \frac{1}{s^2 + 2s + 5}$$

$$= \frac{1}{3}\left[\frac{1}{s^2 + 2s + 2} - \frac{1}{s^2 + 2s + 5}\right] + \frac{1}{s^2 + 2s + 5}$$

$$= \frac{1}{3}\left[\frac{1}{s^2 + 2s + 2} + \frac{2}{s^2 + 2s + 5}\right]$$

$$= \frac{1}{3}\left[\frac{1}{(s+1)^2 + 1} + \frac{2}{(s+1)^2 + 2^2}\right]$$

Taking the inverse Laplace transform of both sides, we get

$$x = \frac{1}{3}L^{-1}\left[\frac{1}{(s+1)^2 + 1} + 2.\frac{1}{(s+1)^2 + 2^2}\right]$$

$$= \frac{1}{3}\left[e^{-t}\sin t + 2.\frac{1}{2}e^{-t}\sin 2t\right]$$

or $$x = \frac{1}{3}e^{-t}(\sin t + \sin 2t)$$ which is the required solution.

Example 4. *Solve the equation*

$$\frac{d^2y}{dt^2} - 3\frac{dy}{dt} + 2y = 4,\ y(0) = 2,\ y'(0) = 3. \qquad \text{(K.U. April 2013)}$$

Sol. The given equation is $y'' - 3y' + 2y = 4$

Taking the Laplace transform of both sides, we get

$$[s^2\bar{y} - sy(0) - y'(0)] - 3[s\bar{y} - y(0)] + 2\bar{y} = 4.\frac{1}{s}$$

Using the given conditions, it reduces to

$$(s^2 - 3s + 2)\bar{y} = \frac{4}{s} + 2s - 3 = \frac{2s^2 - 3s + 4}{s}$$

or

$$\bar{y} = \frac{2s^2 - 3s + 4}{s(s^2 - 3s + 2)} = \frac{2s^2 - 3s + 4}{s(s-1)(s-2)}$$

$$= \frac{2}{s} + \frac{-3}{s-1} + \frac{3}{s-2} \qquad \text{(Partial Fractions)}$$

Taking the inverse Laplace transforms of both sides, we get

$$y = 2\,L^{-1}\left(\frac{1}{s}\right) - 3\,L^{-1}\left(\frac{1}{s-1}\right) + 3\,L^{-1}\left(\frac{1}{s-2}\right)$$

$$= 2 - 3e^t + 3e^{2t}.$$

EXERCISE 4.5

Solve the following equations by the transform method (1 – 17):

1. $(D^2 - 2D + 2)x = 0,\ x = Dx = 1$ at $t = 0$
2. $\frac{d^2x}{dt^2} - 2\frac{dx}{dt} + x = e^t$ with $x = 2,\ \frac{dx}{dt} = -1$ at $t = 0$
3. $y'' + 4y' + 3y = e^{-t},\ y(0) = y'(0) = 1$
4. $x'' - 3x' + 2x = 1 - e^{2t},\ x(0) = 1,\ x'(0) = 0$
5. $y'' + y' - 2y = t,\ y(0) = 1,\ y'(0) = 0$
6. $\frac{dx}{dt} + x = \sin wt,\ x(0) = 2$
7. $(D^2 + 5D + 6)x = 5e^t,\ x(0) = 2,\ x'(0) = 1$
8. $(D^2 - 1)x = a\cosh t,\ x(0) = x'(0) = 0$
9. $\frac{d^2x}{dt^2} + 9x = \cos 2t,\ x(0) = 1,\ x\left(\frac{\pi}{2}\right) = -1$
10. $(D^2 - D - 2)x = 20\sin 2t,\ x(0) = -1,\ x'(0) = 2$
11. $\frac{d^2y}{dt^2} + 2\frac{dy}{dt} + 2y = \frac{17}{2}\sin 5t,\ y = 2$ and $\frac{dy}{dt} = -4$, when $t = 0$
12. $\frac{d^2y}{dt^2} + \frac{dy}{dt} - 2y = 3\cos 3t - 11\sin 3t$ given that $y(0) = 0$ and $y'(0) = 6$
13. $\frac{d^2x}{dt^2} + x = t\cos 2t$ given that $x(0) = x'(0) = 0$
14. $(D^3 + D)x = 2,\ x = 3,\ Dx = 1,\ D^2x = -2$ at $t = 0$
15. $(D^3 + D^2)x = 6t^2 + 4,\ x = D^2x = 0,\ Dx = 2$ at $t = 0$
16. $(D^3 - D)x = 2\cos t,\ x = 3,\ Dx = 2,\ D^2x = 1$, at $t = 0$

17. $(D^3 - 3D^2 + 3D - 1)y = t^2 e^{2t}, y = 1, Dy = 0, D^2y = -2$ at $t = 0$

18. Solve $y'' + y' - 2y = t, y(0) = 1, y'(0) = 0$, using Laplace transform.

19. Solve $y'' + 2y' - 3y = \sin t, y(0) = y'(0) = 0$, using Laplace transform *(K.U. June 2012)*

20. Solve $\frac{d^2y}{dt^2} + 2\frac{dy}{dt} + 6y = 6t\, e^{-t}, y(0) = 2, y'(0) = 5$, using Laplace transform.

Answers

1. $x = e^t \cos t$

2. $x = \left(2 - 3t + \frac{1}{2}t^2\right)e^t$

3. $y = \frac{7}{4}e^{-t} - \frac{3}{4}e^{-3t} - \frac{1}{2} + e^{-t}$

4. $x = \frac{1}{2} + \left(\frac{1}{2} - t\right)e^{2t}$

5. $y = e^t + \frac{1}{4}(e^{-2t} - 1) - \frac{1}{2}t$

6. $x = \left(2 + \frac{w}{w^2+1}\right)e^{-t} + \frac{\sin wt - w\cos wt}{w^2+1}$

7. $x = \frac{5}{12}e^t + \frac{16}{3}e^{-2t} - \frac{15}{4}e^{-3t}$

8. $x = \frac{1}{2}$ at $\sinh t$

9. $x = \frac{1}{5}(4\cos 3t + 4\sin 3t + \cos 2t)$

10. $x = 2e^{2t} - 4e^{-t} + \cos 2t - 3\sin 2t$

11. $y = e^{-t}\left(4\cos 2t - \frac{1}{2}\sin 2t\right) - 2\cos 2t + \frac{1}{2}\sin 2t$

12. $y = e^t - e^{-2t} + \sin 3t$

13. $x = \frac{1}{9}(4\sin 2t - 5\sin t - 3t\cos 2t)$

14. $x = 1 + 2t + 2\cos t - \sin t$

15. $x = \frac{1}{2}t^4 - 2t^3 + 8t^2 - 14t + 16(1 - e^{-t})$

16. $x = 3\sinh t + \cosh t - \sin t + 2$

17. $y = (t^2 - 6t + 12)e^{2t} - \left(\frac{3}{2}t^2 + 7t + 11\right)e^t$

18. $y = -1 + 2t + e^{2t}$

19. $y = \frac{1}{8}e^t - \frac{1}{40}e^{-3t} - \frac{1}{10}(2\sin t + \cos t)$

20. $y = \frac{e^{-t}}{5}\left(6t + 10\cos\sqrt{5}\,t + \frac{29}{\sqrt{5}}\sin\sqrt{5}\,t\right)$

4.13. LAPLACE TRANSFORMS OF SOME OTHER USEFUL FUNCTIONS

(1) **Unit Step Function (or Heaviside's Unit Function)**

Def. The unit step function, $u(t - a)$ is defined as

$$u(t-a) = \begin{cases} 0, & \text{for } t < a \\ 1, & \text{for } t \ge a \end{cases} \quad \text{where, } a \ge 0.$$

As a particular case,

$$u(t) = \begin{cases} 0, & \text{for } t < 0 \\ 1, & \text{for } t \ge 0 \end{cases}$$

The product

$$f(t).\, u(t-a) = \begin{cases} 0, & \text{for } t < a \\ f(t), & \text{for } t \ge a \end{cases}$$

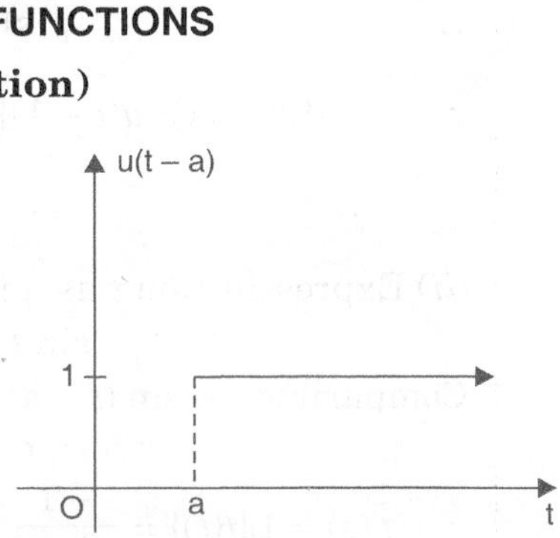

Fig. 4.1

The function $f(t-a).\,u(t-a)$ represents the graph of $f(t)$ shifted through a distance 'a' to the right.

Laplace Transform of Unit Step Function

$$L\{u(t-a)\} = \int_0^\infty e^{-st}\, u(t-a)\,dt$$

$$= \int_0^a e^{-st}.0dt + \int_a^\infty e^{-st}.\,1dt = 0 + \left[\frac{e^{-st}}{-s}\right]_a^\infty = \frac{1}{s}e^{-as}$$

In particular, $L\{u(t)\} = \frac{1}{s}$.

Second Shifting Theorem

If $\qquad L\{f(t)\} = \bar{f}(s)$, then $L\{f(t-a).\,u(t-a)\} = e^{-as}\,\bar{f}(s)$.

$$L\,\{f(t-a).\,u(t-a)\} = \int_0^\infty e^{-st}\, f(t-a).\,u(t-a)\,dt$$

$$= \int_0^\infty e^{-st}\, f(t-a)\,dt$$

$$= \int_0^\infty e^{-s(u+a)}\, f(u)\,du, \text{ where, } u = t-a$$

$$= e^{-as}\int_0^\infty e^{-su}\, f(u)\,du = e^{-as}\,\bar{f}(s)$$

Cor. 1. $\quad L^{-1}\{e^{-as}\,\bar{f}(s)\} = f(t-a).\,u(t-a)$

Cor. 2. If $a = 0$, $L\{f(t)\,u(t)\} = \bar{f}(s) = L\{f(t)\}$.

ILLUSTRATIVE EXAMPLES

Example 1. *Find the Laplace transforms of*

(i) $(t-1)^2\,u(t-1)$ $\qquad$ *(ii)* $\sin t u(t-\pi)$

(iii) $e^{-3t}\,u(t-2)$.

Sol. *(i)* Comparing $(t-1)^2\,u(t-1)$ with $f(t-a)\,u(t-a)$, we have

$$a = 1 \text{ and } f(t) = t^2$$

$$\therefore \quad \bar{f}(s) = L\{f(t)\} = \frac{2}{s^3}$$

$$\therefore \quad L\{(t-1)^2\,u(t-1)\} = e^{-s}\,\bar{f}(s) \qquad \text{(by second shifting property)}$$

$$= \frac{2e^{-s}}{s^3}.$$

(ii) Expressing $\sin t$ as a function of $(t-\pi)$, we have

$$\sin t = \sin\,[(t-\pi)+\pi] = -\sin\,(t-\pi)$$

Comparing $\quad -\sin\,(t-\pi)\,u(t-\pi)$ with $f(t-a)\,u(t-a)$, we get

$$a = \pi \text{ and } f(t) = -\sin t$$

$$\therefore \quad \bar{f}(s) = L\{f(t)\} = \frac{-1}{s^2+1}$$

$$\therefore \quad L\{\sin t\,u(t-\pi)\} = L\{-\sin\,(t-\pi)\,u(t-\pi)\}$$

$$= e^{-\pi s}\,\bar{f}(s) \qquad \text{(by second shifting property)}$$

$$= -\frac{e^{-\pi s}}{s^2+1}.$$

(*iii*) Expressing e^{-3t} as a function of $(t-2)$, we have

$$e^{-3t} = e^{-3[(t-2)+2]} = e^{-6}.\, e^{-3(t-2)}$$

$$\therefore \quad e^{-3t}\, u(t-2) = e^{-6}.\, e^{-3(t-2)}\, u(t-2)$$

Comparing $e^{-3(t-2)}\, u(t-2)$ with $f(t-a)\, u(t-a)$, we get

$$a = 2 \text{ and } f(t) = e^{-3t}$$

$$\therefore \quad \bar{f}(s) = \text{L}\{f(t)\} = \frac{1}{s+3}$$

$$\therefore \quad \text{L}\{e^{-3t}\, u(t-2)\} = e^{-6}\, \text{L}\{e^{-3(t-2)}\, u(t-2)\}$$

$$= e^{-6}.\, e^{-2S}\bar{f}(s) = \frac{e^{-2(s+3)}}{s+3}.$$

Example 2. *Find the inverse Laplace transform of*

(*i*) $\dfrac{e^{-2s}}{s-3}$ (*ii*) $\dfrac{se^{-as}}{s^2-\omega^2},\ a>0$ (*iii*) $\dfrac{s\,e^{-s/2}+\pi\, e^{-s}}{s^2+\pi^2}$

Sol. By second shifting theorem,

$$\text{L}^{-1}\{e^{-as}\,\bar{f}(s)\} = f(t-a).\, u(t-a)$$

(*i*) Let $\bar{f}(s) = \dfrac{1}{s-3}$, then $f(t) = e^{3t}$

$$\therefore \quad \text{L}^{-1}\left\{\frac{e^{-2s}}{s-3}\right\} = \text{L}^{-1}\left\{e^{-2S}\,\bar{f}(s)\right\}$$

$$= f(t-2).\, u(t-2)$$

$$= e^{3(t-2)}.\, u(t-2).$$

(*ii*) Let $\bar{f}(s) = \dfrac{s}{s^2-\omega^2}$, then $f(t) = \cosh \omega t$

$$\therefore \quad \text{L}^{-1}\left\{\frac{s\,e^{-as}}{s^2-\omega^2}\right\} = \text{L}^{-1}\,\{e^{-as}\,\bar{f}(s)\}$$

$$= f(t-a).\, u(t-a) = \cosh \omega(t-a).\, u(t-a)$$

(*iii*) Since $\text{L}^{-1}\left\{\dfrac{s}{s^2+\pi^2}\right\} = \cos \pi\, t$ and $\text{L}^{-1}\left\{\dfrac{\pi}{s^2+\pi^2}\right\} = \sin \pi t$

$$\therefore \quad \text{L}^{-1}\left\{\frac{s\,e^{-s/2}+\pi\, e^{-s}}{s^2+\pi^2}\right\} = \text{L}^{-1}\left\{\frac{s\,e^{-s/2}}{s^2+\pi^2}\right\} + \text{L}^{-1}\left\{\frac{\pi\, e^{-s}}{s^2+\pi^2}\right\}$$

$$= \cos \pi\left(t-\frac{1}{2}\right).\, u\left(t-\frac{1}{2}\right) + \sin \pi(t-1).\, u(t-1)$$

$$= \cos\left(\pi t-\frac{\pi}{2}\right).\, u\left(t\ \ \frac{1}{2}\right) + \sin(\pi t-\pi).\, u(t-1)$$

$$= \sin \pi t . u\left(t - \frac{1}{2}\right) - \sin \pi t . u(t-1)$$

$$= \sin \pi t \left[u\left(t - \frac{1}{2}\right) - u(t-1)\right].$$

(2) Unit Impulse Function (or Dirac–delta Function)

In mechanics, we come across problems where a very large force acts for a very small time. In the study of bending of beams, we have point loads which is equivalent to large pressure acting over a very small area. To deal with such problems, we introduce the unit impulse function or Dirac–delta function.

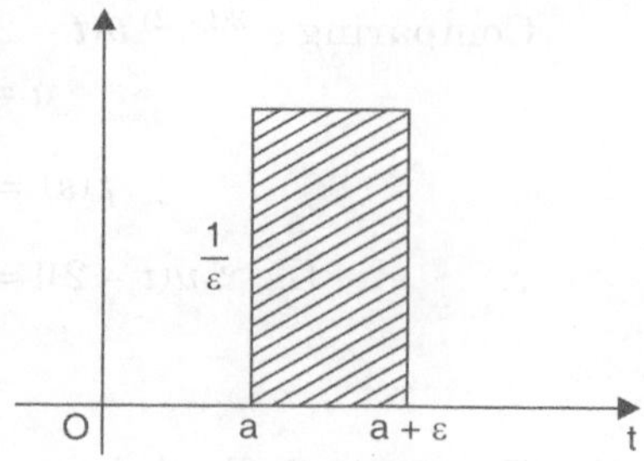

Fig. 4.2

Unit impulse function is considered as the limiting form of the function

$$\delta_\varepsilon (t-a) = \begin{cases} 0, & \text{for} \quad t < a \\ \dfrac{1}{\varepsilon}, & \text{for} \quad a < t < a + \varepsilon \\ 0, & \text{for} \quad t > a \end{cases}$$

This function is represented in the figure. Integrating this function, we get

$$\int_0^\infty \delta_\varepsilon (t-a)\, dt = \int_a^{a+\varepsilon} \frac{1}{\varepsilon}\, dt = \frac{1}{\varepsilon}(a + \varepsilon - a) = 1$$

As $\in \to 0$, the function $\delta_\varepsilon (t-a)$ tends to be infinite at $x = a$ and zero elsewhere, with the characteristic property that its integral across $t = a$ is unity. If $\delta_\varepsilon (t-a)$ represents a force acting for short duration ε at time $t = a$, then the integral

$$\underset{\varepsilon \to 0}{\text{Lt}} \int_a^{a+\varepsilon} \delta_\varepsilon (t-a)\, dt = 1$$

represents unit impulse at $t = a$. Hence the limiting form of $\delta_\varepsilon (t-a)$ as $\varepsilon \to 0$ is expressed as unit impulse function denoted by $\delta(t-a)$.

Thus the unit impulse function $\delta(t-a)$ is defined as follows:

$$\delta(t-a) = \begin{cases} \infty, & \text{for} \quad t = a \\ 0, & \text{for} \quad t \neq a \end{cases}$$

such that $\displaystyle\int_0^\infty \delta(t-a)\, dt = 1.$

Laplace Transform of Unit Impulse Function

If $f(t)$ be a function of t continuous at $t = a$, then

$$\int_0^\infty f(t)\, \delta_\varepsilon (t-a)\, dt = \int_a^{a+\varepsilon} f(t) . \frac{1}{\varepsilon}\, dt$$

$$= (a + \varepsilon - a) f(c) . \frac{1}{\varepsilon} = f(c), \text{ where } a < c < a + \varepsilon$$

(by mean value theorem for integrals)

As $\varepsilon \to 0$, we get $\displaystyle\int_0^\infty f(t)\, \delta(t-a) dt = f(a)$

Cor. 1. $L\{\delta(t-a)\} = \int_0^\infty e^{-st}\,\delta(t-a)\,dt = e^{-sa}$.

Cor. 2. $L\{\delta(t)\} = e^{-s\times 0} = 1$.

(3) Periodic Functions. If $f(t)$ is a periodic function with period T, *i.e.*, $f(t + T) = f(t)$,

then $$L\{f(t)\} = \frac{1}{1-e^{-st}}\int_0^T e^{-st} f(t)\,dt.$$

Proof. $$L\{f(t)\} = \int_0^\infty e^{-st} f(t)\,dt = \int_0^T e^{-st} f(t)\,dt + \int_T^{2T} e^{-st} f(t)\,dt + \int_{2T}^{3T} e^{-st} f(t)\,dt....$$

Putting $t = u$, $t = u + T$, $t = u + 2T$, ... in the successive integrals

$$L\{f(t)\} = \int_0^T e^{-su} f(u)\,du + \int_0^T e^{-s(u+T)} f(u+T)\,du + \int_0^T e^{-s(u+2T)} f(u+2T)\,du + ...$$

Since $f(u) = f(u + T) = f(u + 2T) = ...$, we have

$$L\{f(t)\} = \int_0^T e^{-su} f(u)\,du + e^{-sT}\int_0^T e^{-su} f(u)\,du + e^{-2sT}\int_0^T e^{-su} f(u)\,du + ...$$

$$= (1 + e^{-sT} + e^{-2sT} + ...)\int_0^T e^{-su} f(u)\,du$$

$$= \frac{1}{1-e^{-sT}}\int_0^T e^{-st} f(t)\,dt.$$

ILLUSTRATIVE EXAMPLES

Example 1. *Find the Laplace transform of the triangular wave function of period 2c given by*

$$f(t) = \begin{cases} t, & 0 < t < c \\ 2c - t, & c < t < 2c. \end{cases}$$

Sol. $$L\{f(t)\} = \frac{1}{1-e^{-2cs}}\int_0^{2c} e^{-st} f(t)\,dt$$

$$= \frac{1}{1-e^{-2cs}}\left[\int_0^c e^{-st}.\,t\,dt + \int_c^{2c} e^{-st}(2c-t)\,dt\right]$$

$$= \frac{1}{1-e^{-2cs}}\left[\left\{t\cdot\frac{e^{-st}}{-s} - 1\cdot\frac{e^{-st}}{s^2}\right\}_0^C + \left\{(2c-t)\cdot\frac{e^{-st}}{-s} - (-1)\cdot\frac{e^{-st}}{s^2}\right\}_c^{2c}\right]$$

$$= \frac{1}{1-e^{-2cs}}\left[\left\{-\frac{ce^{-cs}}{-s} - \frac{e^{-cs}}{s^2} + \frac{1}{s^2}\right\} + \left\{\frac{e^{-2cs}}{s^2} + \frac{ce^{-cs}}{s} - \frac{e^{-cs}}{s^2}\right\}\right]$$

$$= \frac{1}{1-e^{-2cs}}\left(\frac{1-2e^{-cs}+e^{-2cs}}{s^2}\right)$$

$$= \frac{1}{s^2} \cdot \frac{(1-e^{-cs})^2}{(1+e^{-cs})(1-e^{-sc})} = \frac{1}{s^2} \cdot \frac{1-e^{-cs}}{1+e^{-cs}}$$

$$= \frac{1}{s^2}\left(\frac{e^{cs/2}-e^{-cs/2}}{e^{cs/2}+e^{-cs/2}}\right) = \frac{1}{s^2}\tanh\left(\frac{cs}{2}\right)$$

Example 2. *Find the Laplace transform of the rectified semi-wave function defined by*

$$f(t) = \begin{cases} \sin \omega t, & 0 < t < \dfrac{\pi}{w} \\ 0, & \dfrac{\pi}{\omega} < t < \dfrac{2\pi}{\omega} \end{cases}.$$

Sol. Here $f(t)$ is a periodic function with period $\dfrac{2\pi}{\omega}$

$$\therefore \quad L\{f(t)\} = \frac{1}{1-e^{-\frac{2\pi s}{\omega}}} \int_0^{\frac{2\pi}{\omega}} e^{-st} f(t)\, dt$$

$$= \frac{1}{1-e^{-\frac{2\pi s}{\omega}}}\left[\int_0^{\frac{\pi}{\omega}} e^{-st} \sin \omega t\, dt + \int_{\pi/\omega}^{2\pi/\omega} e^{-st} . 0\, dt\right]$$

$$= \frac{1}{1-e^{-\frac{2\pi s}{\omega}}}\left[e^{-st}\frac{(-s \sin \omega t - \omega \cos \omega t)}{s^2+\omega^2}\right]_0^{\pi/\omega}$$

$$= \frac{\omega e^{\frac{-\pi s}{\omega}} + \omega}{\left(1-e^{\frac{-\pi s}{\omega}}\right)\left(1+e^{\frac{-\pi s}{\omega}}\right)(s^2+\omega^2)} = \frac{\omega}{(1-e^{-\pi s/\omega})(s^2+\omega^2)}.$$

EXERCISE 4.6

1. Find the Laplace transforms of

   (*i*) $e^{t-2}\, u(t-2)$ (*ii*) $t^2\, u(t-3)$

   (*iii*) $\sin 2t\, u(t-\pi)$ (*iv*) $e^{2t}\, u(t-3)$.

2. Find the inverse Laplace transforms of

   (*i*) $\dfrac{e^{-\pi s}}{s^2+1}$ (*ii*) $\dfrac{e^{-\pi s}}{s^2}$ (*iii*) $\dfrac{se^{-2s}}{s^2-1}$

   (*iv*) $\dfrac{e^{-cs}}{s^2(s+a)}, c > 0$ (*v*) $\dfrac{se^{-s/2}+\pi e^{-s}}{s^2+\pi^2}$ (*vi*) $\dfrac{3}{s} - \dfrac{4e^{-s}}{s^2} + \dfrac{4e^{-3s}}{s^2}$

3. Find the Laplace transforms of the square-wave function of period a defined as

$$f(t) = \begin{cases} 1, & \text{when } 0 < t < \dfrac{a}{2} \\ -1, & \text{when } \dfrac{a}{2} < t < a \end{cases}$$

4. Find the Laplace transform of the saw-toothed wave function of period T, defined by $f(t) = \dfrac{t}{T}$, for $0 < t < T$.

5. Expressing $F(t)$ in terms of unit step functions, obtain $L\{F(t)\}$ where

$$F(t) = \begin{cases} t^2, & 0 < t < 2 \\ t - 1, & 2 < t < 3 \\ 7, & t > 3 \end{cases}$$

6. Expressing the following function in terms of unit step function and find its Laplace transforms

$$f(t) = \begin{cases} 8, & t < 2 \\ 6, & t > 2 \end{cases}$$

7. Find the Laplace transform of the waveform

$$f(t) = \frac{2t}{3}, \; 0 \le t \le 3 \text{ of period } 3.$$ *(K.U. May 2011)*

Answers

1. (i) $\dfrac{e^{-2s}}{s-1}$ (ii) $\dfrac{e^{-3}(2 + 6s + 9s^2)}{s^3}$ (iii) $\dfrac{-2e^{-\pi s}}{s^2 + 4}$

(iv) $\dfrac{e^{-3(s+2)}}{s+2}$

2. (i) $-\sin t.\ u(t - \pi)$ (ii) $(t - \pi).\ u(t - \pi)$ (iii) $\cosh(t - 2).\ u(t - 2)$

(iv) $\dfrac{1}{a^2}\left[a(t - c) - 1 + e^{-a(t-c)}\right].\ u(t - c)$ (v) $\sin \pi t \left[u\left(t - \dfrac{1}{2}\right) - u(t - 1)\right]$

(vi) $3 - 4(t - 1)\, u(t - 1) + 4\,(t - 3)\, u(t - 3)$.

3. $\dfrac{1}{s}\tanh\left(\dfrac{as}{4}\right)$ 4. $\dfrac{\dfrac{1}{s^2 T} - e^{-sT}}{s(1 - e^{-sT})}$

5. $-\left(\dfrac{3}{s} + \dfrac{3}{s^2} + \dfrac{2}{s^3}\right)e^{-2s} + \dfrac{2}{s^3} + \left(\dfrac{5}{s} - \dfrac{1}{s^2}\right)e^{-3s}$ 6. $\dfrac{8}{s} - 2.\ \dfrac{e^{-2s}}{s}$

7. $-\dfrac{2\,e^{-3s}}{s(1 - e^{-3s})} + \dfrac{2}{3s^2}$.

5
Differential Equations

5.1. DEFINITIONS

A **linear differential equation** is the differential equation in which the dependent variable and its derivatives occur only in the first degree and are not multiplied together. Thus, the general linear differential equation of the nth order is of the form $\frac{d^n y}{dx^n} + P_1 \frac{d^{n-1}y}{dx^{n-1}} + P_2 \frac{d^{n-2}y}{dx^{n-2}} + \cdots + P_{n-1}\frac{dy}{dx} + P_n y = X$, where $P_1, P_2, \ldots, P_{n-1}, P_n$ and X are functions of x only.

A **linear differential equation with constant co-efficients** is of the form

$$\frac{d^n y}{dx^n} + a_1 \frac{d^{n-1}y}{dx^{n-1}} + a_2 \frac{d^{n-2}y}{dx^{n-2}} + \cdots + a_{n-1}\frac{dy}{dx} + a_n y = X \qquad \ldots(1)$$

where $a_1, a_2, \ldots, a_{n-1}, a_n$ are constants and X is either a constant or a function of x only.

5.2. THE OPERATOR D

The part $\frac{d}{dx}$ of the symbol $\frac{dy}{dx}$ may be regarded as an operator such that when it operates on y, the result is the derivative of y.

Similarly, $\frac{d^2}{dx^2}, \frac{d^3}{dx^3}, \ldots, \frac{d^n}{dx^n}$ may be regarded as operators.

For brevity, we write $\frac{d}{dx} \equiv D, \frac{d^2}{dx^2} \equiv D^2, \ldots, \frac{d^n}{dx^n} \equiv D^n$

Thus, the symbol D is a **differential operator** or simply an **operator.**

Written in symbolic form, equation (1) becomes

$$(D^n + a_1 D^{n-1} + a_2 D^{n-2} + \cdots + a_{n-1} D + a_n)y = X$$

or $\qquad f(D)y = X$

where $\qquad f(D) = D^n + a_1 D^{n-1} + a_2 D^{n-2} + \cdots + a_{n-1} D + a_n$

i.e., $f(D)$ is a polynomial in D.

The operator D can be treated as an algebraic quantity.

Thus $\qquad D(u + v) = Du + Dv, \quad D(\lambda u) = \lambda\, Du$

$$D^p D^q\, u = D^{p+q}\, u, \qquad D^p D^q u = D^q D^p u$$

The polynomial $f(D)$ can be factorised by ordinary rules of algebra and the factors may be written in any order.

5.3. THEOREMS

Theorem 1. *If $y = y_1, y = y_2, ..., y = y_n$ are n linearly independent solutions of the differential equation*

$$(D^n + a_1 D^{n-1} + a_2 D^{n-2} + \cdots + a_n)y = 0 \quad ...(i)$$

then $u = c_1 y_1 + c_2 y_2 + \cdots + c_n y_n$ is also its solution, where $c_1, c_2, ..., c_n$ are arbitrary constants.

Proof. Since $y = y_1, y = y_2, ..., y = y_n$ are solution of equation (*i*).

$$\therefore \left.\begin{aligned} D^n y_1 + a_1 D^{n-1} y_1 + a_2 D^{n-2} y_1 + \cdots + a_n y_1 &= 0 \\ D^n y_2 + a_1 D^{n-1} y_2 + a_2 D^{n-2} y_2 + \cdots + a_n y_2 &= 0 \\ \cdots \quad \cdots \quad \cdots \quad \cdots \quad & \cdots \\ \cdots \quad \cdots \quad \cdots \quad \cdots \quad & \cdots \\ D^n y_n + a_1 D^{n-1} y_n + a_2 D^{n-2} y_n + \cdots + a_n y_n &= 0 \end{aligned}\right\} \quad ...(ii)$$

Now $D^n u + a_1 D^{n-1} u + a_2 D^{n-2} u + \cdots + a_n u$

$$\begin{aligned} &= D^n (c_1 y_1 + c_2 y_2 + \cdots + c_n y_n) \\ &\quad + a_1 D^{n-1}(c_1 y_1 + c_2 y_2 + \cdots + c_n y_n) + a_2 D^{n-2}(c_1 y_1 + c_2 y_2 + \cdots + c_n y_n) \\ &\quad + \cdots \quad \cdots \quad \cdots \quad \cdots + a_n(c_1 y_1 + c_2 y_2 + \cdots + c_n y_n) \\ &= c_1(D^n y_1 + a_1 D^{n-1} y_1 + a_2 D^{n-2} y_1 + \cdots + a_n y_1) \\ &\quad + c_2(D^n y_2 + a_1 D^{n-1} y_2 + a_2 D^{n-2} y_n + \cdots + a_n y_2) \\ &\quad + \cdots \quad \cdots + c_n(D^n y_n + a_1 D^{n-1} y_n + a_2 D^{n-2} y_n + \cdots + a_n y_n) \\ &= c_1(0) + c_2(0) + \cdots + c_n(0) \qquad \text{[From (ii)]} \\ &= 0 \end{aligned}$$

which shows that $u = c_1 y_1 + c_2 y_2 + \cdots + c_n y_n$ is also the solution of equation (*i*).

Since this solution contains *n* arbitrary constants, it is the general or complete solution of equation (*i*).

Theorem 2. *If $y = u$ is the complete solution of the equation $f(D)y = 0$ and $y = v$ is a particular solution (containing no arbitrary constants) of the equation $f(D)y = X$, then the complete solution of the equation $f(D)y = X$ is $y = u + v$.*

Proof. Since $y = u$ is the complete solution of the equation $f(D)y = 0$ ...(*i*)

$\therefore$ $f(D)u = 0$ ...(*ii*)

Also $y = v$ is a particular solution of the equation $f(D)y = X$ ...(*iii*)

$\therefore$ $f(D)v = X$ ...(*iv*)

Adding (*ii*) and (*iv*), we have $f(D)(u + v) = X$

Thus $y = u + v$ satisfies the equation (*iii*), hence it is the **complete solution (C.S.)** because it contains *n* arbitrary constants.

The part $y = u$ is called the **complementary function (C.F.)** and the part $y = v$ is called the **particular integral (P.I.)** of the equation (*iii*).

$\therefore$ The complete solution of equation (*iii*), is **y = C.F. + P.I.**

Thus in order to solve the equation (iii), we first find the C.F. i.e., the C.S. of equation (i) and then the P.I. i.e., a particular solution of equation (iii).

5.4. AUXILIARY EQUATION (A.E.)

Consider the differential equation $(D^n + a_1 D^{n-1} + a_2 D^{n-2} + \cdots + a_n)y = 0$ ...(*i*)

Let $y = e^{mx}$ be a solution of (*i*), then $Dy = me^{mx}$, $D^2 y = m^2 e^{mx}$,..., $D^{n-2} y = m^{n-2} e^{mx}$

$$D^{n-1}y = m^{n-1}e^{mx},\ D^n y = m^n e^{mx}$$

Substituting the values of y, Dy, D^2y, ..., $D^n y$ in (*i*), we get

$$(m^n + a_1 m^{n-1} + a_2 m^{n-2} + \cdots + a_n)\, e^{mx} = 0$$

or $$m^n + a_1 m^{n-1} + a_2 m^{n-2} + \cdots + a_n = 0, \text{ since } e^{mx} \neq 0 \quad ...(ii)$$

Thus $y = e^{mx}$ will be a solution of equation (*i*) if m satisfies equation (*ii*).

Equation (*ii*) is called the auxiliary equation for the differential equation (*i*).

Replacing m by D in (*ii*), we get $D^n + a_1 D^{n-1} + a_2 D^{n-2} + \cdots + a_n = 0$ ...(*iii*)

Equation (*ii*) gives the same values of m *as* equation (*iii*) gives of D. In practice, we take equation (*iii*) as the auxiliary equation which is obtained by equating to zero the symbolic coefficient of y in equation (*i*).

Definition. The equation obtained by equating to zero the symbolic co-efficient of y is called the **auxiliary equation**, briefly written as **A.E.**

5.5. RULES FOR FINDING THE COMPLEMENTARY FUNCTION

Consider the equation $(D^n + a_1 D^{n-1} + a_2 D^{n-2} + \cdots + a_n)y = 0$ ...(*i*)

where all the a_i's are constant.

Its auxiliary equation is $D^n + a_1 D^{n-1} + a_2 D^{n-2} + \cdots + a_n = 0$ ...(*ii*)

Let $D = m_1, m_2, m_3, ..., m_n$ be the roots of the A.E. The solution of equation (*i*) depends upon the nature of roots of the A.E. The following cases arise :

Case I. *If all the roots of the A.E. are real and distinct,* then equation (*ii*) is equivalent to

$$(D - m_1)(D - m_2) \ldots (D - m_n) = 0 \quad ...(iii)$$

Equation (*iii*) will be satisfied by the solutions of the equations

$$(D - m_1)y = 0,\ (D - m_2)y = 0, ...,\ (D - m_n)y = 0$$

Now, consider the equation $(D - m_1)y = 0$, *i.e.*, $\dfrac{dy}{dx} - m_1 y = 0$

It is a linear equation and I.F. $= e^{\int -m_1 dx} = e^{-m_1 x}$

$\therefore$ its solution is $\quad y \cdot e^{-m_1 x} = \int 0 \cdot e^{-m_1 x}\, dx + c_1 \quad$ or $\quad y = c_1 e^{m_1 x}$

Similarly, the solution of $(D - m_2)y = 0$ is $y = c_2 e^{m_2 x}$

....

the solution of $(D - m_n)y = 0$ is $y = c_n e^{m_n x}$

Hence the complete solution of equation (*i*) is

$$y = c_1 e^{m_1 x} + c_2 e^{m_2 x} + \ldots + c_n e^{m_n x} \quad ...(iv)$$

Case II. *If two roots of the A.E. are equal,* let $m_1 = m_2$.

The solution obtained in equation (*iv*) becomes

$$y = (c_1 + c_2)\, e^{m_1 x} + c_3 e^{m_3 x} + \ldots + c_n e^{m_n x}$$

$$= c e^{m_1 x} + c_3 e^{m_3 x} + \ldots + c_n e^{m_n x}$$

It contains $(n - 1)$ arbitrary constants and is, therefore, not the complete solution of equation (*i*).

The part of the complete solution corresponding to the repeated root is the complete solution of

$$(D - m_1)(D - m_1)y = 0$$

Putting $(D - m_1)y = v$, it becomes $(D - m_1)v = 0$ *i.e.*, $\dfrac{dv}{dx} - m_1 v = 0$

As in case I, its solution is $v = c_1 e^{m_1 x}$

$$\therefore \qquad (D - m_1)y = c_1 e^{m_1 x} \quad \text{or} \quad \frac{dy}{dx} - m_1 y = c_1 e^{m_1 x}$$

which is a linear equation and I.F. $= e^{-m_1 x}$

$$\therefore \quad \text{its solution is} \quad y \cdot e^{-m_1 x} = \int c_1 e^{m_1 x} \cdot e^{-m_1 x}\, dx + c_2 = c_1 x + c_2$$

or
$$y = (c_1 x + c_2)\, e^{m_1 x}$$

Thus, the complete solution of equation (*i*) is

$$y = (c_1 x + c_2)\, e^{m_1 x} + c_3 e^{m_3 x} + \cdots + c_n e^{m_n x}$$

If, however, three roots of the A.E. are equal, say $m_1 = m_2 = m_3$, then proceeding as above, the solution becomes

$$y = (c_1 x^2 + c_2 x + c_3)\, e^{m_1 x} + c_4 e^{m_4 x} + \cdots + c_n e^{m_n x}$$

Case III. *If two roots of the A.E. are imaginary,* let

$$m_1 = \alpha + i\beta \quad \text{and} \quad m_2 = \alpha - i\beta$$

The solution obtained in equation (*iv*) becomes

$$y = c_1 e^{(\alpha + i\beta)x} + c_2 e^{(\alpha - i\beta)x} + c_3 e^{m_3 x} + \cdots + c_n e^{m_n x}$$

$$= e^{\alpha x}(c_1 e^{i\beta x} + c_2 e^{-i\beta x}) + c_3 e^{m_3 x} + \cdots + c_n e^{m_n x}$$

$$= e^{\alpha x}[c_1(\cos \beta x + i \sin \beta x) + c_2 (\cos \beta x - i \sin \beta x)]$$

$$+ c_3 e^{m_3 x} + \cdots + c_n e^{m_n x} \quad [\because \text{ by Euler's Theorem, } e^{i\theta} = \cos\theta + i \sin\theta]$$

$$= e^{\alpha x}[(c_1 + c_2) \cos \beta x + i\,(c_1 - c_2) \sin \beta x] + c_3 e^{m_3 x} + \cdots + c_n e^{m_n x}$$

$$= e^{\alpha x}(C_1 \cos \beta x + C_2 \sin \beta x) + c_3 e^{m_3 x} + \cdots + c_n e^{m_n x}$$

[taking $c_1 + c_2 = C_1$, $i(c_1 - c_2) = C_2$]

Case IV. *If two pairs of imaginary roots be equal,* let

$$m_1 = m_2 = \alpha + i\beta \quad \text{and} \quad m_3 = m_4 = \alpha - i\beta$$

Then by case II, the complete solution is

$$y = e^{\alpha x}\, [(c_1 x + c_2) \cos \beta x + (c_3 x + c_4) \sin \beta x] + c_5 e^{m_5 x} + \cdots + c_n e^{m_n x}.$$

ILLUSTRATIVE EXAMPLES

Example 1. *Solve* $\dfrac{d^3 y}{dx^3} - 7\dfrac{dy}{dx} - 6y = 0.$

Sol. Given equation in symbolic form is $(D^3 - 7D - 6)y = 0$

Its A.E. is $D^3 - 7D - 6 = 0$ or $(D + 1)(D + 2)(D - 3) = 0$

whence $D = -1, -2, 3$

Hence the C.S. is $y = c_1e^{-x} + c_2e^{-2x} + c_3e^{3x}$.

Example 2. *Solve* $(D^3 - 4D^2 + 4D)y = 0$.

Sol. The A.E. is $D^3 - 4D^2 + 4D = 0$ or $D(D^2 - 4D + 4) = 0$

or $D(D - 2)^2 = 0$

whence $D = 0, 2, 2$

Hence, the C.S is $y = c_1e^{0x} + (c_2x + c_3)e^{2x}$ or $y = c_1 + (c_2x + c_3)e^{2x}$.

Example 3. *Solve* $\dfrac{d^4y}{dx^4} + 13\dfrac{d^2y}{dx^2} + 36y = 0$.

Sol. Given equation in symbolic form is $(D^4 + 13D^2 + 36)y = 0$

Its A.E. is $D^4 + 13D^2 + 36 = 0$

or $(D^2 + 4)(D^2 + 9) = 0$ or $D = \pm 2i, \pm 3i$

Hence the C.S. is $y = e^{0x}(c_1 \cos 2x + c_2 \sin 2x) + e^{0x}(c_3 \cos 3x + c_4 \sin 3x)$

or $y = c_1 \cos 2x + c_2 \sin 2x + c_3 \cos 3x + c_4 \sin 3x$.

Example 4. *Solve* $\dfrac{d^4x}{dt^4} + 4x = 0$.

Sol. Given equation in symbolic form is $(D^4 + 4)x = 0$, where $D = \dfrac{d}{dt}$

Its A.E. is $D^4 + 4 = 0$ or $(D^4 + 4D^2 + 4) - 4D^2 = 0$

or $(D^2 + 2)^2 - (2D)^2 = 0$ or $(D^2 + 2D + 2)(D^2 - 2D + 2) = 0$

whence $D = \dfrac{-2 \pm \sqrt{-4}}{2}$ and $\dfrac{2 \pm \sqrt{-4}}{2}$ *i.e.*, $D = -1 \pm i$ and $1 \pm i$

Hence the C.S. is $x = e^{-t}(c_1 \cos t + c_2 \sin t) + e^{t}(c_3 \cos t + c_4 \sin t)$.

EXERCISE 5.1

Solve the following differential equations:

1. $\dfrac{d^2y}{dx^2} - 3\dfrac{dy}{dx} - 4y = 0$.

2. $\dfrac{d^2y}{dx^2} + (a + b)\dfrac{dy}{dx} + aby = 0$.

3. $\dfrac{d^2y}{dx^2} - 4\dfrac{dy}{dx} + y = 0$.

4. $\dfrac{d^2x}{dt^2} + 6\dfrac{dx}{dt} + 9x = 0$.

5. $\dfrac{d^3y}{dx^3} - 3\dfrac{d^2y}{dx^2} + 3\dfrac{dy}{dx} - y = 0$.

6. $\dfrac{d^3y}{dx^3} + 6\dfrac{d^2y}{dx^2} + 11\dfrac{dy}{dx} + 6y = 0$.

7. $\dfrac{d^4y}{dx^4} - 5\dfrac{d^2y}{dx^2} + 4y = 0$.

8. $\dfrac{d^4y}{dx^4} + 8\dfrac{d^2y}{dx^2} + 16y = 0$.

9. $(D^2 + 1)^3(D^2 + D + 1)^2 y = 0$.

10. $\dfrac{d^2x}{dt^2} - 3\dfrac{dx}{dt} + 2x = 0$, given that when $t = 0$, $x = 0$ and $\dfrac{dx}{dt} = 0$.

11. $\frac{d^2y}{dx^2} + 4\frac{dy}{dx} + 29y = 0$, given that when $x = 0, y = 0$ and $\frac{dy}{dx} = 15$.

12. If $\frac{d^4x}{dt^4} = m^4x$, show that $x = c_1 \cos mt + c_2 \sin mt + c_3 \cosh mt + c_4 \sinh mt$.

Answers

1. $y = c_1 e^{4x} + c_2e^{-x}$ **2.** $y = c_1 e^{-ax} + c_2e^{-bx}$ **3.** $y = c_1e^{(2+\sqrt{3})x} + c_2e^{(2-\sqrt{3})x}$

4. $x = (c_1 + c_2t)\, e^{-3t}$ **5.** $y = (c_1 + c_2x + c_3x^2)\, e^x$ **6.** $y = c_1e^{-x} + c_2e^{-2x} + c_3e^{-3x}$

7. $y = c_1e^x + c_2e^{-x} + c_3e^{2x} + c_4e^{-2x}$ **8.** $y = (c_1 + c_2x) \cos 2x + (c_3 + c_4x) \sin 2x$

9. $y = (c_1 + c_2x + c_3x^2) \cos x + (c_4 + c_5x + c_6x^2) \sin x + e^{-\frac{1}{2}x}\left[(c_7 + c_8x) \cos \frac{\sqrt{3}}{2}x + (c_9 + c_{10}x) \sin \frac{\sqrt{3}}{2}x\right]$

10. $x = 0$ **11.** $y = 3e^{-2x} \sin 5x$.

5.6. THE INVERSE OPERATOR $\frac{1}{f(D)}$

Definition. $\frac{1}{f(D)}X$ is that function of x, *free from arbitrary constants,* which when operated upon by $f(D)$ gives X.

Thus $f(D)\left\{\frac{1}{f(D)}X\right\} = X$

$\therefore$ $f(D)$ and $\frac{1}{f(D)}$ are inverse operators.

Theorem 1. $\frac{1}{f(D)}X$ *is the particular integral of* $f(D)y = X$.

Proof. The given equation is $f(D)y = X$ ...(1)

Putting $y = \frac{1}{f(D)}X$ in (1), we have $f(D)\left\{\frac{1}{f(D)}X\right\} = X$ or $X = X$, which is true.

$$\therefore \quad y = \frac{1}{f(D)}X \text{ is a solution of (1).}$$

Since it contains no arbitrary constants, it is the particular integral of $f(D)\, y = X$.

Theorem 2. *Prove that,* $\frac{1}{D}X = \int X\, dx$.

Proof. Let $\frac{1}{D}X = y$

Operating both sides by D, we have $D\left(\frac{1}{D}X\right) = Dy$ or $X = \frac{dy}{dx}$

Integrating both sides w.r.t. x,

$$y = \int X\, dx,$$

no arbitrary constant being added since $y = \frac{1}{D} X$ contains no arbitrary constant.

$$\therefore \qquad \frac{1}{D} X = \int X \, dx.$$

Theorem 3. *Prove that,* $\frac{1}{D-a} X = e^{ax} \int X e^{-ax} \, dx.$

Proof. Let $\qquad \frac{1}{D-a} X = y$

Operating on both sides by $(D - a)$, $(D-a)\left(\frac{1}{D-a} X\right) = (D-a)y$

or $\qquad X = \frac{dy}{dx} - ay \quad$ *i.e.,* $\quad \frac{dy}{dx} - ay = X$

which is a linear equation and I.F. $= e^{\int -a dx} = e^{-ax}$

$\therefore$ Its solution is $ye^{-ax} = \int X e^{-ax} \, dx$, no constant being added or $y = e^{ax} \int X e^{-ax} \, dx$

Hence $$\frac{1}{D-a} X = e^{ax} \int e^{-ax} X \, dx.$$

5.7. RULES FOR FINDING THE PARTICULAR INTEGRAL

Consider the differential equation, $(D^n + a_1 D^{n-1} + a_2 D^{n-2} + \cdots + a_{n-1} D + a_n)y = X$

It can be written as $f(D)y = X$

$$\therefore \qquad \text{P.I.} = \frac{1}{f(D)} X$$

Case I. *When* $X = e^{ax}$

Since
$$D e^{ax} = a e^{ax}$$
$$D^2 e^{ax} = a^2 e^{ax}$$
$$\ldots\ldots\ldots\ldots\ldots\ldots$$
$$\ldots\ldots\ldots\ldots\ldots\ldots$$
$$D^{n-1} e^{ax} = a^{n-1} e^{ax}$$
$$D^n e^{ax} = a^n e^{ax}$$

$$\therefore \quad (D^n + a_1 D^{n-1} + a_2 D^{n-2} + \cdots + a_{n-1} D + a_n) e^{ax} = (a^n + a_1 a^{n-1} + a_2 a^{n-2} + \cdots + a_{n-1} a + a_n) e^{ax}$$

or $$f(D) e^{ax} = f(a) e^{ax}$$

Operating on both sides by $\frac{1}{f(D)}$.

$$\frac{1}{f(D)} (f(D) e^{ax}) = \frac{1}{f(D)} (f(a) e^{ax}) \quad \text{or} \quad e^{ax} = f(a) \frac{1}{f(D)} e^{ax}$$

Dividing both sides by $f(a)$, $\frac{1}{f(a)} e^{ax} = \frac{1}{f(D)} e^{ax}$, provided $f(a) \neq 0$

Hence $$\boldsymbol{\frac{1}{f(D)} e^{ax} = \frac{1}{f(a)} e^{ax},} \qquad \textbf{provided } \boldsymbol{f(a) \neq 0.}$$

Case of failure. If $f(a) = 0$, the above method fails.

Since $f(a) = 0$, $D = a$ is a root of A.E. $f(D) = 0$

$\therefore$ $D - a$ is a factor of $f(D)$.

Let $$f(D) = (D - a)\,\phi(D), \text{ where } \phi(a) \neq 0 \qquad ...(i)$$

Then $$\frac{1}{f(D)}e^{ax} = \frac{1}{(D-a)\,\phi(D)}e^{ax} = \frac{1}{D-a}\cdot\frac{1}{\phi(D)}e^{ax} = \frac{1}{D-a}\cdot\frac{1}{\phi(a)}e^{ax}$$

$$= \frac{1}{\phi(a)}\cdot\frac{1}{D-a}e^{ax} = \frac{1}{\phi(a)}e^{ax}\int e^{ax}\cdot e^{-ax}dx \qquad \text{[by Theorem 3]}$$

$$= \frac{1}{\phi(a)}e^{ax}\int 1\,dx = x\cdot\frac{1}{\phi(a)}e^{ax} \qquad ...(ii)$$

Differentiating both sides of (*i*) w.r.t. D, we have $f'(D) = (D - a)\,\phi'(D) + \phi(D)$

$$\Rightarrow \qquad f'(a) = \phi(a)$$

$\therefore$ From (*ii*), we have $\boldsymbol{\frac{1}{f(D)}e^{ax} = x\cdot\frac{1}{f'(a)}e^{ax}}$, **provided $f'(a) \neq 0$**

If $f'(a) = 0$, then $\frac{1}{f(D)}e^{ax} = x^2\cdot\frac{1}{f''(a)}e^{ax}$, provided $f''(a) \neq 0$

and so on.

ILLUSTRATIVE EXAMPLES

Example 1. *Find the P.I. of $(4D^2 + 4D - 3)\,y = e^{2x}$.*

Sol. $$\text{P.I.} = \frac{1}{4D^2 + 4D - 3}e^{2x} = \frac{1}{4(2)^2 + 4(2) - 3}e^{2x} \qquad \text{(replacing D by 2)}$$

$$= \frac{1}{21}e^{2x}.$$

Example 2. *Find the P.I. of $(D^2 + 3D + 2)y = 5$.*

Sol. $$\text{P.I.} = \frac{1}{D^2 + 3D + 2}(5e^{0x}) \qquad [\because \; e^{0x} = 1]$$

$$= 5\cdot\frac{1}{0+0+2}e^{0x} \qquad \text{(replacing D by 0)}$$

$$= \frac{5}{2}$$

Example 3. *Find the P.I. of $(D^3 - 3D^2 + 4)y = e^{2x}$.*

Sol. $$\text{P.I.} = \frac{1}{D^3 - 3D^2 + 4}e^{2x}.$$

Hence the denominator vanishes when D is replaced by 2. It is a case of failure. We multiply the numerator by x and differentiate the denominator w.r.t. D.

$$\therefore \qquad \text{P.I.} = x\cdot\frac{1}{3D^2 - 6D}e^{2x}$$

It is again a case of failure. We multiply the numerator by x and differentiate the denominator w.r.t. D.

$$\therefore \quad \text{P.I.} = x^2 \cdot \frac{1}{6D-6} e^{2x} = x^2 \cdot \frac{1}{6(2)-6} e^{2x} = \frac{x^2}{6} e^{2x}.$$

Case II. *When* $X = \sin(ax+b)$ *or* $\cos(ax+b)$

$$D \sin(ax+b) = a\cos(ax+b)$$
$$\mathbf{D^2 \sin(ax+b) = (-a^2)\sin(ax+b)}$$
$$D^3 \sin(ax+b) = -a^3\cos(ax+b)$$
$$D^4 \sin(ax+b) = a^4 \sin(ax+b)$$

or $$\mathbf{(D^2)^2 \sin(ax+b) = (-a^2)^2 \sin(ax+b)}$$

....

....

In general, $(D^2)^n \sin(ax+b) = (-a^2)^n \sin(ax+b)$

$\therefore$ $f(D^2)\sin(ax+b) = f(-a^2)\sin(ax+b)$

Operating on both sides by $\dfrac{1}{f(D^2)}$,

$$\frac{1}{f(D^2)}(f(D^2)\sin(ax+b)) = \frac{1}{f(D^2)}(f(-a^2)\sin(ax+b))$$

or $$\sin(ax+b) = f(-a^2)\frac{1}{f(D^2)}\sin(ax+b).$$

Dividing both sides by $f(-a^2)$,

$$\frac{1}{f(-a^2)}\sin(ax+b) = \frac{1}{f(D^2)}\sin(ax+b), \qquad \text{provided } f(-a^2) \neq 0.$$

Hence $$\mathbf{\frac{1}{f(D^2)}\sin(ax+b) = \frac{1}{f(-a^2)}\sin(ax+b),} \qquad \textbf{provided } \mathbf{f(-a^2) \neq 0}$$

Similarly, $$\mathbf{\frac{1}{f(D^2)}\cos(ax+b) = \frac{1}{f(-a^2)}\cos(ax+b),} \qquad \textbf{provided } \mathbf{f(-a^2) \neq 0}$$

Case of failure. If $f(-a^2) = 0$, the above method fails.

Since $\cos(ax+b) + i\sin(ax+b) = e^{i(ax+b)}$ | Euler's Theorem

$$\therefore \quad \frac{1}{f(D^2)}[\cos(ax+b) + i\sin(ax+b)] = \frac{1}{f(D^2)} e^{i(ax+b)}$$

[If we replace D by ia, $f(D^2) = f(-a^2) = 0$, so that it is a case of failure]

$$= x \cdot \frac{1}{f'(D^2)} e^{i(ax+b)} = x \cdot \frac{1}{f'(D^2)}[\cos(ax+b) + i\sin(ax+b)]$$

Equating real parts

$$\mathbf{\frac{1}{f(D^2)}\cos(ax+b) = x \cdot \frac{1}{f'(D^2)}\cos(ax+b),} \qquad \textbf{provided } \mathbf{f'(-a^2) \neq 0}$$

Equating imaginary parts

$$\mathbf{\frac{1}{f(D^2)}\sin(ax+b) = x \cdot \frac{1}{f'(D^2)}\sin(ax+b),} \qquad \textbf{provided } \mathbf{f'(-a^2) \neq 0}$$

If $f'(-a^2) = 0$, then

$$\frac{1}{f(D^2)} \sin(ax+b) = x^2 \cdot \frac{1}{f''(D^2)} \sin(ax+b), \qquad \text{provided } f''(-a^2) \neq 0$$

$$\frac{1}{f(D^2)} \cos(ax+b) = x^2 \cdot \frac{1}{f''(D^2)} \cos(ax+b), \qquad \text{provided } f''(-a^2) \neq 0$$

and so on.

ILLUSTRATIVE EXAMPLES

Example 1. *Find the P.I. of $(D^3 + 1)y = \sin(2x + 3)$.*

Sol. $$\text{P.I.} = \frac{1}{D^3+1} \sin(2x+3) = \frac{1}{D(-2^2)+1} \sin(2x+3)$$

[Putting $D^2 = -2^2$]

$$= \frac{1}{1-4D} \sin(2x+3)$$

Multiplying and dividing by $(1 + 4D)$

$$= \frac{1+4D}{(1-4D)(1+4D)} \sin(2x+3) = \frac{1+4D}{1-16D^2} \sin(2x+3)$$

$$= \frac{1+4D}{1-16(-2^2)} \sin(2x+3) \qquad \text{[Putting } D^2 = -2^2]$$

$$= \frac{1}{65} [\sin(2x+3) + 4D \sin(2x+3)]$$

$$= \frac{1}{65} [\sin(2x+3) + 8\cos(2x+3)] \qquad \left[\because D = \frac{d}{dx}\right]$$

Example 2. *Find the P.I. of $(D^2 + 4)y = \cos 2x$.*

Sol. $$\text{P.I.} = \frac{1}{D^2+4} \cos 2x$$

Here the denominator vanishes when D^2 is replaced by $-2^2 = -4$. It is a case of failure. We multiply the numerator by x and differentiate the denominator w.r.t. D.

$$\therefore \qquad \text{P.I.} = x \cdot \frac{1}{2D} \cos 2x = \frac{x}{2} \int \cos 2x \, dx \qquad \left[\because \frac{1}{D} f(x) = \int f(x) dx\right]$$

$$= \frac{x}{4} \sin 2x.$$

Case III. *When $X = x^m$, m* being a positive integer.

Here $$\text{P.I.} = \frac{1}{f(D)} x^m$$

Take out the lowest degree term from $f(D)$ to make the first term unity (so that Binomial Theorem for a negative index is applicable).

The remaining factor will be of the form $1 + \phi(D)$ or $1 - \phi(D)$

Take this factor in the numerator. It takes the form $[1 + \phi(D)]^{-1}$ or $[1 - \phi(D)]^{-1}$

Expand it in ascending powers of D as far as the term containing D^m, since $D^{m+1}(x^m) = 0$, $D^{m+2}(x^m) = 0$ and so on.

Operate on x^m term by term.

Example. *Find the P.I. of* $(D^2 + 5D + 4)y = x^2 + 7x + 9$.

Sol. $$\text{P.I.} = \frac{1}{D^2 + 5D + 4}(x^2 + 7x + 9) = \frac{1}{4\left(1 + \frac{5D}{4} + \frac{D^2}{4}\right)}(x^2 + 7x + 9)$$

$$= \frac{1}{4}\left[1 + \left(\frac{5D}{4} + \frac{D^2}{4}\right)\right]^{-1}(x^2 + 7x + 9)$$

$$= \frac{1}{4}\left[1 - \left(\frac{5D}{4} + \frac{D^2}{4}\right) + \left(\frac{5D}{4} + \frac{D^2}{4}\right)^2 - \ldots\right](x^2 + 7x + 9)$$

$$= \frac{1}{4}\left(1 - \frac{5D}{4} - \frac{D^2}{4} + \frac{25D^2}{16} \cdots\right)(x^2 + 7x + 9)$$

$$= \frac{1}{4}\left(1 - \frac{5D}{4} + \frac{21D^2}{16} \cdots\right)(x^2 + 7x + 9)$$

$$= \frac{1}{4}\left[(x^2 + 7x + 9) - \frac{5}{4}D(x^2 + 7x + 9) + \frac{21}{16}D^2(x^2 + 7x + 9)\right]$$

$$= \frac{1}{4}\left[(x^2 + 7x + 9) - \frac{5}{4}(2x + 7) + \frac{21}{16}(2)\right] = \frac{1}{4}\left(x^2 + \frac{9}{2}x + \frac{23}{8}\right).$$

Case IV. *When* $X = e^{ax}V$, *where V is a function of x.*

Let u be a function of x, then by successive differentiation, we have

$$D(e^{ax}u) = e^{ax}Du + a\,e^{ax}u = e^{ax}(D + a)u$$

$$D^2(e^{ax}u) = D[e^{ax}(D + a)u] = e^{ax}(D^2 + aD)u + ae^{ax}(D + a)u$$

$$= e^{ax}(D^2 + 2aD + a^2)u = e^{ax}(D + a)^2 u$$

Similarly, $D^3(e^{ax}u) = e^{ax}(D + a)^3 u$

In general, $D^n(e^{ax}u) = e^{ax}(D + a)^n u$

$\therefore$ $f(D)(e^{ax}u) = e^{ax}f(D + a)u$

Operating on both sides by $\frac{1}{f(D)}$,

$$\frac{1}{f(D)}[f(D)(e^{ax}u)] = \frac{1}{f(D)}[e^{ax}f(D + a)u]$$

$$\Rightarrow \quad e^{ax}u = \frac{1}{f(D)}[e^{ax}f(D + a)u] \quad \ldots(i)$$

Now let $f(D + a)u = V$, *i.e.*, $u = \frac{1}{f(D + a)}V$

$\therefore$ From (i), we have $e^{ax}\frac{1}{f(D + a)}V = \frac{1}{f(D)}(e^{ax}V)$

or $$\frac{1}{f(\mathbf{D})}(e^{ax}\,\mathbf{V}) = e^{ax}\,\frac{1}{f(\mathbf{D}+a)}\,\mathbf{V}.$$

Thus e^{ax} which is on the right of $\frac{1}{f(\mathrm{D})}$ may be taken out to the left provided D is replaced by D + a.

Example. *Find the P.I. of* $(D^2 - 4D + 3)y = e^x \cos 2x$.

Sol. $$\text{P.I.} = \frac{1}{\mathrm{D}^2 - 4\mathrm{D} + 3}\, e^x \cos 2x = e^x\, \frac{1}{(\mathrm{D}+1)^2 - 4(\mathrm{D}+1) + 3} \cos 2x$$

$$= e^x\, \frac{1}{\mathrm{D}^2 - 2\mathrm{D}} \cos 2x = e^x\, \frac{1}{-2^2 - 2\mathrm{D}} \cos 2x \qquad [\text{Putting } \mathrm{D}^2 = -\,2^2]$$

$$= -\frac{1}{2} e^x\, \frac{1}{2+\mathrm{D}} \cos 2x = -\frac{1}{2} e^x\, \frac{2-\mathrm{D}}{(2+\mathrm{D})(2-\mathrm{D})} \cos 2x$$

$$= -\frac{1}{2} e^x\, \frac{2-\mathrm{D}}{4-\mathrm{D}^2} \cos 2x = -\frac{1}{2} e^x\, \frac{2-\mathrm{D}}{4-(-2^2)} \cos 2x$$

$$= -\frac{1}{16} e^x (2 \cos 2x - \mathrm{D} \cos 2x) = -\frac{1}{16} e^x (2 \cos 2x + 2 \sin 2x)$$

$$= -\frac{1}{8} e^x (\cos 2x + \sin 2x).$$

Case V. *When X is any other function of x.*

Resolve $f(\mathrm{D})$ into linear factors.

Let $$f(\mathrm{D}) = (\mathrm{D} - m_1)(\mathrm{D} - m_2) \ldots (\mathrm{D} - m_n)\,\mathrm{X}$$

Then $$\text{P.I.} = \frac{1}{f(\mathrm{D})}\mathrm{X} = \frac{1}{(\mathrm{D}-m_1)(\mathrm{D}-m_2)\ldots(\mathrm{D}-m_n)}\mathrm{X}$$

$$= \left(\frac{\mathrm{A}_1}{\mathrm{D}-m_1} + \frac{\mathrm{A}_2}{\mathrm{D}-m_2} + \cdots + \frac{\mathrm{A}_n}{\mathrm{D}-m_n}\right)\mathrm{X} \qquad \text{(Partial Fractions)}$$

$$= \mathrm{A}_1 \frac{1}{\mathrm{D}-m_1}\mathrm{X} + \mathrm{A}_2 \frac{1}{\mathrm{D}-m_2}\mathrm{X} + \cdots + \mathrm{A}_n \frac{1}{\mathrm{D}-m_n}\mathrm{X}$$

$$= \mathrm{A}_1 e^{m_1 x} \int \mathrm{X}\, e^{-m_1 x}\, dx + \mathrm{A}_2 e^{m_2 x} \int \mathrm{X}\, e^{-m_2 x} dx + \cdots + \mathrm{A}_n e^{m_n x} \int \mathrm{X}\, e^{-m_n x} dx$$

$$\left[\because \quad \frac{1}{\mathrm{D}-m}\mathrm{X} = e^{mx} \int \mathrm{X} e^{-mx} dx\right].$$

ILLUSTRATIVE EXAMPLES

Example 1. *Solve* $(D^3 - 6D^2 + 11D - 6)y = e^{-2x} + e^{-3x}$.

Sol. A.E. is $\mathrm{D}^3 - 6\mathrm{D}^2 + 11\mathrm{D} - 6 = 0$ or $(\mathrm{D}-1)(\mathrm{D}-2)(\mathrm{D}-3) = 0$

whence $$\mathrm{D} = 1, 2, 3$$

$\therefore$ $$\text{C.F.} = c_1 e^x + c_2 e^{2x} + c_3 e^{3x}$$

$$\text{P.I.} = \frac{1}{D^3 - 6D^2 + 11D - 6}(e^{-2x} + e^{-3x})$$

$$= \frac{1}{D^3 - 6D^2 + 11D - 6}e^{-2x} + \frac{1}{D^3 - 6D^2 + 11D - 6}e^{-3x}$$

$$= \frac{1}{(-2)^3 - 6(-2)^2 + 11(-2) - 6}e^{-2x} + \frac{1}{(-3)^3 - 6(-3)^2 + 11(-3) - 6}e^{-3x}$$

$$= -\frac{1}{60}e^{-2x} - \frac{1}{120}e^{-3x} = -\frac{1}{120}(2e^{-2x} + e^{-3x})$$

Hence the C.S. is y = C.F. + P.I.

i.e.,
$$y = c_1e^x + c_2e^{2x} + c_3e^{3x} - \frac{1}{120}(2e^{-2x} + e^{-3x}).$$

Example 2. *Solve* $(D - 2)^2y = 8(e^{2x} + \sin 2x + x^2)$.

Sol. A.E. is $(D - 2)^2 = 0$ whence $D = 2, 2$

$\therefore$
$$\text{C.F.} = (c_1 + c_2x)e^{2x}$$

$$\text{P.I.} = \frac{1}{(D-2)^2}[8(e^{2x} + \sin 2x + x^2)]$$

$$= 8\left[\frac{1}{(D-2)^2}e^{2x} + \frac{1}{(D-2)^2}\sin 2x + \frac{1}{(D-2)^2}x^2\right]$$

Now
$$\frac{1}{(D-2)^2}e^{2x} = x \cdot \frac{1}{2(D-2)}e^{2x} \qquad \text{| Case of failure}$$

$$= x^2 \cdot \frac{1}{2}e^{2x} \qquad \text{| Case of failure}$$

$$= \frac{x^2}{2}e^{2x}$$

$$\frac{1}{(D-2)^2}\sin 2x = \frac{1}{D^2 - 4D + 4}\sin 2x = \frac{1}{-2^2 - 4D + 4}\sin 2x \qquad [\text{Putting } D^2 = -2^2]$$

$$= -\frac{1}{4D}\sin 2x = -\frac{1}{4}\int \sin 2x\,dx = -\frac{1}{4}\left(-\frac{\cos 2x}{2}\right) = \frac{1}{8}\cos 2x$$

$$\frac{1}{(D-2)^2}x^2 = \frac{1}{(2-D)^2}x^2 = \frac{1}{4\left(1-\frac{D}{2}\right)^2}x^2 = \frac{1}{4}\left(1-\frac{D}{2}\right)^{-2}x^2$$

$$= \frac{1}{4}\left[1 - 2\left(-\frac{D}{2}\right) + \frac{(-2)(-3)}{2}\left(\frac{D}{2}\right)^2 \cdots\right]x^2$$

$$= \frac{1}{4}\left[1 + D + \frac{3}{4}D^2 + \cdots\right]x^2$$

$$= \frac{1}{4}\left(x^2 + 2x + \frac{3}{2}\right)$$

$$\therefore \qquad \text{P.I.} = 8\left[\frac{x^2}{2}e^{2x} + \frac{1}{8}\cos 2x + \frac{1}{4}\left(x^2 + 2x + \frac{3}{2}\right)\right]$$

$$= 4x^2 e^{2x} + \cos 2x + 2x^2 + 4x + 3$$

Hence the C.S. is $\quad y = (c_1 + c_2x)\, e^{2x} + 4x^2 e^{2x} + \cos 2x + 2x^2 + 4x + 3.$

Example 3. *Solve* $(D + 2)(D - 1)^2\, y = e^{-2x} + 2 \sinh x.$

Sol. A.E. is $(D + 2)(D - 1)^2 = 0$ so that $D = -2, 1, 1$

$$\therefore \qquad \text{C.F.} = c_1e^{-2x} + (c_2 + c_3x)\, e^x$$

$$\text{P.I.} = \frac{1}{(D+2)(D-1)^2}(e^{-2x} + 2 \sinh x)$$

$$= \frac{1}{(D+2)(D-1)^2}(e^{-2x} + e^x - e^{-x}) \qquad \left[\because \sinh x = \frac{e^x - e^{-x}}{2}\right]$$

Now $\dfrac{1}{(D+2)(D-1)^2}e^{-2x} = \dfrac{1}{D+2}\left[\dfrac{1}{(D-1)^2}e^{-2x}\right] = \dfrac{1}{D+2}\left[\dfrac{1}{(-2-1)^2}e^{-2x}\right]$

$$= \frac{1}{9}\cdot\frac{1}{D+2}e^{-2x} \qquad \text{| Case of failure}$$

$$= \frac{1}{9}x\cdot\frac{1}{1}e^{-2x} = \frac{x}{9}e^{-2x}$$

$$\frac{1}{(D+2)(D-1)^2}e^x = \frac{1}{(D-1)^2}\left[\frac{1}{D+2}e^x\right] = \frac{1}{(D-1)^2}\left[\frac{1}{1+2}e^x\right]$$

$$= \frac{1}{3}\cdot\frac{1}{(D-1)^2}e^x \qquad \text{| Case of failure}$$

$$= \frac{1}{3}\cdot x\frac{1}{2(D-1)}e^x \qquad \text{| Case of failure}$$

$$= \frac{1}{3}\cdot x^2\cdot\frac{1}{2}e^x = \frac{1}{6}x^2e^x$$

$$\frac{1}{(D+2)(D-1)^2}e^{-x} = \frac{1}{(-1+2)(-1-1)^2}e^{-x} = \frac{1}{4}e^{-x}$$

$$\therefore \qquad \text{P.I.} = \frac{x}{9}e^{-2x} + \frac{x^2}{6}e^x + \frac{1}{4}e^{-x}$$

Hence the C.S. is $\quad y = c_1e^{-2x} + (c_2 + c_3x)e^x + \dfrac{x}{9}e^{-2x} + \dfrac{x^2}{6}e^x + \dfrac{1}{4}e^{-x}.$

Example 4. *Solve* $\dfrac{d^2y}{dx^2} - 4y = x \sinh x.$

Sol. Given equation in symbolic form is $(D^2 - 4)y = x \sinh x$

A.E. is $\qquad D^2 - 4 = 0 \quad$ so that $\quad D = \pm 2$

$\therefore$ C.F. $= c_1e^{2x} + c_2e^{-2x}$

$$\text{P.I.} = \frac{1}{D^2-4}\, x \sinh x = \frac{1}{D^2-4}\, x\left(\frac{e^x - e^{-x}}{2}\right)$$

$$= \frac{1}{2}\left[\frac{1}{D^2-4} e^x \cdot x - \frac{1}{D^2-4} e^{-x} \cdot x\right]$$

$$= \frac{1}{2}\left[e^x \frac{1}{(D+1)^2-4} x - e^{-x} \frac{1}{(D-1)^2-4} x\right]$$

$$= \frac{1}{2}\left[e^x \frac{1}{D^2+2D-3} x - e^{-x} \frac{1}{D^2-2D-3} x\right]$$

$$= \frac{1}{2}\left[e^x \frac{1}{-3\left(1-\frac{2D}{3}-\frac{D^2}{3}\right)} x - e^{-x} \frac{1}{-3\left(1+\frac{2D}{3}-\frac{D^2}{3}\right)} x\right]$$

$$= -\frac{1}{6}\left[e^x \left\{1-\left(\frac{2D}{3}+\frac{D^2}{3}\right)\right\}^{-1} x - e^{-x}\left\{1+\left(\frac{2D}{3}-\frac{D^2}{3}\right)\right\}^{-1} x\right]$$

$$= -\frac{1}{6}\left[e^x\left(1+\frac{2D}{3}\cdots\right)x - e^{-x}\left(1-\frac{2D}{3}\cdots\right)x\right] = -\frac{1}{6}\left[e^x\left(x+\frac{2}{3}\right) - e^{-x}\left(x-\frac{2}{3}\right)\right]$$

$$= -\frac{x}{3}\left(\frac{e^x - e^{-x}}{2}\right) - \frac{2}{9}\left(\frac{e^x + e^{-x}}{2}\right) = -\frac{x}{3}\sinh x - \frac{2}{9}\cosh x$$

Hence the C.S. is $y = c_1e^{2x} + c_2e^{-2x} - \frac{x}{3}\sinh x - \frac{2}{9}\cosh x.$

Example 5. *Solve* $\frac{d^4y}{dx^4} - y = \cos x \cosh x.$

Sol. Given equation in symbolic form is $(D^4 - 1)y = \cos x \cosh x$

A.E. is $D^4 - 1 = 0$ or $(D^2 - 1)(D^2 + 1) = 0$ so that $D = \pm 1, \pm i$

$\therefore$ C.F. $= c_1e^x + c_2e^{-x} + e^{0x}(c_3 \cos x + c_4 \sin x)$

$= c_1e^x + c_2e^{-x} + c_3 \cos x + c_4 \sin x$

$$\text{P.I.} = \frac{1}{D^4-1} \cos x \cosh x = \frac{1}{D^4-1} \cos x\left(\frac{e^x + e^{-x}}{2}\right)$$

$$= \frac{1}{2}\left[\frac{1}{D^4-1} e^x \cos x + \frac{1}{D^4-1} e^{-x} \cos x\right]$$

$$= \frac{1}{2}\left[e^x \frac{1}{(D+1)^4-1} \cos x + e^{-x} \frac{1}{(D-1)^4-1} \cos x\right]$$

$$= \frac{1}{2}\left[e^x \frac{1}{D^4+4D^3+6D^2+4D} \cos x + e^{-x} \frac{1}{D^4-4D^3+6D^2-4D} \cos x\right]$$

$$= \frac{1}{2}\left[e^x \frac{1}{(-1^2)^2 + 4D(-1^2) + 6(-1^2) + 4D} \cos x \right.$$

$$\left. + e^{-x} \frac{1}{(-1^2)^2 - 4D(-1^2) + 6(-1^2) - 4D} \cos x\right]$$

$$= \frac{1}{2}\left[e^x \frac{1}{-5} \cos x + e^{-x} \frac{1}{-5} \cos x\right] = -\frac{1}{5}\left(\frac{e^x + e^{-x}}{2}\right) \cos x = -\frac{1}{5} \cosh x \cos x$$

Hence the C.S. is $y = c_1e^x + c_2e^{-x} + c_3 \cos x + c_4 \sin x - \frac{1}{5} \cos x \cosh x$.

Example 6. *Solve* $\frac{d^2y}{dx^2} - 2\frac{dy}{dx} + y = xe^x \sin x$.

Sol. Given equation in symbolic form is $(D^2 - 2D + 1)y = xe^x \sin x$

A.E. is $D^2 - 2D + 1 = 0$ or $(D - 1)^2 = 0$ so that $D = 1, 1$

$\therefore$ C.F. $= (c_1 + c_2x)e^x$

$$\text{P.I.} = \frac{1}{(D-1)^2} e^x \cdot x \sin x = e^x \cdot \frac{1}{(D+1-1)^2} x \sin x$$

$$= e^x \frac{1}{D^2} x \sin x = e^x \frac{1}{D}\int x \sin x \, dx$$

Integrating by parts,

$$\text{P.I.} = e^x \frac{1}{D}\left[x(-\cos x) - \int 1(-\cos x)\,dx\right] = e^x \frac{1}{D}(-x \cos x + \sin x)$$

$$= e^x \int(-x \cos x + \sin x)\,dx = e^x\left[-\left\{x \sin x - \int 1\cdot \sin x \, dx\right\} - \cos x\right]$$

$$= e^x[-x \sin x - \cos x - \cos x] = -e^x(x \sin x + 2 \cos x)$$

Hence the C.S. is $y = (c_1 + c_2x)e^x - e^x(x \sin x + 2 \cos x)$.

Example 7. *Solve* $\frac{d^2y}{dx^2} + y = \text{cosec } x$.

Sol. Given equation in symbolic form is $(D^2 + 1)y = \text{cosec } x$

A.E. is $D^2 + 1 = 0 \Rightarrow D = \pm i$

$\therefore$ C.F. $= c_1 \cos x + c_2 \sin x$

$$\text{P.I.} = \frac{1}{D^2+1} \text{cosec } x = \frac{1}{(D+i)(D-i)} \text{cosec } x$$

$$= \frac{1}{2i}\left(\frac{1}{D-i} - \frac{1}{D+i}\right) \text{cosec } x \qquad \text{(Partial Fractions)}$$

$$= \frac{1}{2i}\left(\frac{1}{D-i} \text{cosec } x - \frac{1}{D+i} \text{cosec } x\right)$$

Now $\frac{1}{D-i} \text{cosec } x = e^{ix} \int \text{cosec } x \, e^{-ix}\,dx \qquad \left[\because \frac{1}{D-a} X = e^{ax}\int Xe^{-ax}\,dx\right]$

$$= e^{ix} \int \operatorname{cosec} x \, (\cos x - i \sin x) \, dx = e^{ix} \int (\cot x - i) \, dx$$

$$= e^{ix} (\log \sin x - ix)$$

Changing i to $-i$, we have $\dfrac{1}{D+i} \operatorname{cosec} x = e^{-ix} (\log \sin x + ix)$

$$\therefore \qquad \text{P.I.} = \frac{1}{2i} [e^{ix} (\log \sin x - ix) - e^{-ix} (\log \sin x + ix)]$$

$$= \log \sin x \left(\frac{e^{ix} - e^{-ix}}{2i} \right) - x \left(\frac{e^{ix} + e^{-ix}}{2} \right)$$

$$= \log \sin x \, (\sin x) - x \cos x = \sin x \log \sin x - x \cos x$$

Hence the C.S. is $y = c_1 \cos x + c_2 \sin x + \sin x \log \sin x - x \cos x$.

Example 8. *Solve* $\dfrac{d^2 y}{dx^2} + a^2 y = \tan ax$.

Sol. Given equation in symbolic form is $(D^2 + a^2)y = \tan ax$

A.E. is $\quad D^2 + a^2 = 0 \Rightarrow D = \pm ia$

$$\therefore \qquad \text{C.F.} = c_1 \cos ax + c_2 \sin ax$$

$$\text{P.I.} = \frac{1}{D^2 + a^2} \tan ax = \frac{1}{(D + ia)(D - ia)} \tan ax$$

$$= \frac{1}{2ia} \left[\frac{1}{D - ia} - \frac{1}{D + ia} \right] \tan ax \qquad \text{(Partial Fractions)}$$

$$= \frac{1}{2ia} \left[\frac{1}{D - ia} \tan ax - \frac{1}{D + ia} \tan ax \right]$$

Now $\dfrac{1}{D - ia} \tan ax = e^{iax} \displaystyle\int \tan ax \cdot e^{-iax} \, dx$

$$= e^{iax} \int \tan ax \, (\cos ax - i \sin ax) \, dx = e^{iax} \int \left(\sin ax - i \frac{\sin^2 ax}{\cos ax} \right) dx$$

$$= e^{iax} \int \left(\sin ax - i \frac{1 - \cos^2 ax}{\cos ax} \right) dx = e^{iax} \int [\sin ax - i(\sec ax - \cos ax)] \, dx$$

$$= e^{iax} \left[-\frac{\cos ax}{a} - \frac{i}{a} \log (\sec ax + \tan ax) + i \frac{\sin ax}{a} \right]$$

$$= -\frac{1}{a} e^{iax} [(\cos ax - i \sin ax) + i \log (\sec ax + \tan ax)]$$

$$= -\frac{1}{a} e^{iax} [e^{-iax} + i \log (\sec ax + \tan ax)] = -\frac{1}{a} [1 + ie^{iax} \log (\sec ax + \tan ax)]$$

Changing i to $-i$, we have $\dfrac{1}{D + ia} \tan ax = -\dfrac{1}{a} [1 - ie^{-iax} \log (\sec ax + \tan ax)]$

$$\therefore \quad \text{P.I.} = \frac{1}{2ia}\left[-\frac{1}{a}\left\{1 + ie^{iax}\log(\sec ax + \tan ax)\right\} + \frac{1}{a}\left\{1 - ie^{-iax}\log(\sec ax + \tan ax)\right\}\right]$$

$$= -\frac{1}{a^2}\log(\sec ax + \tan ax)\left(\frac{e^{iax} + e^{-iax}}{2}\right) = -\frac{1}{a^2}\log(\sec ax + \tan ax)\cdot\cos ax$$

Hence the C.S. is $y = c_1 \cos ax + c_2 \sin ax - \frac{1}{a^2}\cos ax \log(\sec ax + \tan ax)$.

EXERCISE 5.2

Solve the following differential equations:

1. $\frac{d^3y}{dx^3} + y = 3 + 5e^x.$

2. $\frac{d^2y}{dx^2} - 4y = (1 + e^x)^2$

3. $\frac{d^2y}{dx^2} + 4\frac{dy}{dx} + 5y = -2\cosh x$

4. $\frac{d^2y}{dx^2} - 2\frac{dy}{dx} + 5y = \sin 3x$

5. $\frac{d^3y}{dx^3} + \frac{d^2y}{dx^2} + \frac{dy}{dx} + y = \sin 2x$

6. $\frac{d^3y}{dx^3} + y = \sin 3x - \cos^2\frac{x}{2}$

7. $(D^2 - 4D + 3)y = \sin 3x \cos 2x$

8. $(D^2 - 3D + 2)y = 6e^{-3x} + \sin 2x$

9. $\frac{d^2y}{dx^2} + 4y = e^x + \sin 2x$

10. $\frac{d^3y}{dx^3} - 2\frac{d^2y}{dx^2} + 4\frac{dy}{dx} = e^{2x} + \sin 2x$

11. $\frac{d^2y}{dx^2} - 4y = x^2$

12. $\frac{d^3y}{dx^3} - \frac{d^2y}{dx^2} - 6\frac{dy}{dx} = 1 + x^2$

13. $\frac{d^2y}{dx^2} + \frac{dy}{dx} = x^2 + 2x + 4$

14. $\frac{d^2y}{dx^2} + y = e^{2x} + \cosh 2x + x^3$

15. $(D^2 - 3D + 2)y = 2e^x \cos\frac{x}{2}$

16. $\frac{d^2y}{dx^2} - 3\frac{dy}{dx} + 2y = xe^{3x} + \sin 2x$

17. $\frac{d^4y}{dx^4} - y = e^x \cos x$

18. $(D^2 - 2D)y = e^x \sin x$

19. $(D^2 + 4D + 8)y = 12e^{-2x}\sin x \sin 3x$

20. $\frac{d^2y}{dx^2} + 2y = x^2e^{3x} + e^x \cos 2x$

21. $(D^3 + 2D^2 + D)y = x^2e^{2x} + \sin^2 x$

22. $(D^2 - 4D + 4)y = 8x^2 e^{2x} \sin 2x$

23. $(D - 1)^2(D + 1)^2y = \sin^2\frac{x}{2} + e^x + x$

24. $\frac{d^2y}{dx^2} + 4y = x \sin x$

25. $(D^2 - 1)y = x \sin x + (1 + x^2)e^x$

26. $\frac{d^2y}{dx^2} + a^2y = \sec ax$

27. $\frac{d^2y}{dx^2} + 4y = 4\tan 2x$

28. $\frac{d^2y}{dx^2} + 3\frac{dy}{dx} + 2y = e^{e^x}.$

Answers

1. $y = c_1e^{-x} + e^{\frac{1}{2}x}\left(c_2 \cos \frac{\sqrt{3}}{2} x + c_3 \sin \frac{\sqrt{3}}{2} x\right) + 3 + \frac{5}{2} e^x$

2. $y = c_1e^{2x} + c_2e^{-2x} - \frac{1}{4} - \frac{2}{3}e^x + \frac{1}{4} xe^{2x}$

3. $y = e^{-2x}(c_1 \cos x + c_2 \sin x) - \frac{1}{10} e^x - \frac{1}{2} e^{-x}$

4. $y = e^x(c_1 \cos 2x + c_2 \sin 2x) + \frac{1}{26} (3 \cos 3x - 2 \sin 3x)$

5. $y = c_1e^{-x} + c_2 \cos x + c_3 \sin x + \frac{1}{15} (2 \cos 2x - \sin 2x)$

6. $y = c_1e^{-x} + e^{\frac{1}{2}x}\left(c_2 \cos \frac{\sqrt{3}}{2} x + c_3 \sin \frac{\sqrt{3}}{2} x\right) + \frac{1}{730} (\sin 3x + 27 \cos 3x) - \frac{1}{2} - \frac{1}{4} (\cos x - \sin x)$

7. $y = c_1e^x + c_2e^{3x} + \frac{1}{884} (10 \cos 5x - 11 \sin 5x) + \frac{1}{20} (\sin x + 2 \cos x)$

8. $y = c_1e^x + c_2e^{2x} + \frac{3}{10} e^{-3x} + \frac{1}{20} (3 \cos 2x - \sin 2x)$

9. $y = c_1 \cos 2x + c_2 \sin 2x + \frac{1}{5} e^x - \frac{x}{4} \cos 2x$

10. $y = c_1 + e^x (c_2 \cos \sqrt{3}\, x + c_3 \sin \sqrt{3}\, x) + \frac{1}{8} (e^{2x} + \sin 2x)$

11. $y = c_1e^{2x} + c_2e^{-2x} - \frac{1}{4}\left(x^2 + \frac{1}{2}\right)$

12. $y = c_1 + c_2e^{3x} + c_3e^{-2x} - \frac{1}{18}\left(x^3 - \frac{x^2}{2} + \frac{25}{6} x\right)$

13. $y = c_1 + c_2e^{-x} + \frac{x^3}{3} + 4x$

14. $y = c_1 \cos x + c_2 \sin x + \frac{1}{5} e^{2x} + \frac{1}{5} \cosh 2x + x^3 - 6x$

15. $y = c_1e^x + c_2e^{2x} - \frac{8}{5} e^x \left(2 \sin \frac{x}{2} + \cos \frac{x}{2}\right)$

16. $y = c_1e^x + c_2e^{2x} + \frac{1}{4} e^{3x} (2x - 3) + \frac{1}{20} (3 \cos 2x - \sin 2x)$

17. $y = c_1e^x + c_2e^{-x} + c_3 \cos x + c_4 \sin x - \frac{1}{5} e^x \cos x$

18. $y = c_1 + c_2e^{2x} - \frac{1}{2} e^x \sin x$

19. $y = e^{-2x} (c_1 \cos 2x + c_2 \sin 2x) + \frac{1}{2} e^{-2x} (3x \sin 2x + \cos 4x)$

20. $y = c_1 \cos \sqrt{2}\, x + c_2 \sin \sqrt{2}\, x + \frac{e^{3x}}{11}\left(x^2 - \frac{12}{11} x + \frac{50}{121}\right) + \frac{e^x}{17} (4 \sin 2x - \cos 2x)$

21. $y = c_1 + (c_2 + c_3x)e^{-x} + \frac{x}{2} + \frac{e^{2x}}{18}\left(x^2 - \frac{7}{3} x + \frac{11}{6}\right) + \frac{1}{100} (3 \sin 2x + 4 \cos 2x)$

22. $y = (c_1 + c_2x)e^{2x} - e^{2x}[4x \cos 2x + (2x^2 - 3) \sin 2x]$

23. $y = (c_1 + c_2x)e^x + (c_3 + c_4x)e^{-x} + \frac{1}{2} - \frac{1}{8} \cos x + \frac{x^2}{8} e^x + x$

24. $y = c_1 \cos 2x + c_2 \sin 2x + \frac{1}{9} (3x \sin x - 2 \cos x)$

25. $y = c_1 e^x + c_2 e^{-x} - \frac{1}{2}(x \sin x + \cos x) + \frac{1}{12} xe^x (2x^2 - 3x + 9)$

26. $y = c_1 \cos ax + c_2 \sin ax + \frac{1}{a}\left(x \sin ax + \cos ax \frac{\log \cos ax}{a}\right)$

27. $y = c_1 \cos 2x + c_2 \sin 2x - \cos 2x \log (\sec 2x + \tan 2x)$

28. $y = c_1 e^{-x} + c_2 e^{-2x} + e^{-2x} . e^{e^x}$.

5.8. METHOD OF VARIATION OF PARAMETERS TO FIND P.I.

Consider the linear equation of **second order** with constant co-efficients

$$\frac{d^2y}{dx^2} + a_1 \frac{dy}{dx} + a_2 y = \text{X} \qquad ...(1)$$

Let its C.F. be $y = c_1 y_1 + c_2 y_2$ so that y_1 and y_2 satisfy the equation

$$\frac{d^2y}{dx^2} + a_1 \frac{dy}{dx} + a_2 y = 0 \qquad ...(2)$$

Now, let us assume that the P.I. of (1) is $y = uy_1 + vy_2$ ...(3)

where u and v are unknown functions of x.

Differentiating (3) w.r.t. x, we have $y' = uy_1' + vy_2' + u'y_1 + v'y_2 = uy_1' + vy_2'$ ...(4)

assuming that u, v satisfy the equation $u'y_1 + v'y_2 = 0$ ...(5)

Differentiating (4) w.r.t. x, we have $y'' = uy_1'' + u'y_1' + vy_2'' + v'y_2'$

Substituting the values of y, y' and y'' in (1), we get

$$(uy_1'' + u'y_1' + vy_2'' + v'y_2') + a_1(uy_1' + vy_2') + a_2(uy_1 + vy_1) = \text{X}$$

or $$u(y_1'' + a_1y_1' + a_2y_1) + v(y_2'' + a_1y_2' + a_2y_2) + u'y_1' + v'y_2' = \text{X}$$

or $$u'y_1' + v'y_2' = \text{X} \qquad ...(6)$$

since y_1 and y_2 satisfy (2).

Solving (5) and (6), we get $u' = \begin{vmatrix} 0 & y_2 \\ \text{X} & y_2' \end{vmatrix} \div \begin{vmatrix} y_1 & y_2 \\ y_1' & y_2' \end{vmatrix} = -\frac{y_2\text{X}}{\text{W}}$

and $v' = \begin{vmatrix} y_1 & 0 \\ y_1' & \text{X} \end{vmatrix} \div \begin{vmatrix} y_1 & y_2 \\ y_1' & y_2' \end{vmatrix} = \frac{y_1\text{X}}{\text{W}}$

where $\text{W} = \begin{vmatrix} y_1 & y_2 \\ y_1' & y_2' \end{vmatrix}$ is called the Wronskian of y_1, y_2.

Integrating, $u = -\int \frac{y_2\text{X}}{\text{W}} dx, \quad v = \int \frac{y_1\text{X}}{\text{W}} dx$

Substituting in (3), the P.I. is known.

Note 1. As the solution is obtained by varying the arbitrary constants c_1, c_2 of the C.F., the method is known as *variation of parameters*.

Note 2. Method of variation of parameters is to be used if instructed to do so.

Example. *Apply the method of variation of parameters to solve* $\frac{d^2y}{dx^2} + 4y = 4 \sec^2 2x$.

Sol. Given equation in symbolic form is $(\text{D}^2 + 4)y = 4 \sec^2 2x$

Its A.E. is $\text{D}^2 + 4 = 0$ so that $\text{D} = \pm 2i$

$\therefore$ C.F. is $y = c_1 \cos 2x + c_2 \sin 2x$

Here $y_1 = \cos 2x$, $y_2 = \sin 2x$ and $\text{X} = 4 \sec^2 2x$

$$\therefore \quad W = \begin{vmatrix} y_1 & y_2 \\ y_1' & y_2' \end{vmatrix} = \begin{vmatrix} \cos 2x & \sin 2x \\ -2\sin 2x & 2\cos 2x \end{vmatrix} = 2$$

$$\text{P.I.} = -y_1 \int \frac{y_2 X}{W}\, dx + y_2 \int \frac{y_1 X}{W}\, dx$$

$$= -\cos 2x \int \frac{\sin 2x \,.\, 4\sec^2 2x}{2}\, dx + \sin 2x \int \frac{\cos 2x \,.\, 4\sec^2 2x}{2}\, dx$$

$$= -2\cos 2x \int \sec 2x \tan 2x\, dx + 2\sin 2x \int \sec 2x\, dx$$

$$= -2\cos 2x \cdot \frac{\sec 2x}{2} + 2\sin 2x \cdot \frac{1}{2}\log(\sec 2x + \tan 2x)$$

$$= -1 + \sin 2x \log(\sec 2x + \tan 2x)$$

Hence the C.S. is $y = c_1 \cos 2x + c_2 \sin 2x - 1 + \sin 2x \log(\sec 2x + \tan 2x)$.

EXERCISE 5.3

Solve by the method of variation of parameters:

1. $\dfrac{d^2y}{dx^2} + y = \operatorname{cosec} x.$
2. $\dfrac{d^2y}{dx^2} + y = \sec x$ *(K.U. May 2010)*
3. $\dfrac{d^2y}{dx^2} + 4y = \tan 2x$ *(K.U. May 2012)*
4. $\dfrac{d^2y}{dx^2} + y = x \sin x$
5. $y'' - 2y' + 2y = e^x \tan x.$

Answers

1. $y = c_1 \cos x + c_2 \sin x - x \cos x + \sin x \log \sin x$
2. $y = c_1 \cos x + c_2 \sin x + \cos x \log(\cos x) + x \sin x$
3. $y = c_1 \cos 2x + c_2 \sin 2x - \dfrac{1}{4}\cos 2x \log(\sec 2x + \tan 2x)$
4. $y = c_1 \cos x + c_2 \sin x + \dfrac{x}{4}\sin x - \dfrac{x^2}{4}\cos x$
5. $y = e^x (c_1 \cos x + c_2 \sin x) - e^x \cos x \log(\sec x + \tan x).$

5.9. CAUCHY'S HOMOGENEOUS LINEAR EQUATION

An equation of the form

$$x^n \frac{d^n y}{dx^n} + a_1 x^{n-1} \frac{d^{n-1}y}{dx^{n-1}} + a_2 x^{n-2} \frac{d^{n-2}y}{dx^{n-2}} + \cdots + a_{n-1} x \frac{dy}{dx} + a_n y = X \qquad ...(i)$$

where a_i's are constants and X is a function of x, is called Cauchy's homogeneous linear equation.

Such equations can be reduced to linear differential equations with constant co-efficients by the substitution $x = e^z$ or $z = \log x$

so that $$\frac{dy}{dx} = \frac{dy}{dz} \cdot \frac{dz}{dx} = \frac{dy}{dz} \cdot \frac{1}{x} \quad \text{or} \quad x\frac{dy}{dx} = \frac{dy}{dz} = \text{D}y, \text{ where D} = \frac{d}{dz}$$

$$\frac{d^2y}{dx^2} = \frac{d}{dx}\left(\frac{1}{x} \cdot \frac{dy}{dz}\right) = -\frac{1}{x^2}\frac{dy}{dz} + \frac{1}{x} \cdot \frac{d^2y}{dz^2} \cdot \frac{dz}{dx}$$

$$= -\frac{1}{x^2}\frac{dy}{dz} + \frac{1}{x^2}\frac{d^2y}{dz^2} \qquad \left(\because \frac{dz}{dx} = \frac{1}{x}\right)$$

or $$x^2\frac{d^2y}{dx^2} = \frac{d^2y}{dz^2} - \frac{dy}{dz} = \text{D}^2y - \text{D}y = \text{D}(\text{D} - 1)y$$

Similarly, $x^3\dfrac{d^3y}{dx^3} = \text{D}(\text{D} - 1)(\text{D} - 2)y$ and so on.

Substituting these values in equation (*i*), we get a linear differential equation with constant co-efficients, which can be solved by the methods already discussed.

ILLUSTRATIVE EXAMPLES

Example 1. *Solve* $x^3\dfrac{d^3y}{dx^3} + 2x^2\dfrac{d^2y}{dx^2} + 2y = 10\left(x + \dfrac{1}{x}\right)$.

Sol. Given equation is a Cauchy's homogeneous linear equation.

Put $x = e^z$ *i.e.*, $z = \log x$

so that $$x\frac{dy}{dx} = \text{D}y, \; x^2\frac{d^2y}{dx^2} = \text{D}(\text{D} - 1)y$$

$$x^3\frac{d^3y}{dx^3} = \text{D}(\text{D} - 1)(\text{D} - 2)y, \text{ where D} \equiv \frac{d}{dz}$$

Substituting these values in the given equation, it reduces to

$$[\text{D}(\text{D} - 1)(\text{D} - 2) + 2\text{D}(\text{D} - 1) + 2]y = 10(e^z + e^{-z})$$

or $$(\text{D}^3 - \text{D}^2 + 2)y = 10(e^z + e^{-z})$$

which is a linear equation with constant co-efficients.

Its A.E. is $\text{D}^3 - \text{D}^2 + 2 = 0$ or $(\text{D} + 1)(\text{D}^2 - 2\text{D} + 2) = 0$

$$\therefore \quad \text{D} = -1, \frac{2 \pm \sqrt{4 - 8}}{2} = -1, 1 \pm i$$

$$\therefore \quad \text{C.F.} = c_1e^{-z} + e^z(c_2 \cos z + c_3 \sin z) = \frac{c_1}{x} + x[c_2 \cos(\log x) + c_3 \sin(\log x)]$$

$$\text{P.I.} = 10\frac{1}{\text{D}^3 - \text{D}^2 + 2}(e^z + e^{-z}) = 10\left(\frac{1}{\text{D}^3 - \text{D}^2 + 2}e^z + \frac{1}{\text{D}^3 - \text{D}^2 + 2}e^{-z}\right)$$

$$= 10\left(\frac{1}{1^3 - 1^2 + 2} e^z + z \cdot \frac{1}{3D^2 - 2D} e^{-z}\right) = 10\left(\frac{1}{2} e^z + z \cdot \frac{1}{3(-1)^2 - 2(-1)} e^{-z}\right)$$

$$= 5e^z + 2ze^{-z} = 5x + \frac{2}{x} \log x$$

Hence the C.S. is $y = \frac{c_1}{x} + x[c_2 \cos(\log x) + c_3 \sin(\log x)] + 5x + \frac{2}{x} \log x.$

Example 2. *Solve* $x^2 \frac{d^2y}{dx^2} - x\frac{dy}{dx} - 3y = x^2 \log x.$

Sol. Given equation is a Cauchy's homogeneous linear equation.

Put $x = e^z$ *i.e.,* $z = \log x$ so that $x\frac{dy}{dx} = Dy,\ x^2 \frac{d^2y}{dx^2} = D(D-1)y,$ where $D \equiv \frac{d}{dz}.$

Substituting these values in the given equation, it reduces to

$$[D(D-1) - D - 3]y = ze^{2z} \quad \text{or} \quad (D^2 - 2D - 3)y = ze^{2z}$$

which is a linear equation with constant co-efficients.

Its A.E. is $D^2 - 2D - 3 = 0$ or $(D-3)(D+1) = 0$

$$\therefore \quad D = 3, -1$$

$$\text{C.F.} = c_1e^{3z} + c_2e^{-z} = c_1x^3 + \frac{c_2}{x}$$

$$\text{P.I.} = \frac{1}{D^2 - 2D - 3}(e^{2z} \cdot z) = e^{2z}\frac{1}{(D+2)^2 - 2(D+2) - 3} z = e^{2z}\frac{1}{D^2 + 2D - 3} z$$

$$= e^{2z}\frac{1}{-3\left(1 - \frac{2D}{3} - \frac{D^2}{3}\right)} z = -\frac{1}{3}e^{2z}\left[1 - \left(\frac{2D}{3} + \frac{D^2}{3}\right)\right]^{-1} z$$

$$= -\frac{1}{3}e^{2z}\left[1 + \left(\frac{2D}{3} + \frac{D^2}{3}\right) + \cdots\right] z = -\frac{1}{3}e^{2z}\left(z + \frac{2}{3}\right) = -\frac{x^2}{3}\left(\log x + \frac{2}{3}\right)$$

Hence the C.S. is $y = c_1x^3 + \frac{c_2}{x} - \frac{x^2}{3}\left(\log x + \frac{2}{3}\right).$

Example 3. *Solve* $x^2 \frac{d^2y}{dx^2} + x\frac{dy}{dx} + y = \log x \sin(\log x).$

Sol. Given equation is a Cauchy's homogeneous linear equation.

Put $x = e^z$ *i.e.,* $z = \log x$ so that $x\frac{dy}{dx} = Dy,\ x^2 \frac{d^2y}{dx^2} = D(D-1)y,$ where $D \equiv \frac{d}{dz}.$

Substituting these values in the given equation, it reduces to

$$[D(D-1) + D + 1]y = z \sin z$$

or

$$(D^2 + 1)y = z \sin z$$

Its A.E. is $D^2 + 1 = 0$ so that $D = \pm i$

$$\text{C.F.} = c_1 \cos z + c_2 \sin z = c_1 \cos(\log x) + c_2 \sin(\log x)$$

$$\text{P.I.} = \frac{1}{D^2+1} z \sin z = \text{Imaginary part of } \frac{1}{D^2+1} z e^{iz}$$

$$= \text{I.P. of } e^{iz} \frac{1}{(D+i)^2+1} z = \text{I.P. of } e^{iz} \frac{1}{D^2+2iD} z$$

$$= \text{I.P. of } e^{iz} \frac{1}{2iD\left(1+\frac{D}{2i}\right)} z = \text{I.P. of } e^{iz} \frac{1}{2iD\left(1-\frac{iD}{2}\right)} z$$

$$= \text{I.P. of } \frac{1}{2i} e^{iz} \frac{1}{D}\left(1-\frac{iD}{2}\right)^{-1} z = \text{I.P. of } \frac{1}{2i} e^{iz} \frac{1}{D}\left(1+\frac{iD}{2}+\cdots\right) z$$

$$= \text{I.P. of } \frac{1}{2i} e^{iz} \frac{1}{D}\left(z+\frac{i}{2}\right) = \text{I.P. of } \frac{1}{2i} e^{iz} \int\left(z+\frac{i}{2}\right) dz$$

$$= \text{I.P of } -\frac{1}{2} e^{iz}\left(\frac{z^2}{2}+\frac{i}{2}z\right) = \text{I.P. of } e^{iz}\left(-\frac{i}{4}z^2+\frac{z}{4}\right)$$

$$= \text{I.P. of } (\cos z + i \sin z)\left(-\frac{i}{4}z^2+\frac{z}{4}\right) = -\frac{z^2}{4}\cos z + \frac{z}{4}\sin z$$

$$= -\frac{1}{4}(\log x)^2 \cos(\log x) + \tfrac{1}{4}\log x \sin(\log x)$$

Hence the C.S. is

$$y = c_1 \cos(\log x) + c_2 \sin(\log x) - \tfrac{1}{4}(\log x)^2 \cos(\log x) + \tfrac{1}{4}\log x \sin(\log x).$$

5.10. LEGENDRE'S LINEAR EQUATION

An equation of the form

$$(a+bx)^n \frac{d^n y}{dx^n} + a_1(a+bx)^{n-1}\frac{d^{n-1}y}{dx^{n-1}} + \cdots + a_{n-1}(a+bx)\frac{dy}{dx} + a_n y = X \quad \ldots(i)$$

where a_i's are constants and X is a function of x, is called Legendre's linear equation.

Such equations can be reduced to linear differential equations with constant co-efficients, by the substitution $a + bx = e^z$ *i.e.*, $z = \log(a+bx)$ so that $\frac{dy}{dx} = \frac{dy}{dz}\cdot\frac{dz}{dx} = \frac{b}{a+bx}\frac{dy}{dz}$

or $$(a+bx)\frac{dy}{dx} = b\frac{dy}{dz} = b\,Dy, \quad \text{where } D = \frac{d}{dz}$$

$$\frac{d^2y}{dx^2} = \frac{d}{dx}\left(\frac{b}{a+bx}\frac{dy}{dz}\right) = -\frac{b^2}{(a+bx)^2}\frac{dy}{dz} + \frac{b}{a+bx}\cdot\frac{d^2y}{dz^2}\cdot\frac{dz}{dx}$$

$$= -\frac{b^2}{(a+bx)^2}\frac{dy}{dz} + \frac{b}{a+bx}\frac{d^2y}{dz^2}\cdot\frac{b}{a+bx} = \frac{b^2}{(a+bx)^2}\left(\frac{d^2y}{dz^2}-\frac{dy}{dz}\right)$$

or $$(a+bx)^2\frac{d^2y}{dx^2} = b^2(D^2y - Dy) = b^2 D(D-1)y$$

Similarly, $(a + bx)^3 \dfrac{d^3y}{dx^3} = b^3 D(D - 1)(D - 2)y$.

Substituting these values in equation (*i*), we get a linear differential equation with constant co-efficients, which can be solved by the methods already discussed.

Example. *Solve* $(3x + 2)^2 \dfrac{d^2y}{dx^2} + 3(3x + 2) \dfrac{dy}{dx} - 36y = 3x^2 + 4x + 1.$

Sol. Given equation is a Legendre's linear equation.

Put $3x + 2 = e^z$ *i.e.*, $z = \log (3x + 2)$ so that $(3x + 2) \dfrac{dy}{dx} = 3Dy$.

$$(3x + 2)^2 \frac{d^2y}{dx^2} = 3^2 D(D - 1)y, \quad \text{where } D \equiv \frac{d}{dz}.$$

Substituting these values in the given equation, it reduces to

$$[3^2 D(D - 1) + 3.3D - 36]y = 3\left(\frac{e^z - 2}{3}\right)^2 + 4\left(\frac{e^z - 2}{3}\right) + 1$$

or $$9(D^2 - 4)y = \frac{1}{3}e^{2z} - \frac{1}{3} \quad \text{or} \quad (D^2 - 4)y = \frac{1}{27}(e^{2z} - 1)$$

which is a linear equation with constant co-efficients.

Its A.E. is $D^2 - 4 = 0 \quad \therefore \quad D = \pm 2$

$$\text{C.F.} = c_1e^{2z} + c_2e^{-2z} = c_1 (3x + 2)^2 + c_2 (3x + 2)^{-2}$$

$$\text{P.I.} = \frac{1}{27} \cdot \frac{1}{D^2 - 4}(e^{2z} - 1) = \frac{1}{27}\left[\frac{1}{D^2 - 4}e^{2z} - \frac{1}{D^2 - 4}e^{0z}\right]$$

$$= \frac{1}{27}\left[z \cdot \frac{1}{2D}e^{2z} - \frac{1}{0 - 4}e^{0z}\right] = \frac{1}{27}\left[\frac{z}{2}\int e^{2z}\, dz + \frac{1}{4}\right]$$

$$= \frac{1}{27}\left[\frac{z}{4}e^{2z} + \frac{1}{4}\right] = \frac{1}{108}(ze^{2z} + 1) = \frac{1}{108}[(3x + 2)^2 \log (3x + 2) + 1]$$

Hence the C.S. is $y = c_1(3x + 2)^2 + c_2(3x + 2)^{-2} + \dfrac{1}{108}[(3x + 2)^2 \log (3x + 2) + 1]$.

EXERCISE 5.4

Solve:

1. $x^2 \dfrac{d^2y}{dx^2} + 9x \dfrac{dy}{dx} + 25y = 50$

2. $x^2 \dfrac{d^2y}{dx^2} - 2y = x^2 + \dfrac{1}{x}$

3. $x^2 \dfrac{d^2y}{dx^2} + 2x \dfrac{dy}{dx} - 20y = (x + 1)^2$

4. $x^2 \dfrac{d^3y}{dx^3} - 4x \dfrac{d^2y}{dx^2} + 6\dfrac{dy}{dx} = 4$

[**Hint.** Multiply throughout by *x*]

5. (*i*) $x^4 \dfrac{d^3y}{dx^3} + 2x^3 \dfrac{d^2y}{dx^2} - x^2 \dfrac{dy}{dx} + xy = 1.$ (*ii*) $x^2\dfrac{d^2y}{dx^2} - 4x \dfrac{dy}{dx} + 6y = x^2.$

6. The radial displacement u in a rotating disc at a distance r from the axis is given by $r^2 \dfrac{d^2u}{dr^2} + r \dfrac{du}{dr} - u + kr^3 = 0$, where k is a constant. Solve the equation under the conditions $u = 0$ when $r = 0$, $u = 0$ when $r = a$.

7. $x^2 \dfrac{d^2y}{dx^2} - x \dfrac{dy}{dx} + y = \log x$

8. $x^2 \dfrac{d^2y}{dx^2} - x \dfrac{dy}{dx} + 2y = x \log x$

9. $x^2 \dfrac{d^2y}{dx^2} + 2x \dfrac{dy}{dx} - 12y = x^3 \log x$

10. $x^2 \dfrac{d^2y}{dx^2} - 2x \dfrac{dy}{dx} - 4y = x^2 + 2 \log x$

11. $x^2 \dfrac{d^2y}{dx^2} - 3x \dfrac{dy}{dx} + 5y = \sin (\log x)$

12. $x^3 \dfrac{d^3y}{dx^3} + 3x^2 \dfrac{d^2y}{dx^2} + x \dfrac{dy}{dx} + 8y = 65 \cos (\log x)$

13. $x^2 \dfrac{d^2y}{dx^2} - 3x \dfrac{dy}{dx} + 5y = x^2 \sin (\log x)$

14. $x^2 \dfrac{d^2y}{dx^2} - 3x \dfrac{dy}{dx} + y = \dfrac{\log x \sin (\log x) + 1}{x}$

15. $\dfrac{d^2y}{dx^2} + \dfrac{1}{x} \cdot \dfrac{dy}{dx} = \dfrac{12 \log x}{x^2}$

16. $(1 + x)^2 \dfrac{d^2y}{dx^2} + (1+x) \dfrac{dy}{dx} + y = 4 \cos \log (1 + x)$

17. $(1 + 2x)^2 \dfrac{d^2y}{dx^2} - 6(1 + 2x) \dfrac{dy}{dx} + 16y = 8(1 + 2x)^2$

18. $(2x + 3)^2 \dfrac{d^2y}{dx^2} - 2(2x + 3) \dfrac{dy}{dx} - 12y = 6x.$

Answers

1. $y = x^{-4} [c_1 \cos (3 \log x) + c_2 \sin (3 \log x)] + 2$

2. $y = c_1x^2 + \dfrac{c_2}{x} + \dfrac{1}{3}\left(x^2 - \dfrac{1}{x}\right) \log x$

3. $y = c_1x^4 + c_2x^{-5} - \dfrac{x^2}{14} - \dfrac{x}{9} - \dfrac{1}{20}$

4. $y = c_1 + c_2x^3 + c_3x^4 + \dfrac{2}{3} x$

5. (*i*) $y = (c_1 + c_2 \log x)x + c_3x^{-1} + \dfrac{1}{4x} \log x$ (*ii*) $y = c_1x^2 + c_2x^3 - x^2 \log x$

6. $u = \dfrac{kr}{8} (a^2 - r^2)$

7. $y = (c_1 + c_2 \log x)x + \log x + 2$

8. $y = x[c_1 \cos (\log x) + c_2 \sin (\log x)] + x \log x$

9. $y = c_1x^3 + c_2x^{-4} + \dfrac{x^3}{98} \log x (7 \log x - 2)$

10. $y = c_1x^{-1} + c_2x^4 - \dfrac{x^2}{6} - \dfrac{1}{2} \log x + \dfrac{3}{8}$

11. $y = x^2[c_1 \cos (\log x) + c_2 \sin (\log x)] + \dfrac{1}{8} [\sin (\log x) + \cos (\log x)]$

12. $y = c_1x^{-2} + x[c_2 \cos (\sqrt{3} \log x) + c_3 \sin (\sqrt{3} \log x)] + 8 \cos (\log x) - \sin (\log x)$

13. $y = x^2[c_1 \cos (\log x) + c_2 \sin (\log x)] - \dfrac{1}{2} x^2 \log x \cos (\log x)$

14. $y = x^2 [c_1 \cosh (\sqrt{3} \log x) + c_2 \sinh (\sqrt{3} \log x)]$

$$+ \frac{1}{61x}\left[\log x \{5 \sin (\log x) + 6 \cos (\log x)\} + \frac{2}{61} \{27 \sin (\log x) + 191 \cos (\log x)\}\right] + \frac{1}{6x}$$

15. $y = c_1 \log x + c_2 + 2 (\log x)^3$

16. $y = c_1 \cos [\log (1 + x)] + c_2 \sin [\log (1 + x)] + 2 \log (1 + x) + \sin [\log (1 + x)]$

17. $y = (1 + 2x)^2[c_1 + c_2 \log (1 + 2x) + \{\log (1 + 2x)\}^2]$

18. $y = c_1(2x + 3)^{-1} + c^2(2x + 3)^3 - \frac{3}{4} (2x + 3) + 3.$

5.11. SIMULTANEOUS LINEAR EQUATIONS WITH CONSTANT CO-EFFICIENTS

Now we discuss differential equations in which there is one independent variable and two or more than two dependent variables. Such equations are called *simultaneous linear equations*. To solve such equations completely, we must have as many simultaneous equations as the number of dependent variables. Here, we shall consider simultaneous linear equations with constant co-efficients only.

Let x, y be the two dependent variables and t the independent variable. Consider the simultaneous equations

$$f_1(D)x + f_2(D)y = T_1 \quad ...(1) \quad \text{and} \quad \phi_1(D)x + \phi_2(D)y = T_2 \quad ...(2)$$

where $D \equiv \frac{d}{dt}$ and T_1, T_2 are functions of t.

To eliminate y, operating on both sides of (1) by $\phi_2(D)$ and on both sides of (2) by $f_2(D)$ and subtracting, we get

$$[f_1(D)\, \phi_2(D) - \phi_1(D)\, f_2(D)]\, x = \phi_2(D)T_1 - f_2(D)T_2 \quad \text{or} \quad f(D)\, x = T$$

which is a linear equation in x and t and can be solved by the methods already discussed. Substituting the value of x in either (1) or (2), we get the value of y.

Note. We can also eliminate x to get a linear equation in y and t.

ILLUSTRATIVE EXAMPLES

Example 1. *Solve* $\frac{dx}{dt} + 4x + 3y = t$

$$\frac{dy}{dt} + 2x + 5y = e^t.$$

Sol. Writing D for $\frac{d}{dt}$, the given equations become $(D + 4)x + 3y = t$ ...(1)

and $2x + (D + 5)y = e^t$ ...(2)

To eliminate y, operating on both sides of (1) by (D + 5) and on both sides of (2) by 3 and subtracting, we get

$$[(D + 4)(D + 5) - 6]x = (D + 5)t - 3e^t$$

or $$(D^2 + 9D + 14)x = 1 + 5t - 3e^t$$

Its A.E. is $$D^2 + 9D + 14 = 0$$

or $$(D + 2)(D + 7) = 0 \quad \therefore \quad D = -2, -7$$

$$\text{C.F.} = c_1e^{-2t} + c_2e^{-7t}$$

$$\text{P.I.} = \frac{1}{D^2 + 9D + 14}(1 + 5t - 3e^t)$$

$$= \frac{1}{D^2 + 9D + 14} e^{0t} + 5\frac{1}{D^2 + 9D + 14} t - 3\frac{1}{D^2 + 9D + 14} e^t$$

$$= \frac{1}{0^2 + 9(0) + 14} e^{0t} + 5 \cdot \frac{1}{14\left(1 + \frac{9D}{14} + \frac{D^2}{14}\right)} t - 3 \frac{1}{1^2 + 9(1) + 14} e^t$$

$$= \frac{1}{14} + \frac{5}{14}\left[1 + \left(\frac{9D}{14} + \frac{D^2}{14}\right)\right]^{-1} t - \frac{1}{8} e^t = \frac{1}{14} + \frac{5}{14}\left[1 - \left(\frac{9D}{14} + \frac{D^2}{14}\right) + \cdots\right] t - \frac{1}{8} e^t$$

$$= \frac{1}{14} + \frac{5}{14}\left(t - \frac{9}{14}\right) - \frac{1}{8} e^t = \frac{1}{14} + \frac{5}{14} t - \frac{45}{196} - \frac{1}{8} e^t = \frac{5}{14} t - \frac{31}{196} - \frac{1}{8} e^t$$

$\therefore \quad x = c_1 e^{-2t} + c_2 e^{-7t} + \frac{5}{14} t - \frac{31}{196} - \frac{1}{8} e^t$

Now $\frac{dx}{dt} = -2c_1 e^{-2t} - 7c_2 e^{-7t} + \frac{5}{14} - \frac{1}{8} e^t$

Substituting the values of x and $\frac{dx}{dt}$ in (1), we have

$$3y = t - \frac{dx}{dt} - 4x$$

$$= t + 2c_1 e^{-2t} + 7c_2 e^{-7t} - \frac{5}{14} + \frac{1}{8} e^t - 4c_1 e^{-2t} - 4c_2 e^{-7t} - \frac{10}{7} t + \frac{31}{49} + \frac{1}{2} e^t$$

$\therefore \quad y = \frac{1}{3}\left[-2c_1 e^{-2t} + 3c_2 e^{-7t} - \frac{3}{7} t + \frac{27}{98} + \frac{5}{8} e^t\right]$

Hence $\quad x = c_1 e^{-2t} + c_2 e^{-7t} + \frac{5}{14} t - \frac{31}{196} - \frac{1}{8} e^t$

$$y = -\frac{2}{3} c_1 e^{-2t} + c_2 e^{-7t} - \frac{1}{7} t + \frac{9}{98} + \frac{5}{24} e^t.$$

Example 2. *Solve* $\frac{d^2x}{dt^2} + 4x + 5y = t^2$

$$\frac{d^2y}{dt^2} + 5x + 4y = t + 1.$$

Sol. Writing D for $\frac{d}{dt}$, the given equations become $(D^2 + 4)x + 5y = t^2$ ...(1)

and $5x + (D^2 + 4)y = t + 1$ ...(2)

To eliminate y, operating on both sides of (1) by $(D^2 + 4)$ and on both sides of (2) by 5 and subtracting, we get

$$[(D^2 + 4)^2 - 25]x = (D^2 + 4)t^2 - 5(t + 1)$$

or $(D^4 + 8D^2 - 9)x = 2 + 4t^2 - 5t - 5$

or $(D^4 + 8D^2 - 9)x = 4t^2 - 5t - 3$

Its A.E. is $D^4 + 8D^2 - 9 = 0$

or $(D^2 + 9)(D^2 - 1) = 0 \quad \therefore \quad D = \pm 1, \pm 3i$

$$\text{C.F.} = c_1 e^t + c_2 e^{-t} + c_3 \cos 3t + c_4 \sin 3t$$

$$\text{P.I.} = \frac{1}{D^4 + 8D^2 - 9}(4t^2 - 5t - 3) = \frac{1}{-9\left(1 - \frac{8D^2}{9} - \frac{D^4}{9}\right)}(4t^2 - 5t - 3)$$

$$= -\frac{1}{9}\left[1 - \left(\frac{8D^2}{9} + \frac{D^4}{9}\right)\right]^{-1}(4t^2 - 5t - 3)$$

$$= -\frac{1}{9}\left[1 + \left(\frac{8D^2}{9} + \frac{D^4}{9}\right) + \cdots\right](4t^2 - 5t - 3)$$

$$= -\frac{1}{9}\left[4t^2 - 5t - 3 + \frac{8}{9}(8)\right] = -\frac{1}{9}\left(4t^2 - 5t + \frac{37}{9}\right)$$

$$\therefore \quad x = c_1e^t + c_2e^{-t} + c_3\cos 3t + c_4\sin 3t - \frac{4}{9}t^2 + \frac{5}{9}t - \frac{37}{81}$$

Now $$\frac{dx}{dt} = c_1e^t - c_2e^{-t} - 3c_3\sin 3t + 3c_4\cos 3t - \frac{8}{9}t + \frac{5}{9}$$

$$\frac{d^2x}{dt^2} = c_1e^t + c_2e^{-t} - 9c_3\cos 3t - 9c_4\sin 3t - \frac{8}{9}$$

Substituting the values of x and $\frac{d^2x}{dt^2}$ in (1), we have

$$5y = t^2 - 4x - \frac{d^2x}{dt^2}$$

$$= t^2 - 4c_1e^t - 4c_2e^{-t} - 4c_3\cos 3t - 4c_4\sin 3t$$

$$+ \frac{16}{9}t^2 - \frac{20}{9}t + \frac{148}{81} - c_1e^t - c_2e^{-t} + 9c_3\cos 3t + 9c_4\sin 3t + \frac{8}{9}$$

$$\therefore \quad y = \frac{1}{5}\left[-5c_1e^t - 5c_2e^{-t} + 5c_3\cos 3t + 5c_4\sin 3t + \frac{25}{9}t^2 - \frac{20}{9}t + \frac{220}{81}\right]$$

Hence $$x = c_1e^t + c_2e^{-t} + c_3\cos 3t + c_4\sin 3t - \frac{1}{9}\left(4t^2 - 5t + \frac{37}{9}\right)$$

$$y = -c_1e^t - c_2e^{-t} + c_3\cos 3t + c_4\sin 3t + \frac{1}{9}\left(5t^2 - 4t + \frac{44}{9}\right).$$

Example 3. *Solve the simultaneous equations:*

$$t\frac{dx}{dt} + y = 0,\ t\frac{dy}{dt} + x = 0 \text{ given } x(1) = 1,\ y(-1) = 0.$$

Sol. The given equations are

$$t\frac{dx}{dt} + y = 0 \qquad \text{...(1)}$$

$$t\frac{dy}{dt} + x = 0 \qquad \text{...(2)}$$

Differentiating (1) w.r.t. t, we have

$$t\frac{d^2x}{dt^2} + \frac{dx}{dt} + \frac{dy}{dt} = 0$$

Multiplying throughout by t

$$t^2 \frac{d^2x}{dt^2} + t\frac{dx}{dt} + t\frac{dy}{dt} = 0$$

or $$t^2 \frac{d^2x}{dt^2} + t\frac{dx}{dt} - x = 0 \quad \text{...(3) [Using (2)]}$$

which is Cauchy's homogeneous linear equation.

Putting $t = e^u$ *i.e.*, $u = \log t$, so that $t\frac{d}{dt} = \frac{d}{du} = \text{D}$, equation (3) becomes

$$[\text{D}(\text{D} - 1) + \text{D} - 1]x = 0 \quad \text{or} \quad (\text{D}^2 - 1)\, x = 0$$

Its A.E. is $\text{D}^2 - 1 = 0$, whence $\text{D} = \pm 1$

$$\therefore \quad x = c_1 e^u + c_2 e^{-u} = c_1 t + \frac{c_2}{t} \quad \text{...(4)}$$

From (1), $$y = -t\frac{dx}{dt} = -t\left(c_1 - \frac{c_2}{t^2}\right) = -c_1 t + \frac{c_2}{t} \quad \text{...(5)}$$

Since $x(1) = 1$, $\therefore$ from (4), we have $1 = c_1 + c_2$

Also $y(-1) = 0$, $\therefore$ from (5), we have $0 = c_1 - c_2$

Solving $$c_1 = c_2 = \frac{1}{2}$$

Hence $$x = \frac{1}{2}\left(t + \frac{1}{t}\right), y = \frac{1}{2}\left(-t + \frac{1}{t}\right).$$

Example 4. *Solve the following simultaneous equations:*

$$\frac{dx}{dt} = 2y, \quad \frac{dy}{dt} = 2z, \quad \frac{dz}{dt} = 2x.$$

Sol. The given equations are

$$\frac{dx}{dt} = 2y \quad \text{...(1)} \qquad \frac{dy}{dt} = 2z \quad \text{...(2)} \qquad \frac{dz}{dt} = 2x \quad \text{...(3)}$$

Differentiating (1) w.r.t. t, $$\frac{d^2x}{dt^2} = 2\frac{dy}{dt} = 2(2z) \quad \text{[Using (2)]}$$

Differentiating again w.r.t. t, $$\frac{d^3x}{dt^3} = 4\frac{dz}{dt} = 4(2x)$$

or $$(\text{D}^3 - 8)x = 0, \qquad \text{where } \text{D} \equiv \frac{d}{dt}$$

Its A.E. is $$\text{D}^3 - 8 = 0 \quad \text{or} \quad (\text{D} - 2)(\text{D}^2 + 2\text{D} + 4) = 0$$

whence $\text{D} = 2, \dfrac{-2 \pm 2i\sqrt{3}}{2}$ or $\text{D} = 2, -1 \pm i\sqrt{3}$

$$\therefore \quad x = c_1 e^{2t} + c_2 e^{-t} \cos(\sqrt{3}\, t - c_3)$$

From (1), $$y = \frac{1}{2}\frac{dx}{dt}$$

$$= \frac{1}{2}[2c_1 e^{2t} - c_2 e^{-t} \cos(\sqrt{3}t - c_3) - c_2\sqrt{3}e^{-t} \sin(\sqrt{3}t - c_3)]$$

$$= c_1 e^{2t} + c_2 e^{-t}\left[\cos\frac{2\pi}{3}\cos(\sqrt{3}t - c_3) - \sin\frac{2\pi}{3}\sin(\sqrt{3}t - c_3)\right]$$

$$= c_1 e^{2t} + c_2 e^{-t}\cos\left(\sqrt{3}t - c_3 + \frac{2\pi}{3}\right)$$

From (2), $z = \dfrac{1}{2}\dfrac{dy}{dt}$

$$= \frac{1}{2}\left[2c_1 e^{2t} - c_2 e^{-t}\cos\left(\sqrt{3}t - c_3 + \frac{2\pi}{3}\right) - c_2\sqrt{3}\, e^{-t}\sin\left(\sqrt{3}t - c_3 + \frac{2\pi}{3}\right)\right]$$

$$= c_1 e^{2t} + c_2 e^{-t}\left[\cos\frac{2\pi}{3}\cos\left(\sqrt{3}t - c_3 + \frac{2\pi}{3}\right) - \sin\frac{2\pi}{3}\sin\left(\sqrt{3}t - c_3 + \frac{2\pi}{3}\right)\right]$$

$$= c_1 e^{2t} + c_2 e^{-t}\cos\left(\sqrt{3}t - c_3 + \frac{4\pi}{3}\right).$$

Note. $c_1 \cos \beta x + c_2 \sin \beta x$ can be replaced by $c_1 \cos(\beta x - c_2)$

or $\quad c_1 \cos(\beta x + c_2)$ or $c_1 \sin(\beta x - c_2)$ or $c_1 \sin(\beta x + c_2)$.

EXERCISE 5.5

Solve the following simultaneous equations:

1. $\dfrac{dx}{dt} = 7x - y, \dfrac{dy}{dt} = 2x + 5y.$
2. $\dfrac{dx}{dt} + y = \sin t, \dfrac{dy}{dt} + x = \cot t$; given that $x = 2$ and $y = 0$ when $t = 0$.
3. $\dfrac{dx}{dt} + 5x - 2y = t, \dfrac{dy}{dt} + 2x + y = 0$; given that $x = y = 0$ when $t = 0$. *(K.U. April, 2013)*
4. $(D + 1)x + (2D + 1)y = e^t, (D - 1)x + (D + 1)y = 1.$
5. $\dfrac{dx}{dt} + 2x + 3y = 0, 3x + \dfrac{dy}{dt} + 2y = 2e^{2t}.$
6. $(D - 1)x + Dy = 2t + 1, (2D + 1)x + 2Dy = t.$
7. $\dfrac{d^2x}{dt^2} - 3x - 4y = 0, \dfrac{d^2y}{dt^2} + x + y = 0.$
8. $\dfrac{d^2x}{dt^2} - \dfrac{dy}{dt} = 2x + 2t, \dfrac{dx}{dt} + 4\dfrac{dy}{dt} = 3y.$
9. $\dfrac{d^2y}{dt^2} + \dfrac{dy}{dt} - 2y = \sin t, \dfrac{dx}{dt} + x - 3y = 0.$
10. A mechanical system with two degrees of freedom satisfies the equations

$$2\frac{d^2x}{dt^2} + 3\frac{dy}{dt} = 4, 2\frac{d^2y}{dt^2} - 3\frac{dx}{dt} = 0$$

Obtain expressions for x and y in terms of t, given $x, y, \dfrac{dx}{dt}, \dfrac{dy}{dt}$ all vanish at $t = 0$.

11. $\dfrac{d^2x}{dt^2} + y = \sin t, \dfrac{d^2y}{dt^2} + x = \cos t$

Answers

1. $x = e^{6t}(c_1 \cos t + c_2 \sin t),\ y = e^{6t}[(c_1 - c_2)\cos t + (c_1 + c_2)\sin t]$
2. $x = e^t + e^{-t},\ y = e^{-t} - e^t + \sin t$
3. $x = -\frac{1}{27}(1 + 6t)\,e^{-3t} + \frac{1}{27}(1 + 3t),\ y = -\frac{2}{27}(2 + 3t)\,e^{-3t} + \frac{2}{27}(2 - 3t)$
4. $x = \frac{1}{2}[e^t - 1 - ac_1e^{at} + bc_2e^{bt}],\ y = c_1e^{at} + c_2e^{bt} + \frac{1}{2}$, where $a = \frac{1}{2}(3 + \sqrt{17}),\ b = \frac{1}{2}(3 - \sqrt{17})$
5. $x = c_1e^t + c_2e^{-5t} - \frac{6}{7}e^{2t},\ y = c_2e^{-5t} - c_1e^t + \frac{8}{7}e^{2t}$
6. $x = -t - \frac{2}{3},\ y = \frac{1}{2}t^2 + \frac{4}{3}t + c$
7. $x = (c_1 + c_2t)e^{-t} + (c_3 + c_4t)\,e^t,\ y = -\frac{1}{2}[c_1 + c_2(1 + t)]e^{-t} + \frac{1}{2}[c_4(1 - t) - c_3]e^t$
8. $x = (c_1 + c_2t)e^t + c_3e^{-\frac{3}{2}t} - t,\ y = [c_2(3 - t) - c_1]e^t - \frac{1}{6}c_3e^{-\frac{3}{2}t}$
9. $x = \frac{3}{2}c_1e^t - 3c_2e^{-2t} + c_3e^{-t} + \frac{3}{10}e^t(\cos t - 2\sin t),\ y = c_1e^t + c_2e^{-2t} - \frac{1}{10}(\cos t + 3\sin t)$
10. $x = \frac{8}{9}\left(1 - \cos\frac{3t}{2}\right),\ y = \frac{4}{3}t - \frac{8}{9}\sin\frac{3t}{2}.$
11. $x = c_1e^t + c_2e^{-t} + c_3\cos t + c_4\sin t + \frac{t}{4}(\sin t - \cos t)$

    $y = -c_1e^t - c_2e^{-t} + c_3\cos t + c_4\sin t + \frac{1}{4}(2 + t)(\sin t - \cos t).$

6
Matrices

6.1. ELEMENTARY TRANSFORMATIONS (OR OPERATIONS)

Any one of the following operations on a matrix is called an elementary transformation (or E-operation).

(*i*) **Interchange of two rows or two columns.**

The interchange of ith and jth rows is denoted by R_{ij}

The interchange of ith and jth columns is denoted by C_{ij}.

(*ii*) **Multiplication of (each element of) a row or column by a non-zero number k.**

The multiplication of ith row by k is denoted by kR_i.

The multiplication of ith column by k is denoted by kC_i.

(*iii*) **Addition of k times the elements of a row (or column) to the corresponding elements of another row (or column), k ≠ 0.**

The addition of k times the jth row to the ith row is denoted by $R_i + kR_j$.

The addition of k times the jth column to the ith column is denoted by $C_i + kC_j$.

If a matrix B is obtained from a matrix A by one or more E-operations, then B is said to be **equivalent** to A.

Two equivalent matrices A and B are written as A ~ B.

6.2. ELEMENTARY MATRICES

The matrix obtained from a unit matrix I by subjecting it to one of the E-operations is called an elementary matrix.

For example, let $I = \begin{bmatrix} 1 & 0 & 0 \\ 0 & 1 & 0 \\ 0 & 0 & 1 \end{bmatrix}$

(*i*) Operating R_{23} or C_{23} on I, we get the same elementary matrix $\begin{bmatrix} 1 & 0 & 0 \\ 0 & 0 & 1 \\ 0 & 1 & 0 \end{bmatrix}$.

It is denoted by E_{23}. Thus, the E-matrix obtained by either of the operations R_{ij} or C_{ij} on I is denoted by E_{ij}.

(*ii*) Operating $5R_2$ or $5C_2$ on I, we get the same elementary matrix $\begin{bmatrix} 1 & 0 & 0 \\ 0 & 5 & 0 \\ 0 & 0 & 1 \end{bmatrix}$.

It is denoted by $5E_2$. Thus, the E-matrix obtained by either of the operations kR_i or kC_i is denoted by kE_i.

(*iii*) Operating $R_2 + 4R_3$ on I, we get the elementary matrix $\begin{bmatrix} 1 & 0 & 0 \\ 0 & 1 & 4 \\ 0 & 0 & 1 \end{bmatrix}$.

It is denoted by $E_{23}(4)$. Thus, the E-matrix obtained by the operation $R_i + kR_j$ is denoted by $E_{ij}(k)$.

(*iv*) Operating $C_2 + 4C_3$ on I, we get the elementary matrix $\begin{bmatrix} 1 & 0 & 0 \\ 0 & 1 & 0 \\ 0 & 4 & 1 \end{bmatrix}$, which is the transpose of $\begin{bmatrix} 1 & 0 & 0 \\ 0 & 1 & 4 \\ 0 & 0 & 1 \end{bmatrix} = E_{23}(4)$ and is, therefore, denoted by $E'_{23}(4)$. Thus, the E-matrix obtained by the operation $C_i + kC_j$ is denoted by $E'_{ij}(k)$.

6.3. THE FOLLOWING THEOREMS ON THE EFFECT OF E-OPERATIONS ON MATRICES HOLD GOOD

(*a*) *Any E-row operation on the product of two matrices is equivalent to the same E-row operation on the pre-factor.*

If the E-row operation is denoted by R, then $R(AB) = R(A) \cdot B$.

(*b*) *Any E-column operation on the product of two matrices is equivalent to the same E-column operation on the post-factor.*

If the E-column operation is denoted by C, then $C(AB) = A \cdot C(B)$.

(*c*) *Every E-row operation on a matrix is equivalent to pre-multiplication by the corresponding E-matrix.*

Thus, the effect of E-row operation R_{ij} on $A = E_{ij} \cdot A$

The effect of E-row operation kR_i on $A = kE_i \cdot A$

The effect of E-row operation $R_i + kR_j$ on $A = E_{ij}(k) \cdot A$.

(*d*) *Every E-column operation on a matrix is equivalent to post-multiplication by the corresponding E-matrix.*

Thus, the effect of E-column operation C_{ij} on $A = A \cdot E_{ij}$

The effect of *E*-column operation kC_i on $A = A \cdot (kE_i)$

The effect of *E*-column operation $C_i + kC_j$ on $A = A \cdot E'_{ij}(k)$.

6.4. INVERSE OF A MATRIX BY E-OPERATIONS (Gauss-Jordan Method)

The elementary row transformations which reduce a square matrix A to the unit matrix, when applied to the unit matrix, give the inverse matrix A^{-1}.

Let A be a non-singular square matrix. Then A = IA

Apply suitable E-row operations to A on the left hand side so that A is reduced to I.

Simultaneously, apply the same E-row operations to the pre-factor I on right hand side. Let I reduce to B, so that

$$I = BA$$

Post-multiplying by A^{-1}, we get

$$IA^{-1} = BAA^{-1} \Rightarrow A^{-1} = B(AA^{-1}) = BI = B$$

$$\therefore \quad B = A^{-1}.$$

Note. In practice, to find the inverse of A by E-row operations, we write A and I side by side and the same operations are performed on both. As soon as A is reduced to I, I will reduce to A^{-1}.

Example 1. *Find the inverse of* $A = \begin{bmatrix} 0 & 1 & 2 \\ 1 & 2 & 3 \\ 3 & 1 & 1 \end{bmatrix}$ *by elementary row operations.*

Sol. Writing the given matrix side by side with unit matrix I_3, we get

$$[A : I_3] = \left[\begin{array}{ccc:ccc} 0 & 1 & 2 & 1 & 0 & 0 \\ 1 & 2 & 3 & 0 & 1 & 0 \\ 3 & 1 & 1 & 0 & 0 & 1 \end{array}\right] \quad \text{operating } R_{12}$$

$$\sim \left[\begin{array}{ccc:ccc} 1 & 2 & 3 & 0 & 1 & 0 \\ 0 & 1 & 2 & 1 & 0 & 0 \\ 3 & 1 & 1 & 0 & 0 & 1 \end{array}\right] \quad \text{operating } R_3 - 3R_1$$

$$\sim \left[\begin{array}{ccc:ccc} 1 & 2 & 3 & 0 & 1 & 0 \\ 0 & 1 & 2 & 1 & 0 & 0 \\ 0 & -5 & -8 & 0 & -3 & 1 \end{array}\right] \quad \text{operating } R_1 - 2R_2,\ R_3 + 5R_2$$

$$\sim \left[\begin{array}{ccc:ccc} 1 & 0 & -1 & -2 & 1 & 0 \\ 0 & 1 & 2 & 1 & 0 & 0 \\ 0 & 0 & 2 & 5 & -3 & 1 \end{array}\right] \quad \text{operating } \tfrac{1}{2}R_3$$

$$\sim \left[\begin{array}{ccc:ccc} 1 & 0 & -1 & -2 & 1 & 0 \\ 0 & 1 & 2 & 1 & 0 & 0 \\ 0 & 0 & 1 & \frac{5}{2} & -\frac{3}{2} & \frac{1}{2} \end{array}\right] \quad \text{operating } R_1 + R_3,\ R_2 - 2R_3$$

$$\sim \left[\begin{array}{ccc:ccc} 1 & 0 & 0 & \frac{1}{2} & -\frac{1}{2} & \frac{1}{2} \\ 0 & 1 & 0 & -4 & 3 & -1 \\ 0 & 0 & 1 & \frac{5}{2} & -\frac{3}{2} & \frac{1}{2} \end{array}\right] = [I_3 : A^{-1}]$$

$$\therefore \qquad A^{-1} = \begin{bmatrix} \frac{1}{2} & -\frac{1}{2} & \frac{1}{2} \\ -4 & 3 & -1 \\ \frac{5}{2} & -\frac{3}{2} & \frac{1}{2} \end{bmatrix}.$$

6.5. RANK OF A MATRIX

Let A be any $m \times n$ matrix. It has square sub-matrices of different orders. The determinants of these square sub-matrices are called minors of A. If **all** minors of order $(r + 1)$ are zero but there **is at least one** non-zero minor of order r, then r is called the rank of A. Symbolically, rank of A is written as $\rho(A)$.

From the definition of the rank of a matrix A, it follows that:

(*i*) If *A* is a null matrix, then $\rho(A) = 0$.

[$\because$ every minor of A has zero value.]

(*ii*) If A is not a null matrix, then $\rho(A) \geq 1$.

(*iii*) If A is a non-singular $n \times n$ matrix, then $\rho(A) = n$

[$\because$ $|A| \neq 0$ is the largest minor of A.]

If I_n is the $n \times n$ unit matrix, then $|I_n| = 1 \neq 0 \Rightarrow \rho(I_n) = n$.

(*iv*) If A is an $m \times n$ matrix, then $\rho(A) \leq$ minimum of m and n.

(*v*) If all minors of order r are equal to zero, then $\rho(A) < r$.

To determine the rank of a matrix A, we adopt the following different methods:

(*i*) Start with the highest order minor (or minors) of A. Let their order be r. **If any one of them** is non-zero, then $\rho(A) = r$.

If **all of them** are zero, start with minors of next lower order $(r - 1)$ and so on till you get a non-zero minor. The order of that minor is the rank of A.

This method **usually** involves a lot of computational work since we have to evaluate several determinants.

(*ii*) If A is an $m \times n$ matrix and by a series of elementary (*row or column or both*) operations, it can be put into one of the following forms (called normal forms):

$$\begin{bmatrix} I_r & \vdots & O \\ \multicolumn{3}{c}{\cdots\cdots\cdots} \\ O & \vdots & O \end{bmatrix}, \begin{bmatrix} I_r \\ \\ O \end{bmatrix}, [I_r \vdots O], [I_r], \text{ where } I_r \text{ is the unit matrix of order } r.$$

Since the rank of a matrix is not changed as a result of elementary transformations, it follows that

$$\rho(A) = r. \qquad [\because \ r\text{th order minor } |I_r| = 1 \neq 0]$$

Note. *For an $m \times n$ matrix A of rank r, to find square matrices P and Q of orders m and n respectively, such that PAQ is in the normal form* $\begin{bmatrix} I_r & O \\ O & O \end{bmatrix}$.

Method. Write A = IAI.

Reduce the matrix on LHS to normal form by affecting elementary row and/or column transformations.

Every elementary row transformation on A must be accompanied by the same transformation on the pre-factor on RHS.

Every elementary column transformation on A must be accompanied by the same transformation on the post-factor on RHS.

Example 2. *Find the rank of the matrix*

(*i*) $\begin{bmatrix} 1 & 2 & 3 & 4 \\ -2 & 0 & 5 & 7 \end{bmatrix}$ (*ii*) $\begin{bmatrix} 1 & 2 & 3 \\ 1 & 4 & 2 \\ 2 & 6 & 5 \end{bmatrix}$

(*iii*) $\begin{bmatrix} 2 & 3 & 4 & -1 \\ 5 & 2 & 0 & -1 \\ -4 & 5 & 12 & -1 \end{bmatrix}$ (*iv*) $\begin{bmatrix} 1 & 2 & 3 & 0 \\ 2 & 4 & 3 & 2 \\ 3 & 2 & 1 & 3 \\ 6 & 8 & 7 & 5 \end{bmatrix}$.

Sol. (*i*) Here $A = \begin{bmatrix} 1 & 2 & 3 & 4 \\ -2 & 0 & 5 & 7 \end{bmatrix}$ is a 2×4 matrix.

$\therefore \quad \rho(A) \leq 2$, the smaller of 2 and 4.

The second order minor $\begin{vmatrix} 1 & 2 \\ -2 & 0 \end{vmatrix} = 4 \neq 0 \qquad \therefore \quad \rho(A) = 2.$

(*ii*) Here $A = \begin{bmatrix} 1 & 2 & 3 \\ 1 & 4 & 2 \\ 2 & 6 & 5 \end{bmatrix}$ is a 3 × 3 matrix. $\therefore \quad \rho(A) \le 3$

Operating $R_2 - R_1, R_3 - 2R_1$

$$A \sim \begin{bmatrix} 1 & 2 & 3 \\ 0 & 2 & -1 \\ 0 & 2 & -1 \end{bmatrix} \quad \text{operating } R_3 - R_2$$

$$\sim \begin{bmatrix} 1 & 2 & 3 \\ 0 & 2 & -1 \\ 0 & 0 & 0 \end{bmatrix}$$

The only third order minor is zero, but the second order minor

$$\begin{vmatrix} 1 & 2 \\ 0 & 2 \end{vmatrix} = 2 \ne 0 \quad \therefore \quad \rho(A) = 2.$$

(*iii*) Here $A = \begin{bmatrix} 2 & 3 & 4 & -1 \\ 5 & 2 & 0 & -1 \\ -4 & 5 & 12 & -1 \end{bmatrix}$ is a 3 × 4 matrix. $\therefore \quad \rho(A) \le 3$

Operating C_{14}

$$A \sim \begin{bmatrix} -1 & 3 & 4 & 2 \\ -1 & 2 & 0 & 5 \\ -1 & 5 & 12 & -4 \end{bmatrix} \quad \text{operating } R_2 - R_1, R_3 - R_1$$

$$\sim \begin{bmatrix} -1 & 3 & 4 & 2 \\ 0 & -1 & -4 & 3 \\ 0 & 2 & 8 & -6 \end{bmatrix} \quad \text{operating } R_1 + 3R_2, R_3 + 2R_2$$

$$\sim \begin{bmatrix} -1 & 0 & -8 & 11 \\ 0 & -1 & -4 & 3 \\ 0 & 0 & 0 & 0 \end{bmatrix}$$

All the third order minors are zero but the second order minor

$$\begin{vmatrix} -1 & 0 \\ 0 & -1 \end{vmatrix} = 1 \ne 0 \quad \therefore \quad \rho(A) = 2.$$

(*iv*) Here $A = \begin{bmatrix} 1 & 2 & 3 & 0 \\ 2 & 4 & 3 & 2 \\ 3 & 2 & 1 & 3 \\ 6 & 8 & 7 & 5 \end{bmatrix}$ is a 4 × 4 matrix $\therefore \quad \rho(A) \le 4$

Operating $R_4 - (R_3 + R_2 + R_1)$

$$A \sim \begin{bmatrix} 1 & 2 & 3 & 0 \\ 2 & 4 & 3 & 2 \\ 3 & 2 & 1 & 3 \\ 0 & 0 & 0 & 0 \end{bmatrix} \quad \text{operating } R_2 - 2R_1, R_3 - 3R_1$$

$$\sim \begin{bmatrix} 1 & 2 & 3 & 0 \\ 0 & 0 & -3 & 2 \\ 0 & -4 & -8 & 3 \\ 0 & 0 & 0 & 0 \end{bmatrix} \quad \text{operating } R_{23}$$

$$\sim \begin{bmatrix} 1 & 2 & 3 & 0 \\ 0 & -4 & -8 & 3 \\ 0 & 0 & -3 & 2 \\ 0 & 0 & 0 & 0 \end{bmatrix}$$

The only fourth order minor is zero. Since the third order minor

$$\begin{vmatrix} 1 & 2 & 3 \\ 0 & -4 & -8 \\ 0 & 0 & -3 \end{vmatrix} = (1)(-4)(-3) = 12 \neq 0 \quad \therefore \quad \rho(A) = 3.$$

Example 3. *Reduce each of the following matrices to normal form and hence, find their ranks:*

$$(i) \begin{bmatrix} 2 & 3 & -1 & -1 \\ 1 & -1 & -2 & -4 \\ 3 & 1 & 3 & -2 \\ 6 & 3 & 0 & -7 \end{bmatrix} \text{ (K.U. May 2011)} \quad (ii) \begin{bmatrix} 8 & 1 & 3 & 6 \\ 0 & 3 & 2 & 2 \\ -8 & -1 & -3 & 4 \end{bmatrix}.$$

Sol. (*i*) Let

$$A = \begin{bmatrix} 2 & 3 & -1 & -1 \\ 1 & -1 & -2 & -4 \\ 3 & 1 & 3 & -2 \\ 6 & 3 & 0 & -7 \end{bmatrix} \quad \text{operating } R_{12}$$

$$A \sim \begin{bmatrix} 1 & -1 & -2 & -4 \\ 2 & 3 & -1 & -1 \\ 3 & 1 & 3 & -2 \\ 6 & 3 & 0 & -7 \end{bmatrix} \quad \text{operating } C_2 + C_1, C_3 + 2C_1, C_4 + 4C_1$$

$$\sim \begin{bmatrix} 1 & 0 & 0 & 0 \\ 2 & 5 & 3 & 7 \\ 3 & 4 & 9 & 10 \\ 6 & 9 & 12 & 17 \end{bmatrix} \quad \text{operating } R_2 - 2R_1, R_3 - 3R_1, R_4 - 6R_1$$

$$\sim \begin{bmatrix} 1 & 0 & 0 & 0 \\ 0 & 5 & 3 & 7 \\ 0 & 4 & 9 & 10 \\ 0 & 9 & 12 & 17 \end{bmatrix} \quad \text{operating } R_2 - R_3, R_4 - 2R_3$$

$$\sim \begin{bmatrix} 1 & 0 & 0 & 0 \\ 0 & 1 & -6 & -3 \\ 0 & 4 & 9 & 10 \\ 0 & 1 & -6 & -3 \end{bmatrix} \quad \text{operating } C_3 + 6C_2, C_4 + 3C_2$$

$$\sim \begin{bmatrix} 1 & 0 & 0 & 0 \\ 0 & 1 & 0 & 0 \\ 0 & 4 & 33 & 22 \\ 0 & 1 & 0 & 0 \end{bmatrix} \quad \text{operating } R_3 - 4R_2, R_4 - R_2$$

$$\sim \begin{bmatrix} 1 & 0 & 0 & 0 \\ 0 & 1 & 0 & 0 \\ 0 & 0 & 33 & 22 \\ 0 & 0 & 0 & 0 \end{bmatrix} \qquad \text{operating } \frac{1}{33} C_3, \frac{1}{22} C_4$$

$$\sim \begin{bmatrix} 1 & 0 & 0 & 0 \\ 0 & 1 & 0 & 0 \\ 0 & 0 & 1 & 1 \\ 0 & 0 & 0 & 0 \end{bmatrix} \qquad \text{operating } C_4 - C_3$$

$$\sim \left[\begin{array}{ccc:c} 1 & 0 & 0 & 0 \\ 0 & 1 & 0 & 0 \\ 0 & 0 & 1 & 0 \\ \hdashline 0 & 0 & 0 & 0 \end{array}\right] = \left[\begin{array}{c:c} I_3 & O \\ \hdashline O & O \end{array}\right] \text{ which is the required normal form.}$$

$\therefore \quad \rho(A) = 3.$

(*ii*) Let

$$A = \begin{bmatrix} 8 & 1 & 3 & 6 \\ 0 & 3 & 2 & 2 \\ -8 & -1 & -3 & 4 \end{bmatrix} \qquad \text{operating } \frac{1}{8} C_1$$

$$\sim \begin{bmatrix} 1 & 1 & 3 & 6 \\ 0 & 3 & 2 & 2 \\ -1 & -1 & -3 & 4 \end{bmatrix} \qquad \text{operating } R_3 + R_1$$

$$\sim \begin{bmatrix} 1 & 1 & 3 & 6 \\ 0 & 3 & 2 & 2 \\ 0 & 0 & 0 & 10 \end{bmatrix} \qquad \text{operating } C_2 - C_1, C_3 - 3C_1, C_4 - 6C_1$$

$$\sim \begin{bmatrix} 1 & 0 & 0 & 0 \\ 0 & 3 & 2 & 2 \\ 0 & 0 & 0 & 10 \end{bmatrix} \qquad \text{operating } \frac{1}{3} C_2$$

$$\sim \begin{bmatrix} 1 & 0 & 0 & 0 \\ 0 & 1 & 2 & 2 \\ 0 & 0 & 0 & 10 \end{bmatrix} \qquad \text{operating } C_3 - 2C_2, C_4 - 2C_2$$

$$\sim \begin{bmatrix} 1 & 0 & 0 & 0 \\ 0 & 1 & 0 & 0 \\ 0 & 0 & 0 & 10 \end{bmatrix} \qquad \text{operating } \frac{1}{10} C_4$$

$$\sim \begin{bmatrix} 1 & 0 & 0 & 0 \\ 0 & 1 & 0 & 0 \\ 0 & 0 & 0 & 1 \end{bmatrix} \qquad \text{operating } C_{34}$$

$$\sim \left[\begin{array}{ccc:c} 1 & 0 & 0 & 0 \\ 0 & 1 & 0 & 0 \\ 0 & 0 & 1 & 0 \end{array}\right] = [I_3 : O]$$

which is the required normal form

$\rho(A) = 3.$

Example 4. *For the matrix* $A = \begin{bmatrix} 1 & 1 & 2 \\ 1 & 2 & 3 \\ 0 & -1 & -1 \end{bmatrix}$, *find non-singular matrices P and Q such that PAQ is in the normal form.*

Sol. We write A = IAI

$$\Rightarrow \quad \begin{bmatrix} 1 & 1 & 2 \\ 1 & 2 & 3 \\ 0 & -1 & -1 \end{bmatrix} = \begin{bmatrix} 1 & 0 & 0 \\ 0 & 1 & 0 \\ 0 & 0 & 1 \end{bmatrix} A \begin{bmatrix} 1 & 0 & 0 \\ 0 & 1 & 0 \\ 0 & 0 & 1 \end{bmatrix}$$

Operating $C_2 - C_1$, $C_3 - 2C_1$ (subjecting the *post-factor* on RHS to same operations)

$$\begin{bmatrix} 1 & 0 & 0 \\ 1 & 1 & 1 \\ 0 & -1 & -1 \end{bmatrix} = \begin{bmatrix} 1 & 0 & 0 \\ 0 & 1 & 0 \\ 0 & 0 & 1 \end{bmatrix} A \begin{bmatrix} 1 & -1 & -2 \\ 0 & 1 & 0 \\ 0 & 0 & 1 \end{bmatrix}$$

Operating $R_2 - R_1$ (subjecting the *pre-factor* on RHS to same operation)

$$\begin{bmatrix} 1 & 0 & 0 \\ 0 & 1 & 1 \\ 0 & -1 & -1 \end{bmatrix} = \begin{bmatrix} 1 & 0 & 0 \\ -1 & 1 & 0 \\ 0 & 0 & 1 \end{bmatrix} A \begin{bmatrix} 1 & -1 & -2 \\ 0 & 1 & 0 \\ 0 & 0 & 1 \end{bmatrix}$$

Operating $C_3 - C_2$ (subjecting the *post-factor* on RHS to same operation)

$$\begin{bmatrix} 1 & 0 & 0 \\ 0 & 1 & 0 \\ 0 & -1 & 0 \end{bmatrix} = \begin{bmatrix} 1 & 0 & 0 \\ -1 & 1 & 0 \\ 0 & 0 & 1 \end{bmatrix} A \begin{bmatrix} 1 & -1 & -1 \\ 0 & 1 & -1 \\ 0 & 0 & 1 \end{bmatrix}$$

Operating $R_3 + R_2$ (subjecting the *pre-factor* on RHS to same operation)

$$\begin{bmatrix} 1 & 0 & 0 \\ 0 & 1 & 0 \\ 0 & 0 & 0 \end{bmatrix} = \begin{bmatrix} 1 & 0 & 0 \\ -1 & 1 & 0 \\ -1 & 1 & 0 \end{bmatrix} A \begin{bmatrix} 1 & -1 & -1 \\ 0 & 1 & -1 \\ 0 & 0 & 1 \end{bmatrix}$$

$$\Rightarrow \quad \begin{bmatrix} I_2 & O \\ O & O \end{bmatrix} = PAQ, \quad \text{where} \quad P = \begin{bmatrix} 1 & 0 & 0 \\ -1 & 1 & 0 \\ -1 & 1 & 1 \end{bmatrix} \text{ and } Q = \begin{bmatrix} 1 & -1 & -1 \\ 0 & 1 & -1 \\ 0 & 0 & 1 \end{bmatrix}.$$

Example 5. *Reduce the matrix* $\begin{bmatrix} 1 & 2 & 3 & -2 \\ 2 & -2 & 1 & 1 \\ 3 & 0 & 4 & 3 \end{bmatrix}$ *to its normal form and find its rank.*

Sol. Let $A = \begin{bmatrix} 1 & 2 & 3 & -2 \\ 2 & -2 & 1 & 1 \\ 3 & 0 & 4 & 3 \end{bmatrix}$, operating $R_2 - 2R_1$, $R_3 - 3R_1$

$$\sim \begin{bmatrix} 1 & 2 & 3 & -2 \\ 0 & -6 & -5 & 5 \\ 0 & -6 & -5 & 9 \end{bmatrix}, \text{ operating } R_1 + \frac{1}{3}R_2, \frac{1}{-6}R_2, \frac{1}{4}R_3 - \frac{1}{4}R_2$$

$$\sim \begin{bmatrix} 1 & 0 & \frac{4}{3} & \frac{-1}{3} \\ 0 & 1 & \frac{5}{6} & \frac{-5}{6} \\ 0 & 0 & 0 & 1 \end{bmatrix}, \text{ operating } C_3 + C_4$$

$$\sim \begin{bmatrix} 1 & 0 & 1 & \frac{-1}{3} \\ 0 & 1 & 0 & \frac{-5}{6} \\ 0 & 0 & 1 & 1 \end{bmatrix}, \text{ operating } C_3 - C_1$$

$$\sim \begin{bmatrix} 1 & 0 & 0 & \frac{-1}{3} \\ 0 & 1 & 0 & \frac{-5}{6} \\ 0 & 0 & 1 & 1 \end{bmatrix}, C_4 + \frac{1}{3}C_1$$

$$\sim \begin{bmatrix} 1 & 0 & 0 & 0 \\ 0 & 1 & 0 & \frac{-5}{6} \\ 0 & 0 & 1 & 1 \end{bmatrix}, C_4 + \frac{5}{6}C_2 \sim \begin{bmatrix} 1 & 0 & 0 & 0 \\ 0 & 1 & 0 & 0 \\ 0 & 0 & 1 & 1 \end{bmatrix}, C_4 - C_3 \sim \begin{bmatrix} 1 & 0 & 0 & 0 \\ 0 & 1 & 0 & 0 \\ 0 & 1 & 1 & 0 \end{bmatrix}$$

$$\sim \left[\begin{array}{ccc:c} 1 & 0 & 0 & 0 \\ 0 & 1 & 0 & 0 \\ 0 & 0 & 1 & 0 \end{array}\right]$$

$$\sim [I_3 : 0]$$

which is the required normal form. The rank of the matrix is 3.

6.6. ECHELON FORM OF A MATRIX

A matrix is said to be an Echelon matrix if it satisfies the following conditions:

(*i*) The first non-zero number in any row is 1 (one).

(*ii*) Rows are arranged so that the column containing the first non-zero number is to the left of the column containing the first non-zero number of the next row.

(*iii*) All non zero rows are above any rows of all zeros.

For example:

$$(a)\ A = \begin{bmatrix} 1 & -3 & 4 & 2 \\ 0 & 1 & 3 & 1 \\ 0 & 0 & 1 & -1 \end{bmatrix} \qquad (b)\ B = \begin{bmatrix} 1 & 2 & -1 & 3 \\ 0 & 1 & 4 & 5 \\ 0 & 0 & 0 & 0 \end{bmatrix}$$

The rank of any matrix can be easily computed by reducing the matrix into Echelon form using elementary transformations.

Example 1. *Find the rank of the matrix,* $A = \begin{bmatrix} 2 & 4 & 1 & 5 \\ 6 & 3 & 2 & 9 \end{bmatrix}$, *by reducing it into an Echelon matrix using elementary operations.*

Sol.
$$A = \begin{bmatrix} 2 & 4 & 1 & 5 \\ 6 & 3 & 2 & 9 \end{bmatrix}$$

$$\sim \begin{bmatrix} 1 & 4 & 2 & 5 \\ 2 & 3 & 6 & 9 \end{bmatrix} \qquad C_1 \leftrightarrow C_2$$

$$\sim \begin{bmatrix} 1 & 4 & 2 & 5 \\ 0 & 5 & -2 & 1 \end{bmatrix} \qquad R_1 \rightarrow 2R_1 - R_2$$

$$\sim \begin{bmatrix} 1 & 5 & 2 & 4 \\ 0 & 1 & -2 & 5 \end{bmatrix} \qquad C_2 \leftrightarrow C_4$$

The rank of the matrix A is same as the rank of the last revised matrix, which is in Echelon form. The largest square submatrix $\begin{bmatrix} 1 & 5 \\ 1 & 1 \end{bmatrix}$ is having a non zero determinant value; and its order is 2. Obviously, the rank of A is 2.

Example 2. *Find the rank of the matrix A, by reducing it into Echelon form using elementary row transformation where the matrix is*

$$A = \begin{bmatrix} 1 & 5 & 2 \\ 4 & 5 & 3 \\ 9 & -3 & 6 \\ 7 & 5 & 4 \end{bmatrix}.$$

Sol.
$$A = \begin{bmatrix} 1 & 5 & 2 \\ 4 & 5 & 3 \\ 9 & -3 & 6 \\ 7 & 5 & 4 \end{bmatrix}$$

$$\sim \begin{bmatrix} 1 & 5 & 2 \\ 0 & 15 & 5 \\ 0 & 48 & 12 \\ 0 & 30 & 10 \end{bmatrix} \qquad \begin{aligned} &R_1 \to R_1 \\ &R_2 \to 4R_1 - R_2 \\ &R_3 \to 9R_1 - R_3 \\ &R_4 \to 7R_1 \end{aligned}$$

$$\sim \begin{bmatrix} 1 & 5 & 2 \\ 0 & 3 & 1 \\ 0 & 4 & 1 \\ 0 & 3 & 1 \end{bmatrix} \qquad \begin{aligned} &R_1 \to R_1 \\ &R_2 \to \frac{1}{5} R_2 \\ &R_3 \to \frac{1}{12} R_3 \\ &R_4 \to \frac{1}{10} R_4 \end{aligned}$$

$$\sim \begin{bmatrix} 1 & 5 & 2 \\ 0 & 3 & 1 \\ 0 & 1 & 0 \\ 0 & 0 & 0 \end{bmatrix} \qquad \begin{aligned} &R_1 \to R_1 \\ &R_2 \to R_2 \\ &R_3 \to R_3 - R_2 \\ &R_4 \to R_4 - R_2 \end{aligned}$$

$$\sim \begin{bmatrix} 1 & 5 & 2 \\ 0 & 1 & 0 \\ 0 & 3 & 1 \\ 0 & 0 & 0 \end{bmatrix} \qquad \begin{aligned} &R_1 \to R_1 \\ &R_2 \leftrightarrow R_3 \\ &R_4 \to R_4 \end{aligned}$$

$$\sim \begin{bmatrix} 1 & 5 & 2 \\ 0 & 1 & 0 \\ 0 & 0 & 1 \\ 0 & 0 & 0 \end{bmatrix} \qquad \begin{aligned} &R_1 \to R_1 \\ &R_2 \to R_2 \\ &R_3 \to R_3 - 3R_2 \\ &R_4 \to R_4 \end{aligned}$$

Obviously the rank of A is 3. Hence the result.

EXERCISE 6.1

1. Reduce to triangular form $\begin{bmatrix} 3 & -4 & -5 \\ -9 & 1 & 4 \\ -5 & 3 & 1 \end{bmatrix}$.

2. Use Gauss-Jordan method to find the inverse of each of the following matrices:

(i) $\begin{bmatrix} 8 & 4 & 3 \\ 2 & 1 & 1 \\ 1 & 2 & 1 \end{bmatrix}$ (ii) $\begin{bmatrix} 2 & 0 & -1 \\ 5 & 1 & 0 \\ 0 & 1 & 3 \end{bmatrix}$ (iii) $\begin{bmatrix} 2 & 1 & -1 \\ 0 & 2 & 1 \\ 5 & 2 & -3 \end{bmatrix}$

(iv) $\begin{bmatrix} 1 & 1 & 3 \\ 1 & 3 & -3 \\ -2 & -4 & -4 \end{bmatrix}$ (v) $\begin{bmatrix} 3 & -3 & 4 \\ 2 & -3 & 4 \\ 0 & -1 & 1 \end{bmatrix}$.

3. Find the ranks of the matrices

(i) $\begin{bmatrix} 2 & -1 & 0 & 5 \\ 0 & 3 & 1 & 4 \end{bmatrix}$ (ii) $\begin{bmatrix} 1 & 2 & 3 \\ 2 & 4 & 7 \\ 3 & 6 & 10 \end{bmatrix}$ (iii) $\begin{bmatrix} 1 & 3 & 4 & 5 \\ 1 & 2 & 6 & 7 \\ 1 & 5 & 0 & 10 \end{bmatrix}$

(iv) $\begin{bmatrix} 1 & 3 & 4 & 3 \\ 3 & 9 & 12 & 3 \\ 1 & 3 & 4 & 1 \end{bmatrix}$.

4. Determine the rank of the matrix $A = \begin{bmatrix} 6 & 1 & 3 & 8 \\ 4 & 2 & 6 & -1 \\ 10 & 3 & 9 & 7 \\ 16 & 4 & 12 & 15 \end{bmatrix}$ by reducing it to the normal form.

5. Find the ranks of the following matrices:

(i) $\begin{bmatrix} 3 & -1 & 2 \\ -6 & 2 & 4 \\ -3 & 1 & 2 \end{bmatrix}$ (ii) $\begin{bmatrix} 1 & 2 & 3 & 2 \\ 2 & 3 & 5 & 1 \\ 1 & 3 & 4 & 5 \end{bmatrix}$ (iii) $\begin{bmatrix} 1 & -1 & 2 & -3 \\ 4 & 1 & 0 & 2 \\ 0 & 3 & 1 & 4 \\ 0 & 1 & 0 & 2 \end{bmatrix}$

(iv) $\begin{bmatrix} 0 & 1 & -3 & -1 \\ 1 & 0 & 1 & 1 \\ 3 & 1 & 0 & 2 \\ 1 & 1 & -2 & 0 \end{bmatrix}$ (v) $\begin{bmatrix} 5 & 6 & 7 & 8 \\ 6 & 7 & 8 & 9 \\ 11 & 12 & 13 & 14 \\ 16 & 17 & 18 & 19 \end{bmatrix}$.

6. Find the ranks of the following matrices:

(i) $\begin{bmatrix} 3 & -4 & -1 & 2 \\ 1 & 7 & 3 & 1 \\ 5 & -2 & 5 & 4 \\ 9 & -3 & 7 & 7 \end{bmatrix}$ (ii) $\begin{bmatrix} -1 & 2 & 3 & -2 \\ 2 & -5 & 1 & 2 \\ 3 & -8 & 5 & 2 \\ 5 & -12 & -1 & 6 \end{bmatrix}$

(iii) $\begin{bmatrix} 2 & 4 & 3 & -2 \\ -3 & -2 & -1 & 4 \\ 6 & -1 & 7 & 2 \end{bmatrix}$ (iv) $\begin{bmatrix} 1 & 2 & 3 & 4 \\ 2 & 3 & 4 & 5 \\ 3 & 4 & 5 & 6 \\ 4 & 5 & 6 & 7 \end{bmatrix}$.

7. Reduce each of the following matrices to normal form and hence, find their ranks:

(i) $\begin{bmatrix} 0 & 2 & 3 & 4 \\ 2 & 3 & 5 & 4 \\ 4 & 8 & 13 & 12 \end{bmatrix}$ (ii) $\begin{bmatrix} 1 & 2 & 0 & -1 \\ 3 & 4 & 1 & 2 \\ -2 & 3 & 2 & 5 \end{bmatrix}$ (iii) $\begin{bmatrix} 1 & 2 & 1 & 0 \\ 3 & 2 & 1 & 2 \\ 2 & -1 & 2 & 5 \\ 5 & 6 & 3 & 2 \\ 1 & 3 & -1 & -3 \end{bmatrix}$.

8. Find non-singular matrices P and Q such that PAQ is in the normal form for the matrices:

$(i)\ A = \begin{bmatrix} 1 & -1 & -1 \\ 1 & 1 & 1 \\ 3 & 1 & 1 \end{bmatrix}$ $\quad (ii)\ A = \begin{bmatrix} 1 & 2 & 3 \\ 3 & 2 & 1 \\ 1 & 3 & 2 \\ 2 & 1 & 3 \end{bmatrix}$ $\quad (iii)\ A = \begin{bmatrix} 1 & 2 & 3 & -2 \\ 2 & -2 & 1 & 3 \\ 3 & 0 & 4 & 1 \end{bmatrix}$.

9. Find the rank of the matrix $\begin{bmatrix} 3 & 2 & 5 & 7 & 12 \\ 1 & 1 & 2 & 3 & 5 \\ 3 & 3 & 6 & 9 & 15 \end{bmatrix}$.

10. Find the inverse of the matrix $\begin{bmatrix} 1 & -1 & 1 \\ 4 & 1 & 0 \\ 8 & 1 & 1 \end{bmatrix}$ using elementary transformations.

11. Find the inverse of the matrix $\begin{bmatrix} 3 & -1 & 2 \\ 2 & 1 & -2 \\ 1 & 3 & -5 \end{bmatrix}$ using elementary transformations.

12. If $A = \begin{bmatrix} 3 & -3 & 4 \\ 2 & -3 & 4 \\ 0 & -1 & 1 \end{bmatrix}$, find A^{-1} and show that $A^3 = A^{-1}$

13. Determine the rank of the following matrix

$$S = \begin{bmatrix} 1 & 2 & 3 & 4 & 5 \\ 3 & 4 & 5 & 6 & 7 \\ -2 & -1 & 1 & 2 & 3 \\ 5 & 6 & 7 & 8 & 4 \end{bmatrix}$$

Answers

1. $\begin{bmatrix} 3 & -4 & -5 \\ 0 & 1 & 1 \\ 0 & 0 & 11 \end{bmatrix}$

2. $(i)\ \dfrac{1}{21}\begin{bmatrix} 1 & 10 & -7 \\ 1 & -11 & 14 \\ -3 & 12 & 0 \end{bmatrix}$ $\quad (ii)\ \begin{bmatrix} 3 & -1 & 1 \\ -15 & 6 & -5 \\ 5 & -2 & 2 \end{bmatrix}$ $\quad (iii)\ \begin{bmatrix} 8 & -1 & -3 \\ -5 & 1 & 2 \\ 10 & -1 & -4 \end{bmatrix}$

$(iv)\ \begin{bmatrix} 3 & 1 & \frac{3}{2} \\ -\frac{5}{4} & -\frac{1}{4} & -\frac{3}{4} \\ -\frac{1}{4} & -\frac{1}{4} & -\frac{1}{4} \end{bmatrix}$ $\quad (v)\ \begin{bmatrix} 1 & -1 & 0 \\ -2 & 3 & -4 \\ -2 & 3 & -3 \end{bmatrix}$.

3. (*i*) 2 (*ii*) 2 (*iii*) 3 (*iv*) 2

4. 2 **5.** (*i*) 2 (*ii*) 3 (*iii*) 4

(*iv*) 2 (*v*) 2

6. (*i*) 3 (*ii*) 2 (*iii*) 3 (*iv*) 2

7. $(i)\ \begin{bmatrix} I_2 & O \\ O & O \end{bmatrix}, 2$ $\quad (ii)\ [I_3\ \ O], 3$ $\quad (iii)\ \begin{bmatrix} I_3 & O \\ O & O \end{bmatrix}, 3$

8. $(i)\ P = \begin{bmatrix} 1 & 0 & 0 \\ -\frac{1}{2} & \frac{1}{2} & 0 \\ \frac{1}{2} & 1 & -\frac{1}{2} \end{bmatrix}, Q = \begin{bmatrix} 1 & 1 & -1 \\ 0 & 1 & 0 \\ 0 & 0 & 1 \end{bmatrix}$ $\quad (ii)\ P = \begin{bmatrix} 1 & 0 & 0 & 0 \\ \frac{3}{4} & -\frac{1}{4} & 0 & 0 \\ \frac{7}{12} & -\frac{1}{12} & -\frac{4}{12} & 0 \\ -\frac{3}{2} & -\frac{1}{2} & 1 & 1 \end{bmatrix}, Q = \begin{bmatrix} 1 & -2 & -1 \\ 0 & 1 & -2 \\ 0 & 0 & 1 \end{bmatrix}$

(*iii*) $P = \begin{bmatrix} 1 & 0 & 0 \\ -2 & 1 & 0 \\ -1 & -1 & 1 \end{bmatrix}$, $Q = \begin{bmatrix} 1 & \frac{1}{3} & -\frac{4}{3} & -\frac{1}{3} \\ 0 & -\frac{1}{6} & -\frac{5}{6} & \frac{7}{6} \\ 0 & 0 & 1 & 0 \\ 0 & 0 & 0 & 1 \end{bmatrix}$.

[**Note.** The matrices P and Q are not unique.]

9. 2 **10.** $\begin{bmatrix} 1 & 2 & -1 \\ -4 & -7 & 4 \\ -4 & -9 & 5 \end{bmatrix}$ **11.** $\frac{1}{5}\begin{bmatrix} 2 & -1 & 1 \\ -9 & 17 & -7 \\ -5 & 10 & -5 \end{bmatrix}$

12. $A^{-1} = \begin{bmatrix} 1 & -1 & 0 \\ -2 & 3 & -4 \\ -2 & 3 & -3 \end{bmatrix}$ **13.** Rank = 4.

6.7. SOLUTION OF A SYSTEM OF LINEAR EQUATIONS

Consider the system of equations $\left.\begin{matrix} a_1x + b_1y + c_1z = d_1 \\ a_2x + b_2y + c_2z = d_2 \\ a_3x + b_3y + c_3z = d_3 \end{matrix}\right\}$ (3 equations in 3 unknowns)

In matrix notation, these equations can be written as

$$\begin{bmatrix} a_1x + b_1y + c_1z \\ a_2x + b_2y + c_2z \\ a_3x + b_3y + c_3z \end{bmatrix} = \begin{bmatrix} d_1 \\ d_2 \\ d_3 \end{bmatrix} \quad \text{or} \quad \begin{bmatrix} a_1 & b_1 & c_1 \\ a_2 & b_2 & c_2 \\ a_3 & b_3 & c_3 \end{bmatrix}\begin{bmatrix} x \\ y \\ z \end{bmatrix} = \begin{bmatrix} d_1 \\ d_2 \\ d_3 \end{bmatrix}$$

or

$$AX = B$$

where $A = \begin{bmatrix} a_1 & b_1 & c_1 \\ a_2 & b_2 & c_2 \\ a_3 & b_3 & c_3 \end{bmatrix}$ is called the co-efficient matrix,

$X = \begin{bmatrix} x \\ y \\ z \end{bmatrix}$ is the column matrix of unknowns

$B = \begin{bmatrix} d_1 \\ d_2 \\ d_3 \end{bmatrix}$ is the column matrix of constants.

If $d_1 = d_2 = d_3 = 0$, then B = O and the matrix equation AX = B reduces to AX = O.
Such a system of equations is called a system of **homogeneous linear equations.**
If at least one of d_1, d_2, d_3 is non-zero, then B ≠ O.
Such a system of equations is called a system of **non-homogeneous linear equations.**

Solving the matrix equation AX = B means finding X, *i.e.*, finding a column matrix $\begin{bmatrix} \alpha \\ \beta \\ \gamma \end{bmatrix}$

such that

$$X = \begin{bmatrix} x \\ y \\ z \end{bmatrix} = \begin{bmatrix} \alpha \\ \beta \\ \gamma \end{bmatrix}. \text{ Then } x = \alpha, y = \beta, z = \gamma.$$

The matrix equation AX = B need not always have a solution. It may have no solution or a unique solution or an infinite number of solutions.

A system of equations having **no solution** is called an **inconsistent** system of equations.

A system of equations having **one or more solutions** is called **a consistent** system of equations.

For a system of non-homogeneous linear equations AX = B

(*i*) if $\rho[A : B] \neq \rho(A)$, the system is inconsistent.

(*ii*) if $\rho[A : B] = \rho(A)$ = number of unknowns, the system has a unique solution.

(*iii*) $\rho[A : B] = \rho(A)$ < number of unknowns, the system has an infinite number of solutions.

The matrix [A : B] in which the elements of A and B are written side by side is called the **augmented matrix.**

For a system of homogeneous linear equations AX = O

(*i*) X = O is always a solution. This solution in which each unknown has the value zero is called the **Null Solution** or the **Trivial Solution.** Thus a homogeneous system is always consistent.

A system of homogeneous linear equations has either the trivial solution or an infinite number of solutions.

(*ii*) if $\rho(A)$ = number of unknowns, the system has only the trivial solution.

(*iii*) if $\rho(A)$ < number of unknowns, the system has an infinite number of non-trivial solutions.

6.8. IF A IS A NON-SINGULAR MATRIX, THEN THE MATRIX EQUATION AX = B HAS A UNIQUE SOLUTION

The given equation is $\qquad AX = B \qquad$ ...(1)

$\because$ A is a non-singular matrix, $\therefore$ A^{-1} exists.

Pre-multiplying both sides of (1) by A^{-1}, we get

$$A^{-1}AX = A^{-1}B \quad \text{or} \quad (A^{-1}A)X = A^{-1}B$$

or

$$IX = A^{-1}B \quad \text{or} \quad X = A^{-1}B$$

which is the required unique solution (since A^{-1} is unique).

Another Method

Write the augmented matrix [A : B]. By **E-row operations** on A and B, reduce A to a diagonal matrix, thus getting

$$[A : B] \sim \begin{bmatrix} p_1 & 0 & 0 & \vdots & q_1 \\ 0 & p_2 & 0 & \vdots & q_2 \\ 0 & 0 & p_3 & \vdots & q_3 \end{bmatrix}$$

Then $\qquad p_1x = q_1, p_2y = q_2, p_3z = q_3.$

ILLUSTRATIVE EXAMPLES

Example 1. *Solve, with the help of matrices, the simultaneous equations:*

$$x + y + z = 3, \quad x + 2y + 3z = 4, \quad x + 4y + 9z = 6.$$

Sol. [In this question, there is no restriction that the solution must be obtained by finding A^{-1}].

$$\text{Augmented matrix } [A : B] = \left[\begin{array}{ccc|c} 1 & 1 & 1 & 3 \\ 1 & 2 & 3 & 4 \\ 1 & 4 & 9 & 6 \end{array}\right] \quad \text{operating } R_2 - R_1, R_3 - R_1$$

$$\sim \left[\begin{array}{ccc|c} 1 & 1 & 1 & 3 \\ 0 & 1 & 2 & 1 \\ 0 & 3 & 8 & 3 \end{array}\right] \quad \text{operating } R_3 - 3R_2$$

$$\sim \left[\begin{array}{ccc|c} 1 & 1 & 1 & 3 \\ 0 & 1 & 2 & 1 \\ 0 & 0 & 2 & 0 \end{array}\right] \quad \text{operating } \tfrac{1}{2} R_3$$

$$\left[\begin{array}{ccc|c} 1 & 1 & 1 & 3 \\ 0 & 1 & 2 & 1 \\ 0 & 0 & 1 & 0 \end{array}\right] \quad \text{operating } R_1 - R_3, R_2 - 2R_3$$

$$\sim \left[\begin{array}{ccc|c} 1 & 1 & 0 & 3 \\ 0 & 1 & 0 & 1 \\ 0 & 0 & 1 & 0 \end{array}\right] \quad \text{operating } R_1 - R_2$$

$$\sim \left[\begin{array}{ccc|c} 1 & 0 & 0 & 2 \\ 0 & 1 & 0 & 1 \\ 0 & 0 & 1 & 0 \end{array}\right]$$

$\therefore \quad x = 2, y = 1, z = 0.$

Example 2. *Solve the system of equations:*

$$2x_1 + x_2 + 2x_3 + x_4 = 6, \qquad 6x_1 - 6x_2 + 6x_3 + 12x_4 = 36,$$

$$4x_1 + 3x_2 + 3x_3 - 3x_4 = -1, \qquad 2x_1 + 2x_2 - x_3 + x_4 = 10.$$

Sol. In matrix notation, the given system of equations can be written as AX = B

$$\text{where} \quad A = \begin{bmatrix} 2 & 1 & 2 & 1 \\ 6 & -6 & 6 & 12 \\ 4 & 3 & 3 & -3 \\ 2 & 2 & -1 & 1 \end{bmatrix}, X = \begin{bmatrix} x_1 \\ x_2 \\ x_3 \\ x_4 \end{bmatrix}, B = \begin{bmatrix} 6 \\ 36 \\ -1 \\ 10 \end{bmatrix}$$

Augmented matrix

$$[A : B] = \left[\begin{array}{cccc|c} 2 & 1 & 2 & 1 & 6 \\ 6 & -6 & 6 & 12 & 36 \\ 4 & 3 & 3 & -3 & -1 \\ 2 & 2 & -1 & 1 & 10 \end{array}\right] \quad \text{operating } R_2 - 3R_1, R_3 - 2R_1, R_4 - R_1$$

$$\sim \begin{bmatrix} 2 & 1 & 2 & 1 & : & 6 \\ 0 & -9 & 0 & 9 & : & 18 \\ 0 & 1 & -1 & -5 & : & -13 \\ 0 & 1 & -3 & 0 & : & 4 \end{bmatrix} \quad \text{operating } -\frac{1}{9} R_2$$

$$\sim \begin{bmatrix} 2 & 1 & 2 & 1 & : & 6 \\ 0 & 1 & 0 & -1 & : & -2 \\ 0 & 1 & -1 & -5 & : & -13 \\ 0 & 1 & -3 & 0 & : & 4 \end{bmatrix} \quad \text{operating } R_1 - R_2, R_3 - R_2, R_4 - R_2$$

$$\sim \begin{bmatrix} 2 & 0 & 2 & 2 & : & 8 \\ 0 & 1 & 0 & -1 & : & -2 \\ 0 & 0 & -1 & -4 & : & -11 \\ 0 & 0 & -3 & 1 & : & 6 \end{bmatrix} \quad \text{operating } R_4 - 3R_3, \tfrac{1}{2} R_1$$

$$\sim \begin{bmatrix} 1 & 0 & 1 & 1 & : & 4 \\ 0 & 1 & 0 & -1 & : & -2 \\ 0 & 0 & -1 & -4 & : & -11 \\ 0 & 0 & 0 & 13 & : & 39 \end{bmatrix} \quad \text{operating } R_1 + R_3, \frac{1}{13} R_4$$

$$\sim \begin{bmatrix} 1 & 0 & 0 & -3 & : & 7 \\ 0 & 1 & 0 & -1 & : & -2 \\ 0 & 0 & -1 & -4 & : & -11 \\ 0 & 0 & 0 & 1 & : & 3 \end{bmatrix} \quad \text{operating } R_1 + 3R_4, R_2 + R_4, R_3 + 4R_4$$

$$\sim \begin{bmatrix} 1 & 0 & 0 & 0 & : & 2 \\ 0 & 1 & 0 & 0 & : & 1 \\ 0 & 0 & -1 & 0 & : & 1 \\ 0 & 0 & 0 & 1 & : & 3 \end{bmatrix} \quad \text{operating } (-1) R_3$$

$$\sim \begin{bmatrix} 1 & 0 & 0 & 0 & : & 2 \\ 0 & 1 & 0 & 0 & : & 1 \\ 0 & 0 & 1 & 0 & : & -1 \\ 0 & 0 & 0 & 1 & : & 3 \end{bmatrix}$$

Hence $x_1 = 2, x_2 = 1, x_3 = -1, x_4 = 3.$

Example 3. *Using matrix method, show that the equations*

$$3x + 3y + 2z = 1,\ x + 2y = 4,\ 10y + 3z = -2,\ 2x - 3y - z = 5$$

are consistent and hence obtain the solutions for x, y and z.

Sol. In matrix notation, the given system of equations can be written as AX = B

where $A = \begin{bmatrix} 3 & 3 & 2 \\ 1 & 2 & 0 \\ 0 & 10 & 3 \\ 2 & -3 & -1 \end{bmatrix}, X = \begin{bmatrix} x \\ y \\ z \end{bmatrix}, B = \begin{bmatrix} 1 \\ 4 \\ -2 \\ 5 \end{bmatrix}$

Augmented matrix

$$[A : B] = \begin{bmatrix} 3 & 3 & 2 & : & 1 \\ 1 & 2 & 0 & : & 4 \\ 0 & 10 & 3 & : & -2 \\ 2 & -3 & 1 & : & 5 \end{bmatrix} \quad \text{operating } R_{12}$$

$$\sim \begin{bmatrix} 1 & 2 & 0 & : & 4 \\ 3 & 3 & 2 & : & 1 \\ 0 & 10 & 3 & : & -2 \\ 2 & -3 & -1 & : & 5 \end{bmatrix} \quad \text{operating } R_2 - 3R_1,\ R_4 - 2R_1$$

$$\sim \begin{bmatrix} 1 & 2 & 0 & : & 4 \\ 0 & -3 & 2 & : & -11 \\ 0 & 10 & 3 & : & -2 \\ 0 & -7 & -1 & : & -3 \end{bmatrix} \quad \text{operating } R_3 + 3R_2,\ R_4 - 2R_2$$

$$\sim \begin{bmatrix} 1 & 2 & 0 & : & 4 \\ 0 & -3 & 2 & : & -11 \\ 0 & 1 & 9 & : & -35 \\ 0 & -1 & -5 & : & 19 \end{bmatrix} \quad \text{operating } R_1 - 2R_3,\ R_2 + 3R_3,\ R_4 + R_3$$

$$\sim \begin{bmatrix} 1 & 0 & -18 & : & 74 \\ 0 & 0 & 29 & : & -116 \\ 0 & 1 & 9 & : & -35 \\ 0 & 0 & 4 & : & -16 \end{bmatrix} \quad \text{operating } R_{23},\ \tfrac{1}{4}R_4$$

$$\sim \begin{bmatrix} 1 & 0 & -18 & : & 74 \\ 0 & 1 & 9 & : & -35 \\ 0 & 0 & 29 & : & -116 \\ 0 & 0 & 1 & : & -4 \end{bmatrix} \quad \text{operating } R_1 + 18R_4,\ R_2 - 9R_4,\ R_3 - 29R_4$$

$$\sim \begin{bmatrix} 1 & 0 & 0 & : & 2 \\ 0 & 1 & 0 & : & 1 \\ 0 & 0 & 0 & : & 0 \\ 0 & 0 & 1 & : & -4 \end{bmatrix} \quad \text{operating } R_{34}$$

$$\sim \begin{bmatrix} 1 & 0 & 0 & : & 2 \\ 0 & 1 & 0 & : & 1 \\ 0 & 0 & 1 & : & -4 \\ 0 & 0 & 0 & : & 0 \end{bmatrix}$$

$\rho(A) = \rho(A : B) = 3$ = number of unknowns

$\Rightarrow$ The given system of equations is consistent and the unique solution is

$$x = 2,\ y = 1,\ z = -4.$$

Example 4. *For what values of λ and μ do the system of equations*

$$x + y + z = 6,\ x + 2y + 3z = 10,\ x + 2y + \lambda z = \mu$$

have (i) no solution (ii) unique solution (iii) more than one solution? [K.U. May 2010]

Sol. In matrix notation, the given system of equations can be written as AX = B

where $A = \begin{bmatrix} 1 & 1 & 1 \\ 1 & 2 & 3 \\ 1 & 2 & \lambda \end{bmatrix},\ X = \begin{bmatrix} x \\ y \\ z \end{bmatrix},\ B = \begin{bmatrix} 6 \\ 10 \\ \mu \end{bmatrix}$

$$\text{Augmented matrix } [A : B] = \begin{bmatrix} 1 & 1 & 1 & : & 6 \\ 1 & 2 & 3 & : & 10 \\ 1 & 2 & \lambda & : & \mu \end{bmatrix} \quad \text{operating } R_2 - R_1, R_3 - R_1$$

$$\sim \begin{bmatrix} 1 & 1 & 1 & : & 6 \\ 0 & 1 & 2 & : & 4 \\ 0 & 1 & \lambda - 1 & : & \mu - 6 \end{bmatrix} \quad \text{operating } R_1 - R_2, R_3 - R_2$$

$$\sim \begin{bmatrix} 1 & 0 & -1 & : & 2 \\ 0 & 0 & 2 & : & 4 \\ 0 & 0 & \lambda - 3 & : & \mu - 10 \end{bmatrix}$$

Case I. If $\lambda = 3, \mu \neq 10$

$$\rho(A) = 2, \rho(A : B) = 3$$

$\because$ $\rho(A) \neq \rho(A : B)$

$\therefore$ The system has no solution.

Case II. If $\lambda \neq 3$, μ may have any value

$$\rho(A) = \rho(A : B) = 3 = \text{number of unknowns}$$

$\therefore$ The system has unique solution.

Case III. If $\lambda = 3, \mu = 10$

$$\rho(A) = \rho(A : B) = 2 < \text{number of unknowns}$$

$\therefore$ The system has an infinite number of solutions.

Example 5. *Solve the equations :*

$x_1 + 3x_2 + 2x_3 = 0, 2x_1 - x_2 + 3x_3 = 0, 3x_1 - 5x_2 + 4x_3 = 0, x_1 + 17x_2 + 4x_3 = 0.$

Sol. In matrix notation, the given system of equations can be written as AX = O

$$\text{where } \quad A = \begin{bmatrix} 1 & 3 & 2 \\ 2 & -1 & 3 \\ 3 & -5 & 4 \\ 1 & 17 & 4 \end{bmatrix}, X = \begin{bmatrix} x_1 \\ x_2 \\ x_3 \end{bmatrix}$$

Operating $R_2 - 2R_1, R_3 - 3R_1, R_4 - R_1$

$$A \sim \begin{bmatrix} 1 & 3 & 2 \\ 0 & -7 & -1 \\ 0 & -14 & -2 \\ 0 & 14 & 2 \end{bmatrix} \quad \text{operating } R_3 - 2R_2, R_4 + 2R_2$$

$$\sim \begin{bmatrix} 1 & 3 & 2 \\ 0 & -7 & -1 \\ 0 & 0 & 0 \\ 0 & 0 & 0 \end{bmatrix} \quad \text{operating } R_1 + 2R_2$$

$$\sim \begin{bmatrix} 1 & -11 & 0 \\ 0 & -7 & -1 \\ 0 & 0 & 0 \\ 0 & 0 & 0 \end{bmatrix}$$

$\therefore \quad \rho(A) = 2 <$ number of unknowns

$\Rightarrow$ The system has an infinite number of non-trivial solutions given by

$$x_1 - 11x_2 = 0, \quad -7x_2 - x_3 = 0$$

i.e., $x_1 = 11k, \quad x_2 = k, \quad x_3 = -7k$, where k is arbitrary.

Different values of k give different solutions.

Example 6. *Show that the system of equations:*

$$2x_1 - 2x_2 + x_3 = \lambda x_1$$
$$2x_1 - 3x_2 + 2x_3 = \lambda x_2$$
$$-x_1 + 2x_2 = \lambda x_3$$

can possess a non-trivial solution only if $\lambda = 1$ *or* -3. *Obtain the general solution in each case.*

Sol. The given system of equation is

$$(2 - \lambda)x_1 - 2x_2 + x_3 = 0$$
$$2x_1 - (3 + \lambda)x_2 + 2x_3 = 0$$
$$-x_1 + 2x_2 - \lambda x_3 = 0$$

In matrix notation, it can be written as AX = O

$$\text{where} \quad A = \begin{bmatrix} 2-\lambda & -2 & 1 \\ 2 & -(3+\lambda) & 2 \\ -1 & 2 & -\lambda \end{bmatrix}, X = \begin{bmatrix} x_1 \\ x_2 \\ x_3 \end{bmatrix}$$

For non-trivial solution, $|\ A\ | = 0$

$$\Rightarrow \quad \begin{vmatrix} 2-\lambda & -2 & 1 \\ 2 & -(3+\lambda) & 2 \\ -1 & 2 & -\lambda \end{vmatrix} = 0$$

$$\Rightarrow \quad (2 - \lambda)[\lambda(3 + \lambda) - 4] + 2(-2\lambda + 2) + [4 - (3 + \lambda)] = 0$$

$$\Rightarrow \quad \lambda^3 + \lambda^2 - 5\lambda + 3 = 0$$

$$\Rightarrow \quad (\lambda - 1)^2 (\lambda + 3) = 0$$

$$\therefore \quad \lambda = 1 \quad \text{or} \quad 3.$$

When $\lambda = 1$, the equations become

$$x_1 - 2x_2 + x_3 = 0$$
$$2x_1 - 4x_2 + 2x_3 = 0$$
$$-x_1 + 2x_2 - x_3 = 0$$

which are identical.

The given system is equivalent to a single equation

$$x_1 - 2x_2 + x_3 = 0$$

Taking $x_2 = t, x_3 = s$, we get $x_1 = 2t - s$

$$\therefore \quad x_1 = 2t - s, x_2 = t, x_3 = s$$

which give an infinite number of non-trivial solutions, t and s being the parameters.

When $\lambda = -3$, the equations become

$$5x_1 - 2x_2 + x_3 = 0$$
$$2x_1 + 2x_3 = 0$$
$$-x_1 + 2x_2 + 3x_3 = 0$$

Solving the first two, we have

$$\frac{x_1}{-4} = \frac{x_2}{2-10} = \frac{x_3}{4} \quad \text{or} \quad x_1 = \frac{x_2}{2} = \frac{x_3}{-1}$$

$\therefore \qquad x_1 = t, \quad x_2 = 2t, \quad x_3 = -t$

which give an infinite number of non-trivial solutions, t being the parameter.

EXERCISE 6.2

1. Write the following equations in matrix form AX = B and solve for X:

(*i*) $2x + 5y + 3z = 1$, $-x + 2y + z = 2$, $x + y + z = 0$

(*ii*) $2x_1 - x_2 + x_3 = 4$, $x_1 + x_2 + x_3 = 1$, $x_1 - 3x_2 - 2x_3 = 2$.

2. Using the loop current method on a circuit, the following equations were obtained:

$$7i_1 - 4i_2 = 12, \; -4i_1 + 12i_2 - 6i_3 = 0, \; -6i_2 + 14i_3 = 0.$$

By matrix method, solve for i_1, i_2 and i_3.

3. Solve the following systems of equations by matrix method:

(*i*) $x + y + z = 8$, $\quad x - y + 2z = 6$, $\quad 3x + 5y - 7z = 14$

(*ii*) $x + y + z = 6$, $\quad x - y + 2z = 5$, $\quad 3x + y + z = 8$

(*iii*) $x + 2y + 3z = 1$, $\quad 2x + 3y + 2z = 2$, $\quad 3x + 3y + 4z = 1$.

4. Show that the equations $x + 2y - z = 3$, $3x - y + 2z = 1$, $2x - 2y + 3z = 2$, $x - y + z = -1$ are consistent and solve them.

5. (*a*) Test for consistency the equations $2x - 3y + 7z = 5$, $3x + y - 3z = 13$, $2x + 19y - 47z = 32$.

(*b*) Examine if the following equations are consistent and solve them if they are consistent:

$$2x + 6y + 11 = 0, \quad 6x + 20y - 6z + 3 = 0, \quad 6y - 18z + 1 = 0$$

(*c*) Test for consistency and solve the equations :

$x + 2y + z = 3$, $\quad 2x + 3y + 2z = 5$,

$3x - 5y + 5z = 2$, $\quad 3x + 9y - z = 4$ (*K.U. May 2009*)

6. Solve the equations $x + 3y + 2z = 0$, $2x - y + 3z = 0$, $3x - 5y + 4z = 0$, $x + 17y + 4z = 0$.

7. For what values of a and b do the equations $x + 2y + 3z = 6$, $x + 3y + 5z = 9$, $2x + 5y + az = b$ have (*i*) no solution (*ii*) a unique solution (*iii*) more than one solution?

8. Find the value of k so that the equations $x + y + 3z = 0$, $4x + 3y + kz = 0$, $2x + y + 2z = 0$ have a non-trivial solution.

9. Show that the equations $3x + 4y + 5z = a$, $4x + 5y + 6z = b$, $5x + 6y + 7z = c$ do not have a solution unless $a + c = 2b$.

10. Investigate the values of λ and μ so that the equations $2x + 3y + 5z = 9$, $7x + 3y - 2z = 8$, $2x + 3y + \lambda z = \mu$ have (*i*) no solution, (*ii*) a unique solution and (*iii*) an infinite number of solutions.

11. Determine the values of λ for which the following system of equations may possess non-trivial solution. $3x_1 + x_2 - \lambda x_3 = 0$, $4x_1 - 2x_2 - 3x_3 = 0$, $2\lambda x_1 + 4x_2 + \lambda x_3 = 0$. For each permissible value of λ, determine the general solution.

12. Investigate for consistency the following equations and if possible, find the solutions:

$4x - 2y + 6z = 8$, $x + y - 2z = -1$, $15x - 3y + 9z = 21$. (*K.U. May 2011*)

13. For what values of k the equations $x + y + z = 1$, $2x + y + 4z = k$, $4x + y + 10z = k^2$ have a solution and solve them completely in each case. (*K.U. April 2013*)

14. Show that if $\lambda \neq -5$, the system of equations $3x - y + 4z = 3$, $x + 2y - 3z = -2$, $6x + 5y + \lambda z = -3$ have a unique solution. If $\lambda = -5$, show that the equations are consistent. Determine the solutions in each case.

15. Find the values of a and b for which the equations $x + ay + z = 3$, $x + 2y + 2z = b$, $x + 5y + 3z = 9$ are consistent. When will these equations have a unique solution?

16. Solve the equations

$$x + y + z + w = 0, \quad x + 3y + 2z + w = 0, \quad 2x + z - w = 0.$$

17. Find the values of λ for which the equations

$$(2 - \lambda)x + 2y + 3 = 0, \quad 2x + (4 - \lambda)y + 7 = 0, \quad 2x + 5y + (6 - \lambda) = 0$$

are consistent and find the values of x and y corresponding to each of these values of λ.

18. Find the values of λ for which the equations

$$(\lambda - 1)x + (3\lambda + 1)y + 2\lambda z = 0$$
$$(\lambda - 1)x + (4\lambda - 2)y + (\lambda + 3)z = 0$$
$$2x + (3\lambda + 1)y + 3(\lambda - 1)z = 0$$

have non-trivial solution and find the ratios $x : y : z$, when λ has the smallest of these values. What happens when λ has the greatest of these values?

19. Using matrix method solve the following system of linear equations

$$x - y + z = 3$$
$$2x + z = 1$$
$$3x + 2y + z = 4$$

20. Test the following system for consistency

$$x + y + z = -3$$
$$3x + y - 2z = -2$$
$$2x + 4y + 7z = 7$$

21. Determine the values of λ so that the equations

$$2x + y + 2z = 10$$
$$x + y + 3z = 10$$
$$4x + 3y + \lambda z = 10$$

are consistent ; hence solve the system.

Answers

1. (*i*) $x = -1, y = 0, z = 1$ (*ii*) $x_1 = 1, x_2 = -1, x_3 = 1$ **2.** $i_1 = \frac{396}{175}, i_2 = \frac{24}{25}, i_3 = \frac{72}{175}$

3. (*i*) $x = 5, y = \frac{5}{3}, z = \frac{4}{3}$ (*ii*) $x = 1, y = 2, z = 3$ (*iii*) $x = -\frac{3}{7}, y = \frac{8}{7}, z = -\frac{2}{7}$

4. $x = -1, y = 4, z = 4$ **5.** (*a*) Inconsistent (*b*) Inconsistent (*c*) $x = -1, y = 1, z = 2$

6. $x = 11k, y = k, z = -7k$, where k is arbitrary

7. (*i*) $a = 8, b \neq 15$ (*ii*) $a \neq 8$, b may have any value (*iii*) $a = 8, b = 15$

8. $k = 8$ **10.** (*i*) $\lambda = 5, \mu \neq 9$ (*ii*) $\lambda \neq 5$, μ arbitrary (*iii*) $\lambda = 5, \mu = 9$

11. $\lambda = 1, -9$. For $\lambda = 1$, solution is $x = k, y = -k, z = 2k$. For $\lambda = -9$, solution is $x = 3k, y = 9k, z = -2k$.

12. Consistent ; $x = 1, y = 3k - 2, z = k$, where k is arbitrary

13. $k = 1, x = -3\lambda, y = 1 + 2\lambda, z = \lambda$; $k = 2, x = 1 - 3\lambda, y = 2\lambda, z = \lambda$, where λ is arbitrary

14. $\lambda \neq -5, x = \frac{4}{7}, y = -\frac{9}{7}, z = 0$; $\lambda = -5, x = \frac{4 - 5k}{7}, y = \frac{13k - 9}{7}, z = k$, where k is arbitrary

15. If $a = -1, b = 6$, equations will be consistent and have infinitely many solutions
If $a = -1, b \neq 6$, equations will be inconsistent
If $a \neq -1$, b is arbitrary, equations will be consistent and have a unique solution

16. $x = \frac{1}{2}(\mu - \lambda), y = -\frac{1}{2}(\lambda + 3\mu), z = \lambda, w = \mu$

17. $\lambda = -1, 1, 12 ; x = -\frac{1}{11}, y = -\frac{15}{11} ; x = -5, y = 1 ; x = \frac{1}{2}, y = 1$

18. $\lambda = 0, 3 ; x : y : z = 1 : 1 : 1$; equations become identical

19. $x = -1,\ y = 2,\ z = 3$

20. Not consistent

21. Consistent when $\lambda \neq 8$. $x = \frac{20}{8-\lambda},\ y = \frac{-10\lambda}{8-\lambda},\ z = \frac{20}{8-\lambda}$

6.9. LU DECOMPOSITION METHOD

This method can be used to solve a system of equations or to find the inverse of a matrix. In this method, the matrix A is decomposed or factorized as the product of a lower triangular matrix L and an upper triangular matrix U.

Consider the system of equations

$$\left.\begin{array}{l} a_{11}x_1 + a_{12}x_2 + \cdots + a_{1n}x_n = b_1 \\ a_{21}x_1 + a_{22}x_2 + \cdots + a_{2n}x_n = b_2 \\ \cdots\cdots\cdots\cdots\cdots\cdots\cdots\cdots \\ \cdots\cdots\cdots\cdots\cdots\cdots\cdots\cdots \\ a_{n1}x_1 + a_{n2}x_2 + \cdots + a_{nn}x_n = b_n \end{array}\right\} \quad \ldots(1)$$

In matrix notation, equation (1) can be written as

$$\begin{bmatrix} a_{11} & a_{12} & \cdots & a_{1n} \\ a_{21} & a_{22} & \cdots & a_{2n} \\ \cdots & \cdots & \cdots & \cdots \\ a_{n1} & a_{12} & \cdots & a_{nn} \end{bmatrix} \begin{bmatrix} x_1 \\ x_2 \\ \vdots \\ x_n \end{bmatrix} = \begin{bmatrix} b_1 \\ b_2 \\ \vdots \\ b_n \end{bmatrix}$$

i.e.,

$$AX = B \quad \ldots(2)$$

Let

$$A = LU$$

where $L = \begin{bmatrix} l_{11} & 0 & 0 & \cdots & 0 \\ l_{21} & l_{22} & 0 & \cdots & 0 \\ l_{31} & l_{32} & l_{33} & \cdots & 0 \\ \cdots & \cdots & \cdots & \cdots & \cdots \\ l_{n1} & l_{n2} & l_{n3} & \cdots & l_{nn} \end{bmatrix}$ and $U = \begin{bmatrix} u_{11} & u_{12} & u_{13} & \cdots & u_{1n} \\ 0 & u_{22} & u_{23} & \cdots & u_{2n} \\ 0 & 0 & u_{33} & \cdots & u_{3n} \\ \cdots & \cdots & \cdots & \cdots & \cdots \\ 0 & 0 & 0 & \cdots & u_{nn} \end{bmatrix}$

Hence equation (2) becomes

$$LUX = B \quad \ldots(3)$$

We rewrite this system as

$$UX = Y \quad \ldots(4)$$

and

$$LY = B \quad \ldots(5)$$

We first solve Y from equation (5) using forward substitution and then X from equation (4) using the back substitution.

Alternately from equation (5), we have

$$Y = L^{-1} B \text{ and}$$

from equation (4), we have

$$X = U^{-1}Y$$

The inverse of A can be obtained from

$$A^{-1} = U^{-1}L^{-1} \qquad ...(6)$$

Note. This method is also called the factorization method or triangularization method.

Remark 1. If U_{ii} are chosen as $U_{ii} = 1, i = 1, 2, ..., n$, then the first column of L is $l_{i1} = a_{i1}$, $i = 1, 2, ..., n$ and the first row of U is

$$U_{ij} = \frac{a_{ij}}{a_{11}}, j = 2, 3, ..., n.$$

If l_{ii} are chosen as $l_{ii} = 1, i = 1, 2, ..., n$ then the first row of u is $u_{ij} = a_{ij}, j = 1, 2, ..., n$ and the first column of L is $l_{i1} = \dfrac{a_{i1}}{a_{11}}, i = 1, 2, ..., n.$

Remark. 2 L^{-1} can be computed from $LL^{-1} = I$, using forward substitution and U^{-1} can be computed from $UU^{-1} = I$, using back substitution. *i.e.*, the inverse of a lower triangular matrix is a lower triangular matrix and inverse of an upper triangular matrix is an upper triangular matrix.

ILLUSTRATIVE EXAMPLES

Example 1. *Solve the equations*

$$2x + 3y + z = 9, \quad x + 2y + 3z = 6, \quad 3x + y + 2z = 8.$$

by the LU decomposition method.

Sol. In matrix form, AX = B ...(1)

$$\begin{bmatrix} 2 & 3 & 1 \\ 1 & 2 & 3 \\ 3 & 1 & 2 \end{bmatrix} \begin{bmatrix} x \\ y \\ z \end{bmatrix} = \begin{bmatrix} 9 \\ 6 \\ 8 \end{bmatrix}$$

Here $$A = \begin{bmatrix} 2 & 3 & 1 \\ 1 & 2 & 3 \\ 3 & 1 & 2 \end{bmatrix}$$

Let $$A = LU \qquad ...(2)$$

$$\begin{bmatrix} 2 & 3 & 1 \\ 1 & 2 & 3 \\ 3 & 1 & 2 \end{bmatrix} = \begin{bmatrix} 1 & 0 & 0 \\ l_{21} & 1 & 0 \\ l_{31} & l_{32} & 1 \end{bmatrix} \begin{bmatrix} u_{11} & u_{12} & u_{13} \\ 0 & u_{22} & u_{23} \\ 0 & 0 & u_{33} \end{bmatrix}$$

$$= \begin{bmatrix} u_{11} & u_{12} & u_{13} \\ l_{21}u_{11} & l_{21}u_{12} + u_{22} & l_{21}u_{13} + u_{23} \\ l_{31}u_{11} & l_{31}u_{12} + l_{32}u_{22} & l_{31}u_{13} + l_{32}u_{23} + u_{33} \end{bmatrix}$$

Comparing the elements on both sides, we get

$$u_{11} = 2, u_{12} = 3, u_{13} = 1$$

$$l_{21}\,u_{11} = 1 \;\Rightarrow\; l_{21} = \frac{1}{u_{11}} = \frac{1}{2}$$

$$l_{31}\,u_{11} = 3 \;\Rightarrow\; l_{31} = \frac{3}{u_{11}} = \frac{3}{2}$$

$$l_{21}u_{12} + u_{22} = 2 \;\Rightarrow\; u_{22} = 2 - l_{21}\,u_{12} = 2 - \frac{1}{2} \times 3 = \frac{1}{2}$$

$$l_{21}u_{13} + u_{23} = 3 \;\Rightarrow\; u_{23} = 3 - l_{21}\,u_{13} = 3 - \frac{1}{2} \times 1 = \frac{5}{2}$$

$$l_{31}u_{12} + l_{32}\,u_{22} = 1 \;\Rightarrow\; l_{32} = \frac{1}{u_{22}}[1 - l_{31}u_{12}]$$

$$= \frac{1 - \frac{3}{2} \times 3}{\left(\frac{1}{2}\right)} = -7$$

$$l_{31}u_{13} + l_{32}u_{23} + u_{33} = 2 \;\Rightarrow\; u_{33} = 2 - l_{31}u_{13} - l_{32}\,u_{23} = 2 - \frac{3}{2} \times 1 - (-7) \times \frac{5}{2} = 18$$

$$\therefore \qquad L = \begin{bmatrix} 1 & 0 & 0 \\ \frac{1}{2} & 1 & 0 \\ \frac{3}{2} & -7 & 1 \end{bmatrix} \quad \text{and} \quad U = \begin{bmatrix} 2 & 3 & 1 \\ 0 & \frac{1}{2} & \frac{5}{2} \\ 0 & 0 & 18 \end{bmatrix}$$

From equation (2), we get

$$A = \begin{bmatrix} 1 & 0 & 0 \\ \frac{1}{2} & 1 & 0 \\ \frac{3}{2} & -7 & 1 \end{bmatrix} \begin{bmatrix} 2 & 3 & 1 \\ 0 & \frac{1}{2} & \frac{5}{2} \\ 0 & 0 & 18 \end{bmatrix}$$

Hence equation (1) becomes, LUX = B ...(3)

$$\text{i.e.,} \qquad \begin{bmatrix} 1 & 0 & 0 \\ \frac{1}{2} & 1 & 0 \\ \frac{3}{2} & -7 & 1 \end{bmatrix} \begin{bmatrix} 2 & 3 & 1 \\ 0 & \frac{1}{2} & \frac{5}{2} \\ 0 & 0 & 18 \end{bmatrix} \begin{bmatrix} x \\ y \\ z \end{bmatrix} = \begin{bmatrix} 9 \\ 6 \\ 8 \end{bmatrix}$$

Substituting UX = Y ...(4) in equation (3), we get

$$LY = B$$

$$\text{i.e.,} \qquad \begin{bmatrix} 1 & 0 & 0 \\ \frac{1}{2} & 1 & 0 \\ \frac{3}{2} & -7 & 1 \end{bmatrix} \begin{bmatrix} y_1 \\ y_2 \\ y_3 \end{bmatrix} = \begin{bmatrix} 9 \\ 6 \\ 8 \end{bmatrix}$$

Solving this system by forward substitution, we get,

$$y_1 = 9$$

$$\frac{1}{2}y_1 + y_2 = 6 \Rightarrow y_2 = \frac{3}{2}$$

and $$\frac{3}{2}y_1 - 7y_2 + y_3 = 8 \Rightarrow y_3 = 5$$

Now we can find the solution of the original system, obtained from equation (4) as

$$\begin{bmatrix} 2 & 3 & 1 \\ 0 & \frac{1}{2} & \frac{5}{2} \\ 0 & 0 & 18 \end{bmatrix} \begin{bmatrix} x \\ y \\ z \end{bmatrix} = \begin{bmatrix} 9 \\ \frac{3}{2} \\ 5 \end{bmatrix}$$

Solving this by back substitution, we get

$$18z = 5 \Rightarrow z = \frac{5}{18}$$

$$\frac{1}{2}y + \frac{5}{2}z = \frac{3}{2} \Rightarrow y = 2\left(\frac{3}{2} - \frac{5}{2}z\right)$$

$$= 2\left(\frac{3}{2} - \frac{5}{2} \times \frac{5}{18}\right) = \frac{29}{18}$$

and $$2x + 3y + z = 9 \Rightarrow x = \frac{1}{2}(9 - 3y - z)$$

$$= \frac{1}{2}\left(9 - 3 \times \frac{29}{18} - \frac{5}{18}\right)$$

$$= \frac{35}{18}$$

$\therefore$ The solution of the given system of equation is

$$X = \left(\frac{35}{18}, \frac{29}{18}, \frac{5}{18}\right)^T$$

Example 2. *Factorise the matrix* $A = \begin{bmatrix} 5 & -2 & 1 \\ 7 & 1 & -5 \\ 3 & 7 & 4 \end{bmatrix}$ *into the form LU, where L is unit lower triangular and U is upper triangular, and hence solve the system of equations*

$$5x - 2y + z = 4,\ 7x + y - 5z = 8,\ 3x + 7y + 4z = 10$$

Determine also L^{-1} *and* U^{-1} *and hence find* A^{-1}.

Sol. The given system of equations in matrix form is AX = B ...(1)

where $$A = \begin{bmatrix} 5 & -2 & 1 \\ 7 & 1 & -5 \\ 3 & 7 & 4 \end{bmatrix}$$

Let A = LU ...(2)

i.e.,
$$\begin{bmatrix} 1 & 0 & 0 \\ l_{21} & 1 & 0 \\ l_{31} & l_{32} & 1 \end{bmatrix}\begin{bmatrix} u_{11} & u_{12} & u_{13} \\ 0 & u_{22} & u_{23} \\ 0 & 0 & u_{33} \end{bmatrix} = \begin{bmatrix} 5 & -2 & 1 \\ 7 & 1 & -5 \\ 3 & 7 & 4 \end{bmatrix}$$

Comparing the elements on both sides and solving by forward substitution, we get

$$L = \begin{bmatrix} 1 & 0 & 0 \\ \frac{7}{5} & 1 & 0 \\ \frac{3}{5} & \frac{41}{19} & 1 \end{bmatrix} \text{ and } U = \begin{bmatrix} 5 & -2 & 1 \\ 0 & \frac{19}{5} & \frac{-32}{5} \\ 0 & 0 & \frac{327}{19} \end{bmatrix}$$

Using equation (2), the given system becomes

$$LUX = B \qquad ...(3)$$

i.e.,
$$\begin{bmatrix} 1 & 0 & 0 \\ \frac{7}{5} & 1 & 0 \\ \frac{3}{5} & \frac{41}{19} & 1 \end{bmatrix}\begin{bmatrix} 5 & -2 & 1 \\ 0 & \frac{19}{5} & \frac{-32}{5} \\ 0 & 0 & \frac{327}{19} \end{bmatrix}\begin{bmatrix} x \\ y \\ z \end{bmatrix} = \begin{bmatrix} 4 \\ 8 \\ 10 \end{bmatrix}$$

Substituting UX = Y ...(4), in equation (3), we get

$$LY = B$$

i.e.,
$$\begin{bmatrix} 1 & 0 & 0 \\ \frac{7}{5} & 1 & 0 \\ \frac{3}{5} & \frac{41}{19} & 1 \end{bmatrix}\begin{bmatrix} y_1 \\ y_2 \\ y_3 \end{bmatrix} = \begin{bmatrix} 4 \\ 8 \\ 10 \end{bmatrix}$$

Solving this system, by forward substitution, we get

$$Y = \left(4, \frac{12}{5}, \frac{46}{19}\right)^T$$

Now we can find this solution of the original system, obtained from equation (4) as

$$\begin{bmatrix} 5 & -2 & 1 \\ 0 & \frac{19}{5} & \frac{-32}{5} \\ 0 & 0 & \frac{327}{19} \end{bmatrix}\begin{bmatrix} x \\ y \\ z \end{bmatrix} = \begin{bmatrix} 4 \\ \frac{12}{5} \\ \frac{46}{19} \end{bmatrix}$$

Solving this by back substitution, we get

$$X = \left(\frac{122}{109}, \frac{284}{327}, \frac{46}{327}\right)^T \text{ From } LL^{-1} = I,$$

i.e.,
$$\begin{bmatrix} 1 & 0 & 0 \\ \frac{7}{5} & 1 & 0 \\ \frac{3}{5} & \frac{41}{19} & 1 \end{bmatrix}\begin{bmatrix} l_{11}^* & 0 & 0 \\ l_{21}^* & l_{22}^* & 0 \\ l_{31}^* & l_{32}^* & l_{33}^* \end{bmatrix} = \begin{bmatrix} 1 & 0 & 0 \\ 0 & 1 & 0 \\ 0 & 0 & 1 \end{bmatrix}$$

Multiplying the matrices on the left and equating the corresponding elements on both sides, we get by forward substitution,

$$L^{-1} = \begin{bmatrix} 1 & 0 & 0 \\ \frac{-7}{5} & 1 & 0 \\ \frac{46}{19} & \frac{-41}{19} & 1 \end{bmatrix}$$

From $UU^{-1} = I$,

i.e.,
$$\begin{bmatrix} 5 & -2 & 1 \\ 0 & \frac{19}{5} & \frac{-32}{5} \\ 0 & 0 & \frac{327}{19} \end{bmatrix} \begin{bmatrix} u_{11}^* & u_{12}^* & u_{13}^* \\ 0 & u_{22}^* & u_{23}^* \\ 0 & 0 & u_{33}^* \end{bmatrix} = \begin{bmatrix} 1 & 0 & 0 \\ 0 & 1 & 0 \\ 0 & 0 & 1 \end{bmatrix}$$

Comparing the elements on both sides and solving by back substitution, we get

$$U^{-1} = \begin{bmatrix} \frac{1}{5} & \frac{2}{19} & \frac{9}{327} \\ 0 & \frac{5}{19} & \frac{32}{327} \\ 0 & 0 & \frac{19}{327} \end{bmatrix}$$

Now
$$A^{-1} = U^{-1}L^{-1}$$

$$= \begin{pmatrix} \frac{1}{5} & \frac{2}{19} & \frac{9}{327} \\ 0 & \frac{5}{19} & \frac{32}{327} \\ 0 & 0 & \frac{19}{327} \end{pmatrix} \begin{pmatrix} 1 & 0 & 0 \\ \frac{-7}{5} & 1 & 0 \\ \frac{46}{19} & \frac{-41}{19} & 1 \end{pmatrix}$$

$$= \frac{1}{6213} \begin{bmatrix} 741 & 285 & 171 \\ -817 & 327 & 608 \\ 874 & -779 & 361 \end{bmatrix}$$

EXERCISE 6.3

1. Solve the following system of equations, using LU decomposition method

$$4x_1 + x_2 + x_3 = 4$$
$$x_1 + 4x_2 - 2x_3 = 4$$
$$3x_1 + 2x_2 - 4x_3 = 6$$

2. Using the decomposition method, solve the system of equations

$$\begin{bmatrix} 2 & -4 & 3 \\ 4 & 1 & -6 \\ 5 & 8 & -4 \end{bmatrix} \begin{bmatrix} x \\ y \\ z \end{bmatrix} = \begin{bmatrix} 1 \\ -1 \\ 9 \end{bmatrix}$$

3. Find the inverse of the matrix using LU decomposition method,

(*i*) $\begin{bmatrix} 2 & 3 & 1 \\ 1 & 2 & 3 \\ 3 & 1 & 2 \end{bmatrix}$ (*ii*) $\begin{bmatrix} 1 & 1 & 1 \\ 3 & 1 & -3 \\ 1 & -2 & -5 \end{bmatrix}$

Answers

1. $L = \begin{bmatrix} 1 & 0 & 0 \\ \frac{1}{4} & 1 & 0 \\ \frac{3}{4} & \frac{1}{3} & 1 \end{bmatrix}, U = \begin{bmatrix} 4 & 1 & 1 \\ 0 & \frac{15}{4} & \frac{-9}{4} \\ 0 & 0 & -4 \end{bmatrix}$

$X = \left(1, \frac{1}{2}, \frac{-1}{2}\right)^T$

2. $X = (1, 1, 1)^T$

3. (*i*) $L^{-1} = \begin{bmatrix} 1 & 0 & 0 \\ -\frac{1}{2} & 1 & 0 \\ -5 & 7 & 1 \end{bmatrix}, U^{-1} = \frac{1}{18}\begin{bmatrix} 9 & -54 & 7 \\ 0 & 36 & -5 \\ 0 & 0 & 1 \end{bmatrix}.$

$A^{-1} = U^{-1}L^{-1} = \frac{1}{18}\begin{bmatrix} 1 & -5 & 7 \\ 7 & 1 & -5 \\ -5 & 7 & 1 \end{bmatrix}.$

(*ii*) $L^{-1} = \begin{bmatrix} 1 & 0 & 0 \\ -3 & 1 & 0 \\ \frac{7}{2} & \frac{-3}{2} & 1 \end{bmatrix}, U^{-1} = \frac{1}{6}\begin{bmatrix} 6 & 3 & 4 \\ 0 & -3 & -6 \\ 0 & 0 & 2 \end{bmatrix},$

$A^{-1} = U^{-1} L^{-1} = \frac{1}{6}\begin{bmatrix} 11 & -3 & 4 \\ -12 & 6 & -6 \\ 7 & -3 & 2 \end{bmatrix}.$

6.10. VECTORS

Any ordered n-tuple of numbers is called an n-vector. By an ordered n-tuple, we mean a set consisting of n numbers in which the place of each number is fixed. If $x_1, x_2, ..., x_n$ be any n numbers then the ordered n-tuple $X = (x_1, x_2, ..., x_n)$ is called an n-vector. Thus the coordinates of a point in space can be represented by a 3-vector (x, y, z). Similarly (1, 0, 2, – 1) and (2, 7, 5, – 3) are 4-vectors. The n numbers $x_1, x_2, ..., x_n$ are called the components of the n-vector X $= (x_1, x_2, ..., x_n)$. A vector may be written either as a *row vector* or as a *column vector*. If A be a matrix of order $m \times n$, then each row of A will be an n-vector and each column of A will be an m-vector. A vector whose components are all zero is called a zero vector and is denoted by O. Thus O = (0, 0, 0, ..., 0).

Let $X = (x_1, x_2, ..., x_n)$ and $Y = (y_1, y_2, ..., y_n)$ be two vectors.

Then X = Y if and only if their corresponding components are equal.

i.e., if $x_i = y_i$, for $i = 1, 2, ..., n$

$$X + Y = (x_1 + y_1, x_2 + y_2, ..., x_n + y_n)$$

If k be a scalar, then $kX = (kx_1, kx_2, ..., kx_n)$.

6.11. LINEAR DEPENDENCE AND LINEAR INDEPENDENCE OF VECTORS

A set of r n-vectors $X_1, X_2, ..., X_r$ is said to be *linearly dependent* if there exist r scalars (numbers) $k_1, k_2, ..., k_r$ not all zero, such that

$$k_1X_1 + k_2X_2 + \cdots + k_rX_r = O$$

If $k_i \neq 0$, then $k_iX_i = -k_1X_1 - k_2X_2 \cdots - k_rX_r$

$$\Rightarrow \quad X_i = -\frac{k_1}{k_i}X_1 - \frac{k_2}{k_i}X_2 - \cdots - \frac{k_r}{k_i}X_r$$

or

$$X_i = \lambda_1X_1 + \lambda_2X_2 + \cdots + \lambda_rX_r$$

The vector X_i is called a linear combination of the remaining vectors $X_1, X_2, ..., X_r$.

If a set of vectors is linearly dependent, then at least one member of the set can be expressed as a linear combination of the remaining vectors.

A set of r n-vectors $X_1, X_2, ..., X_r$ is said to be *linearly independent* if every relation of the type

$$k_1X_1 + k_2X_2 + \cdots + k_rX_r = O \text{ implies } k_1 = k_2 = \cdots = k_r = 0$$

To test the linear dependence of r given vectors, write them as row vectors. Add suitable multiples of one vector to the others so that the resulting $(r - 1)$ vectors have their first component zero. Choose any one of these $(r - 1)$ vectors and add its multiples to the others so that the resulting $(r - 2)$ vectors have their second component zero. In this way continue, reducing the successive components to zero. If the final reduction gives a vector all of whose components are zero, then the original vectors are linearly dependent. However, if the final reduction gives a vector all of whose components are not zero, then the original vectors are linearly independent.

Example. *Show that the vectors $x_1 = (1, 2, 4)$, $x_2 = (2, -1, 3)$, $x_3 = (0, 1, 2)$ and $x_4 = (-3, 7, 2)$ are linearly dependent and find the relation between them.*

Sol. Adding suitable multiples of x_1 to x_2 and x_4 so that the first component reduces to zero, we have

$$x_2 - 2x_1 = (2, -1, 3) - (2, 4, 8) = (0, -5, -5)$$

$$x_4 + 3x_1 = (-3, 7, 2) + (3, 6, 12) = (0, 13, 14)$$

Also $x_3 = (0, 1, 2)$.

Adding suitable multiples of x_3 to the above vectors so that the second component reduces to zero, we have

$$(x_2 - 2x_1) + 5x_3 = (0, -5, -5) + (0, 5, 10) = (0, 0, 5)$$

$$(x_4 + 3x_1) - 13x_3 = (0, 13, 14) - (0, 13, 26) = (0, 0, -12)$$

To reduce the third component to zero, multiplying the above vectors by 12 and 5 respectively and adding, we have

$$12(x_2 - 2x_1 + 5x_3) + 5(x_4 + 3x_1 - 13x_3) = (0, 0, 60) + (0, 0, -60)$$

$$\Rightarrow \quad -9x_1 + 12x_2 - 5x_3 + 5x_4 = (0, 0, 0)$$

$$\Rightarrow \quad 9x_1 - 12x_2 + 5x_3 - 5x_4 = O \quad ...(1)$$

Thus, there exist numbers $k_1 = 9, k_2 = -12, k_3 = 5, k_4 = -5$ which are not all zero such that

$$k_1x_1 + k_2x_2 + k_3x_3 + k_4x_4 = O$$

Hence the vectors x_1, x_2, x_3 and x_4 are linearly dependent. Also, (1) is the relation between them.

Second Method

Consider the relation $k_1x_1 + k_2x_2 + k_3x_3 + k_4x_4 = 0$

i.e., $k_1(1, 2, 4) + k_2(2, -1, 3) + k_3(0, 1, 2) + k_4(-3, 7, 2) = 0$

or $(k_1, 2k_1, 4k_1) + (2k_2, -k_2, 3k_2) + (0, k_3, 2k_3) + (-3k_4, 7k_4, 2k_4) = 0$

or $(k_1 + 2k_2 - 3k_4, 2k_1 - k_2 + k_3 + 7k_4, 4k_1 + 3k_2 + 2k_3 + 2k_4) = (0, 0, 0)$

Equating corresponding components on both sides, we have

$$k_1 + 2k_2 - 3k_4 = 0 \quad ...(1)$$
$$2k_1 - k_2 + k_3 + 7k_4 = 0 \quad ...(2)$$
$$4k_1 + 3k_2 + 2k_3 + 2k_4 = 0 \quad ...(3)$$

We have a homogeneous system of 3 linear equations in 4 unknowns. The coefficient matrix

$$A = \begin{bmatrix} 1 & 2 & 0 & -3 \\ 2 & -1 & 1 & 7 \\ 4 & 3 & 2 & 2 \end{bmatrix} \text{ is } 3 \times 4.$$

$\rho(A) \le \min\{3, 4\} = 3$. Since rank of A is less than the number of unknowns, the homogeneous system has infinitely many non-zero solution. Thus there exist scalars k_1, k_2, k_3, k_4, not all zero, such that

$$k_1x_1 + k_2x_2 + k_3x_3 + k_4x_4 = 0$$

$\therefore$ The vectors x_1, x_2, x_3, x_4 are linearly dependent.

Now multiplying (2) by 2 and subtracting from (3), we have

$$5k_2 - 12k_4 = 0 \Rightarrow k_2 = \frac{12}{5}k_4$$

From (1), $k_1 = 3k_4 - 2k_2 = 3k_4 - \frac{24}{5}k_4 = -\frac{9}{5}k_4$

From (2), $k_3 = -2k_1 + k_2 - 7k_4 = \frac{18}{5}k_4 + \frac{12}{5}k_4 - 7k_4 = -k_4$

Thus $k_1 = -\frac{9}{5}k_4, \quad k_2 = \frac{12}{5}k_4, \quad k_3 = -k_4$ and k_4 is arbitrary.

Taking $k_4 = -5$, we have $k_1 = 9, k_2 = -12, k_3 = 5$

$\therefore$ $9x_1 - 12x_2 + 5x_3 - 5x_4 = 0$

Since k_1, k_2, k_3, k_4 are all non-zero, we can express each of the given vectors as a linear combination of the remaining three.

e.g., $9x_1 = 12x_2 - 5x_3 + 5x_4$

$\therefore$ $x_1 = \frac{4}{3}x_2 - \frac{5}{9}x_3 + \frac{5}{9}x_4.$

Note. The elements in the four columns of A are the components of the four given vectors.

6.12. LINEAR TRANSFORMATIONS

Let a point P(x, y) in a plane transform to the point P$'(x', y')$ under reflection in the co-ordinate axes, or reflection in the line $y = x \tan\theta$ or rotation of OP through an angle θ about the origin or rotation of axes through an angle θ etc. Then the co-ordinates of P$'$ can be expressed in terms of those of P by the linear relations of the form

$$\left.\begin{aligned} x' &= a_1x + b_1y \\ y' &= a_2x + b_2y \end{aligned}\right]$$

which in matrix notation is $\begin{bmatrix} x' \\ y' \end{bmatrix} = \begin{bmatrix} a_1 & b_1 \\ a_2 & b_2 \end{bmatrix} \begin{bmatrix} x \\ y \end{bmatrix}$ or $X' = AX$

Such transformations are called linear transformations in two dimensions.

Similarly, relations of the form $\left.\begin{aligned} x' &= a_1x + b_1y + c_1z \\ y' &= a_2x + b_2y + c_2z \\ z' &= a_3x + b_3y + c_3z \end{aligned}\right]$

which in matrix notation is $\begin{bmatrix} x' \\ y' \\ z' \end{bmatrix} = \begin{bmatrix} a_1 & b_1 & c_1 \\ a_2 & b_2 & c_2 \\ a_3 & b_3 & c_3 \end{bmatrix} \begin{bmatrix} x \\ y \\ z \end{bmatrix}$ or $X' = AX$ gives a linear transformation $(x, y, z) \longrightarrow (x', y', z')$ in three dimensions.

In general, the relation Y = AX, where $Y = \begin{bmatrix} y_1 \\ y_2 \\ \vdots \\ y_n \end{bmatrix}$, $A = \begin{bmatrix} a_{11} & a_{12} & \cdots & a_{1n} \\ a_{21} & a_{22} & \cdots & a_{2n} \\ \ldots & \ldots & \ldots & \ldots \\ a_{n1} & a_{n2} & \cdots & a_{nn} \end{bmatrix}$, $X = \begin{bmatrix} x_1 \\ x_2 \\ \vdots \\ x_n \end{bmatrix}$

defines a linear transformation which carries any vector X into another vector Y over the matrix A which is called the linear operator of the transformation.

This transformation is called linear because $Y_1 = AX_1$ and $Y_2 = AX_2$ implies $aY_1 + bY_2 = A(aX_1 + bX_2)$ for all values of a and b.

Thus, if $X = \begin{bmatrix} 2 \\ -3 \end{bmatrix}$ and $A = \begin{bmatrix} 1 & -1 \\ 2 & 3 \end{bmatrix}$, then $Y = \begin{bmatrix} y_1 \\ y_2 \end{bmatrix} = \begin{bmatrix} 1 & -1 \\ 2 & 3 \end{bmatrix} \begin{bmatrix} 2 \\ -3 \end{bmatrix} = \begin{bmatrix} 5 \\ -5 \end{bmatrix}$

so that $(2, -3) \rightarrow (5, -5)$ under the transformation defined by A.

If the transformation matrix A is non-singular, *i.e.*, if $|A| \neq 0$, then the linear transformation is called *non-singular* or *regular*.

If the transformation matrix A is singular, *i.e.*, if $|A| = 0$, then the linear transformation is also called *singular*.

For a non-singular transformation Y = AX, since A is non-singular, A^{-1} exists and we can write the inverse transformation, which carries the vector Y back into the vector X, as $X = A^{-1}Y$.

Note. If a transformation from $(x_1, x_2, ..., x_n)$ to $(y_1, y_2, ..., y_n)$ is given by Y = AX and another transformation from $(y_1, y_2, ..., y_n)$ to $(z_1, z_2, ..., z_n)$ is given by Z = BY, then the transformation from $(x_1, x_2, ..., x_n)$ to $(z_1, z_2, ..., z_n)$ is given by Z = BY = B(AX) = (BA)X.

6.13. ORTHOGONAL TRANSFORMATION

The linear transformation Y = AX, where

$$Y = \begin{bmatrix} y_1 \\ y_2 \\ \vdots \\ y_n \end{bmatrix}, A = \begin{bmatrix} a_{11} & a_{12} & \cdots & a_{1n} \\ a_{21} & a_{22} & \cdots & a_{2n} \\ \ldots & \ldots & \ldots & \ldots \\ a_{n1} & a_{n2} & \cdots & a_{nn} \end{bmatrix}, X = \begin{bmatrix} x_1 \\ x_2 \\ \vdots \\ x_n \end{bmatrix}$$

is said to be *orthogonal* if it transforms $y_1^2 + y_2^2 + \cdots + y_n^2$ into $x_1^2 + x_2^2 + \cdots + x_n^2$.

The matrix A of this transformation is called an *orthogonal matrix.*

Now $$X'X = [x_1\, x_2 \,\dots\, x_n] \begin{bmatrix} x_1 \\ x_2 \\ \vdots \\ x_n \end{bmatrix} = x_1^2 + x_2^2 + \cdots + x_n^2$$

and similarly $$Y'Y = y_1^2 + y_2^2 + \cdots + y_n^2$$

$\therefore$ If Y = AX is an orthogonal transformation, then

$$X'X = x_1^2 + x_2^2 + \cdots + x_n^2 = y_1^2 + y_2^2 + \cdots + y_n^2$$
$$= Y'Y = (AX)'(AX) = (X'A')(AX) \qquad [\because \; (AB)' = B'A']$$
$$= X'(A'A)X$$

which holds only when $A'A = I$ or when $A'A = A^{-1}A$ $\qquad [\because \; A^{-1}A = I]$

or when $A' = A^{-1}$.

Hence *a square matrix A is said to be orthogonal if* $AA' = A'A = I$

Also, *for an orthogonal matrix A,* $A' = A^{-1}$.

ILLUSTRATIVE EXAMPLES

Example 1. *Show that the transformation* $y_1 = 2x_1 + x_2 + x_3, y_2 = x_1 + x_2 + 2x_3, y_3 = x_1 - 2x_3$ *is regular. Write down the inverse transformation.*

Sol. In matrix notation, the given transformation is Y = AX,

where $$Y = \begin{bmatrix} y_1 \\ y_2 \\ y_3 \end{bmatrix}, A = \begin{bmatrix} 2 & 1 & 1 \\ 1 & 1 & 2 \\ 1 & 0 & -2 \end{bmatrix}, X = \begin{bmatrix} x_1 \\ x_2 \\ x_3 \end{bmatrix}$$

Since $$|A| = \begin{vmatrix} 2 & 1 & 1 \\ 1 & 1 & 2 \\ 1 & 0 & -2 \end{vmatrix} = -1 \neq 0,$$

the matrix A is non-singular and hence, the given transformation is non-singular or regular.

As usual, $$A^{-1} = \begin{bmatrix} 2 & -2 & -1 \\ -4 & 5 & 3 \\ 1 & -1 & -1 \end{bmatrix}$$

The inverse transformation is given by $X = A^{-1}Y$

$$\Rightarrow \quad \begin{bmatrix} x_1 \\ x_2 \\ x_3 \end{bmatrix} = \begin{bmatrix} 2 & -2 & -1 \\ -4 & 5 & 3 \\ 1 & -1 & -1 \end{bmatrix} \begin{bmatrix} y_1 \\ y_2 \\ y_3 \end{bmatrix}$$

$$\Rightarrow \quad x_1 = 2y_1 - 2y_2 - y_3,\; x_2 = -4y_1 + 5y_2 + 3y_3,\; x_3 = y_1 - y_2 - y_3.$$

Example 2. *Prove that the following matrix is orthogonal* $\dfrac{1}{3}\begin{bmatrix} 1 & 2 & 2 \\ 2 & 1 & -2 \\ 2 & -2 & 1 \end{bmatrix}$.

Sol. Denoting the given matrix by A, we have

$$A' = \frac{1}{3}\begin{bmatrix} 1 & 2 & 2 \\ 2 & 1 & -2 \\ 2 & -2 & 1 \end{bmatrix}$$

Now $$AA' = \frac{1}{3}\begin{bmatrix} 1 & 2 & 2 \\ 2 & 1 & -2 \\ 2 & -2 & 1 \end{bmatrix} \times \frac{1}{3}\begin{bmatrix} 1 & 2 & 2 \\ 2 & 1 & -2 \\ 2 & -2 & 1 \end{bmatrix}$$

$$= \frac{1}{9}\begin{bmatrix} 9 & 0 & 0 \\ 0 & 9 & 0 \\ 0 & 0 & 9 \end{bmatrix} = \begin{bmatrix} 1 & 0 & 0 \\ 0 & 1 & 0 \\ 0 & 0 & 1 \end{bmatrix} = I$$

Since AA′ = I, A is an orthogonal matrix.

Example 3. *If A is an orthogonal matrix, prove that | A | = ± 1.*

Sol. A is an orthogonal matrix

$\Rightarrow$ $AA' = I$ $\Rightarrow$ $|AA'| = |I|$

$\Rightarrow$ $|A||A'| = 1$ $\Rightarrow$ $|A||A| = 1$ $[\because \; |A'| = |A|]$

$\Rightarrow$ $|A|^2 = 1$ $\therefore$ $|A| = \pm 1.$

Example 4. *Prove that the inverse of an orthogonal matrix is orthogonal and its transpose is also orthogonal.*

Sol. (*i*) A is orthogonal $\Rightarrow$ $AA' = I$

$\Rightarrow$ $(AA')^{-1} = I^{-1}$ $\Rightarrow$ $(A')^{-1} A^{-1} = I$ $[\because \; (AB)^{-1} = B^{-1}A^{-1}]$

$\Rightarrow$ $(A^{-1})' A^{-1} = I$ $[\because \; (A')^{-1} = (A^{-1})']$

$\Rightarrow$ Product of A^{-1} and its transpose $(A^{-1})'$ is equal to I

$\Rightarrow$ A^{-1} is also orthogonal.

(*ii*) A is orthogonal $\Rightarrow$ $AA' = I$ $\Rightarrow$ $(AA')' = I'$ $\Rightarrow$ $(A')' A' = I$

$\Rightarrow$ Product of A′ and its transpose (A′)′ is equal to I.

$\Rightarrow$ A′ is also orthogonal.

EXERCISE 6.4

1. Are the following vectors linearly dependent? If so, find a relation between them.
   (*i*) $x_1 = (1, 2, 1), x_2 = (2, 1, 4), x_3 = (4, 5, 6)$
   (*ii*) $x_1 = (2, -1, 4), x_2 = (0, 1, 2), x_3 = (6, -1, 16)$
   (*iii*) $x_1 = (2, -1, 3, 2), x_2 = (1, 3, 4, 2), x_3 = (3, -5, 2, 2)$
   (*iv*) $x_1 = (2, 3, 1, -1), x_2 = (2, 3, 1, -2), x_3 = (4, 6, 2, 1).$
   (*v*) $x_1 = (1, 1, 1, 3), x_2 = (1, 2, 3, 4), x_3 = (2, 3, 4, 9)$
2. Show that the transformation $y_1 = x_1 - x_2 + x_3, y_2 = 3x_1 - x_2 + 2x_3, y_3 = 2x_1 - 2x_2 + 3x_3$ is non-singular. Find the inverse transformation.
3. Represent each of the transformations $x_1 = 3y_1 + 2y_2, x_2 = -y_1 + 4y_2, y_1 = z_1 + 2z_2$ and $y_2 = 3z_1$ by the use of matrices and find the composite transformation which expresses x_1, x_2 in terms of z_1, z_2.
4. A transformation from the variables x_1, x_2, x_3 to y_1, y_2, y_3 is given by Y = AX, and another transformation from y_1, y_2, y_3 to z_1, z_2, z_3 is given by Z = BY, where $A = \begin{bmatrix} 2 & 1 & 0 \\ 0 & 1 & 2 \\ -1 & 2 & 1 \end{bmatrix}, B = \begin{bmatrix} 1 & 1 & 1 \\ 1 & 2 & -3 \\ 1 & 3 & 5 \end{bmatrix}.$

   Obtain the transformation from x_1, x_2, x_3 to z_1, z_2, z_3.
5. Which of the following matrices are orthogonal?

   (*i*) $\frac{1}{9}\begin{bmatrix} -8 & 4 & 1 \\ 1 & 4 & -8 \\ 4 & 7 & 4 \end{bmatrix}$ (*ii*) $\begin{bmatrix} 2 & -3 & 1 \\ 4 & 3 & 1 \\ -3 & 1 & 9 \end{bmatrix}.$

6. Prove that the matrix $\begin{bmatrix} -\frac{2}{3} & \frac{1}{3} & \frac{2}{3} \\ \frac{2}{3} & \frac{2}{3} & \frac{1}{3} \\ \frac{1}{3} & -\frac{2}{3} & \frac{2}{3} \end{bmatrix}$ is orthogonal.

7. Verify that the matrix $\begin{bmatrix} \cos\theta & 0 & \sin\theta \\ 0 & 1 & 0 \\ -\sin\theta & 0 & \cos\theta \end{bmatrix}$ is orthogonal

8. If A and B are orthogonal matrices, prove that AB is also orthogonal.

Answers

1. (*i*) Yes; $x_3 = 2x_1 + x_2$ (*ii*) Yes; $x_3 = 3x_1 + 2x_2$
(*iii*) Yes; $2x_1 - x_2 - x_3 = 0$ (*iv*) Yes; $5x_1 - 3x_2 - x_3 = 0$ (*v*) No

2. $x_1 = \frac{1}{2}(y_1 + y_2 - y_3),\ x_2 = \frac{1}{2}(-5y_1 + y_2 + y_3),\ x_3 = -4y_1 + 2y_3$

3. $\begin{bmatrix} x_1 \\ x_2 \end{bmatrix} = \begin{bmatrix} 9 & 6 \\ 11 & -2 \end{bmatrix}\begin{bmatrix} z_1 \\ z_2 \end{bmatrix}$

4. Z = (BA)X, where $BA = \begin{bmatrix} 1 & 4 & -1 \\ -1 & 9 & -1 \\ -3 & 14 & -1 \end{bmatrix}$

5. (*i*) Orthogonal (*ii*) Not orthogonal.

6.14. CHARACTERISTIC EQUATION

If A is a square matrix of order n, we can form the matrix $A - \lambda I$, where λ is a scalar and I is the unit matrix of order n. The determinant of this matrix equated to zero, *i.e.*,

$$|A - \lambda I| = \begin{vmatrix} a_{11}-\lambda & a_{12} & \cdots a_{1n} \\ a_{21} & a_{22}-\lambda & \cdots a_{2n} \\ \cdots & \cdots & \cdots \\ a_{n1} & a_{n2} & \cdots a_{nn}-\lambda \end{vmatrix} = 0$$ is called the *characteristic equation* of A.

On expanding the determinant, the characteristic equation can be written as a polynomial equation of degree n in λ of the form $(-1)^n \lambda^n + k_1\lambda^{n-1} + k_2\lambda^{n-2} + \cdots + k_n = 0$.

The roots of this equation are called the *characteristic roots* or *latent roots* or *eigen values* of A.

6.15. EIGEN VECTORS

Consider the linear transformation Y = AX ...(1)
which transforms the column vector X into the column vector Y. In practice, we are often required to find those vectors X which transform into scalar multiples of themselves.

Let X be such a vector which transforms into λX (λ being a non-zero scalar) by the transformation (1).

Then $Y = \lambda X$ ...(2)

From (1) and (2), $AX = \lambda X \Rightarrow AX - \lambda IX = O \Rightarrow (A - \lambda I)X = O$ ...(3)

This matrix equation gives n homogeneous linear equations

$$\left.\begin{array}{l} (a_{11}-\lambda)x_1 + a_{12}x_2 + \cdots + a_{1n}x_n = 0 \\ a_{21}x_1 + (a_{22}-\lambda)x_2 + \cdots + a_{2n}x_n = 0 \\ \cdots\cdots\cdots\cdots\cdots\cdots\cdots\cdots \\ a_{n1}x_1 + a_{n2}x_2 + \cdots + (a_{nn}-\lambda)x_n = 0 \end{array}\right] \quad \ldots(4)$$

These equations will have a non-trivial solution only if the co-efficient matrix $A - \lambda I$ is singular

i.e., if $$|A - \lambda I| = 0 \quad ...(5)$$

This is the characteristic equation of the matrix A and has n roots which are the eigen values of A. Corresponding to each root of (5), the homogeneous system (3) has a non-zero solution

$$X = \begin{bmatrix} x_1 \\ x_2 \\ \vdots \\ x_n \end{bmatrix}$$ which is called an *eigen vector* or *latent vector*.

Note. If X is a solution of (3), then so is kX, where k is an arbitrary constant. Thus, the eigen vector corresponding to an eigen value is not unique.

ILLUSTRATIVE EXAMPLES

Example 1. *Find the eigen values and eigen vectors of the matrix* $A = \begin{bmatrix} 1 & -2 \\ -5 & 4 \end{bmatrix}$.

Sol. The characteristic equation of the given matrix is

$$|A - \lambda I| = 0 \quad \text{or} \quad \begin{vmatrix} 1-\lambda & -2 \\ -5 & 4-\lambda \end{vmatrix} = 0$$

or $$(1-\lambda)(4-\lambda) - 10 = 0 \quad \text{or} \quad \lambda^2 - 5\lambda - 6 = 0$$

or $$(\lambda - 6)(\lambda + 1) = 0 \quad \therefore \quad \lambda = 6, -1.$$

Thus, the eigen values of A are 6, –1.

Corresponding to $\lambda = 6$, the eigen vectors are given by $(A - 6I)X = O$

or $$\begin{bmatrix} 1-6 & -2 \\ -5 & 4-6 \end{bmatrix}\begin{bmatrix} x_1 \\ x_2 \end{bmatrix} = O \quad \text{or} \quad \begin{bmatrix} -5 & -2 \\ -5 & -2 \end{bmatrix}\begin{bmatrix} x_1 \\ x_2 \end{bmatrix} = O$$

We get only one independent equation $-5x_1 - 2x_2 = 0$, since rank of co-efficient matrix is 1.

$\therefore \quad \dfrac{x_1}{2} = \dfrac{x_2}{-5}$ giving the eigen vector (2, – 5).

Corresponding to $\lambda = -1$, the eigen vectors are given by $\begin{bmatrix} 2 & -2 \\ -5 & 5 \end{bmatrix}\begin{bmatrix} x_1 \\ x_2 \end{bmatrix} = O$

We get only one independent equation $2x_1 - 2x_2 = 0$, since rank of co-efficient matrix is 1.

$\therefore \quad x_1 = x_2$ giving the eigen vector (1, 1).

Example 2. *Find the eigen values and eigen vectors of the matrix* $A = \begin{bmatrix} -2 & 2 & -3 \\ 2 & 1 & -6 \\ -1 & -2 & 0 \end{bmatrix}$.

Sol. The characteristic equation of the given matrix is $|A - \lambda I| = 0$

or $$\begin{vmatrix} -2-\lambda & 2 & -3 \\ 2 & 1-\lambda & -6 \\ -1 & -2 & -\lambda \end{vmatrix} = 0$$

or $\quad (-2-\lambda)[-\lambda(1-\lambda)-12]-2[-2\lambda-6]-3[-4+1(1-\lambda)]=0$

or $\quad \lambda^3+\lambda^2-21\lambda-45=0$

By trial, $\lambda = -3$ satisfies it.

$\therefore \quad (\lambda+3)(\lambda^2-2\lambda-15)=0 \Rightarrow (\lambda+3)(\lambda+3)(\lambda-5)=0 \Rightarrow \lambda=-3,-3,5$

Thus, the eigen values of A are –3, –3, 5.

Corresponding to $\lambda = -3$, the eigen vectors are given by

$$(A+3I)X = O \qquad \text{or} \qquad \begin{bmatrix} 1 & 2 & -3 \\ 2 & 4 & -6 \\ -1 & -2 & 3 \end{bmatrix}\begin{bmatrix} x_1 \\ x_2 \\ x_3 \end{bmatrix} = O$$

We get only one independent equation $x_1 + 2x_2 - 3x_3 = 0$, since rank of co-efficient matrix is 1.

Choosing $x_2 = 0$, we have $x_1 - 3x_3 = 0$

$\therefore \quad \dfrac{x_1}{3} = \dfrac{x_2}{0} = \dfrac{x_3}{1}$ giving the eigen vector (3, 0, 1)

Choosing $x_3 = 0$, we have $x_1 + 2x_2 = 0$

$\therefore \quad \dfrac{x_1}{2} = \dfrac{x_2}{-1} = \dfrac{x_3}{0}$ giving the eigen vector (2, –1, 0)

Any other eigen vector corresponding to $\lambda = -3$ will be a linear combination of these two.

Corresponding to $\lambda = 5$, the eigen vectors are given by $\begin{bmatrix} -7 & 2 & -3 \\ 2 & -4 & -6 \\ -1 & -2 & -5 \end{bmatrix}\begin{bmatrix} x_1 \\ x_2 \\ x_3 \end{bmatrix} = O$

$$\Rightarrow \quad \begin{aligned} -7x_1 + 2x_2 - 3x_3 &= 0 \\ 2x_1 - 4x_2 - 6x_3 &= 0 \\ -x_1 - 2x_2 - 5x_3 &= 0 \end{aligned}$$

Since only two of them are independent, we can omit one of them.

From first two equations, we have $\dfrac{x_1}{-12-12} = \dfrac{x_2}{-6-42} = \dfrac{x_3}{28-4}$

or $\quad \dfrac{x_1}{1} = \dfrac{x_2}{2} = \dfrac{x_3}{-1}$ giving the eigen vector (1, 2, –1).

Example 3. *Find the eigen values of* $A = \begin{bmatrix} 2 & 2 & 1 \\ 1 & 3 & 1 \\ 1 & 2 & 2 \end{bmatrix}$ *and find eigen vector corresponding to the largest eigen value.*

Sol. The characteristic equation is

$$\begin{vmatrix} 2-\lambda & 2 & 1 \\ 1 & 3-\lambda & 1 \\ 1 & 2 & 2-\lambda \end{vmatrix} = 0$$

i.e., $\quad (\lambda-1)^2(\lambda-5)=0$

So the eigen values are $\lambda = 1, 1, 5$. The largest eigen value is 5. Eigen vector corresponding to $\lambda = 5$ is obtained by finding the non trivial solution of the system.

$$\begin{bmatrix} -3 & 2 & 1 \\ 1 & -2 & 1 \\ 1 & 2 & -3 \end{bmatrix} \begin{bmatrix} x \\ y \\ z \end{bmatrix} = \begin{bmatrix} 0 \\ 0 \\ 0 \end{bmatrix}$$

The augmented matrix of the above system is

$$\sim \begin{bmatrix} -3 & 2 & 1 & 0 \\ 1 & -2 & 1 & 0 \\ 1 & 2 & -3 & 0 \end{bmatrix}, \text{ operating } R_{13},$$

$$\sim \begin{bmatrix} 1 & 2 & -3 & 0 \\ 1 & -2 & 1 & 0 \\ -3 & 2 & 1 & 0 \end{bmatrix}, \text{ operating } R_2 - R_1, R_3 + 3R_1$$

$$\sim \begin{bmatrix} 1 & 2 & -3 & 0 \\ 0 & 4 & -4 & 0 \\ 0 & 8 & -8 & 0 \end{bmatrix}, \text{ operating } \frac{1}{4} R_2, R_3 - 2R_2$$

$$\sim \begin{bmatrix} 1 & 2 & -3 & 0 \\ 0 & 1 & -1 & 0 \\ 0 & 0 & 0 & 0 \end{bmatrix}$$

This leads to the conclusion that the rank of the augmented matrix is two and any one of the variables can be equated to an arbitrary value α. Note that the number of unknowns is three and the difference between the number of unknown and the rank of the augmented matrix is one.

Thus, we have

$$x + 2y - z = 0$$

and $$y - z = 0 \text{ with } y = \alpha$$

This gives that $x = y = z = \alpha$, a nontrivial general solution. *i.e.*, (α, α, α) is an eigen vector. In particular $(1, 1, 1)$ is an eigen vector corresponding the eigen value $\lambda = 5$; where $\alpha = 1$.

Hence the result.

6.16. PROPERTIES OF EIGEN VALUES

(*a*) *The eigen values of a square matrix A and its transpose A′ are the same.*

Let $$A = \begin{bmatrix} a_{11} & a_{12} & \cdots\cdots & a_{1n} \\ a_{21} & a_{22} & \cdots\cdots & a_{2n} \\ \cdots & \cdots & \cdots & \cdots \\ a_{n1} & a_{n2} & \cdots\cdots & a_{nn} \end{bmatrix}$$

then $$A' = \begin{bmatrix} a_{11} & a_{21} & \cdots\cdots & a_{n1} \\ a_{12} & a_{22} & \cdots\cdots & a_{n2} \\ \cdots & \cdots & \cdots & \cdots \\ a_{1n} & a_{2n} & \cdots\cdots & a_{nn} \end{bmatrix}$$

The characteristic equation of A is given by

$$|A - \lambda I| = \begin{vmatrix} a_{11}-\lambda & a_{12} & \cdots\cdots & a_{1n} \\ a_{21} & a_{22}-\lambda & \cdots\cdots & a_{2n} \\ \cdots & \cdots & \cdots & \cdots \\ a_{n1} & a_{n2} & \cdots\cdots & a_{nn}-\lambda \end{vmatrix} = 0 \qquad ...(1)$$

Also, the characteristic equation of A′ is given by

$$|A' - \lambda I| = \begin{vmatrix} a_{11}-\lambda & a_{21} & \cdots\cdots & a_{n1} \\ a_{12} & a_{22}-\lambda & \cdots\cdots & a_{n2} \\ \cdots & \cdots & \cdots & \cdots \\ a_{1n} & a_{2n} & \cdots\cdots & a_{nn}-\lambda \end{vmatrix} = 0 \qquad ...(2)$$

Since the value of a determinant remains unchanged when its rows and columns are interchanged, therefore, the characteristic equations (1) and (2) are same. Hence the eigen values of A and A′ are the same.

(*b*) *The sum of the eigen values of a matrix is the sum of the elements on the principal diagonal.*

We prove the property for a matrix of order 3. However, the result is true for a matrix of any order.

Let $$A = \begin{bmatrix} a_{11} & a_{12} & a_{13} \\ a_{21} & a_{22} & a_{23} \\ a_{31} & a_{32} & a_{33} \end{bmatrix}$$

so that $$|A - \lambda I| = \begin{vmatrix} a_{11}-\lambda & a_{12} & a_{13} \\ a_{21} & a_{22}-\lambda & a_{23} \\ a_{31} & a_{32} & a_{33}-\lambda \end{vmatrix}$$

$$= -\lambda^3 + \lambda^2(a_{11} + a_{22} + a_{33}) - \cdots \qquad ...(1)$$

If $\lambda_1, \lambda_2, \lambda_3$ be the eigen values of A, then

$$|A - \lambda I| = (-1)^3 (\lambda - \lambda_1)(\lambda - \lambda_2)(\lambda - \lambda_3)$$

$$= -\lambda^3 + \lambda^2(\lambda_1 + \lambda_2 + \lambda_3) - \cdots \qquad ...(2)$$

From (1) and (2), we have

$$-\lambda^3 + \lambda^2(a_{11} + a_{22} + a_{33}) \cdots = -\lambda^3 + \lambda^2(\lambda_1 + \lambda_2 + \lambda_3) \cdots$$

Comparing co-efficients of λ^2, we get

$$\lambda_1 + \lambda_2 + \lambda_3 = a_{11} + a_{22} + a_{33}$$

= The sum of the elements on the principal diagonal.

(*c*) *The product of the eigen values of a matrix A is equal to* |A|.

If $\lambda_1, \lambda_2, \ldots, \lambda_n$ be the eigen values of a square matrix A of order n, then

$$|A - \lambda I| = (-1)^n (\lambda - \lambda_1)(\lambda - \lambda_2) \ldots (\lambda - \lambda_n)$$

Putting $\lambda = 0$, we get

$$|A| = (-1)^n \cdot (-1)^n \lambda_1\lambda_2 \ldots \lambda_n$$

$$\Rightarrow \quad \lambda_1\lambda_2 \ldots \lambda_n = |A| \qquad [\because \ (-1)^{2n} = 1]$$

(d) If λ is an eigen value of a non-singular matrix A, then $\frac{1}{\lambda}$ is an eigen value of A^{-1}.

λ is an eigen value of A (K.U. May 2009)

$\Rightarrow$ There exists a non-zero vector X such that $AX = \lambda X$

Pre-multiplying both sides by A^{-1}, we get

$$A^{-1}(AX) = A^{-1}(\lambda X)$$

$$\Rightarrow \quad (A^{-1}A)X = \lambda(A^{-1}X) \Rightarrow X = \lambda(A^{-1}X)$$

$$\Rightarrow \quad \frac{1}{\lambda}X = A^{-1}X \quad \Rightarrow \quad A^{-1}X = \frac{1}{\lambda}X$$

$\Rightarrow$ $\frac{1}{\lambda}$ is an eigen value of A^{-1}.

(e) If λ is an eigen value of an orthogonal matrix, then $\frac{1}{\lambda}$ is also its eigen value. (K.U. May 2010)

Let λ be an eigen value of an orthogonal matrix A, then $\frac{1}{\lambda}$ is an eigen value of A^{-1}.

[Part (*d*) above]

But $\quad A^{-1} = A'$ $\quad$ ($\because$ A is an orthogonal matrix)

$\therefore$ $\frac{1}{\lambda}$ is an eigen value of A'.

But the matrices A and A' have the same eigen values. [Part (*a*) above]

$\therefore$ $\frac{1}{\lambda}$ is also an eigen value of A.

(f) If $\lambda_1, \lambda_2, ..., \lambda_n$ are the eigen values of a matrix A, then A^m has the eigen values $\lambda_1^m, \lambda_2^m, ..., \lambda_n^m$ (m being a positive integer). (K.U. May 2012)

Let λ be an eigen value of a matrix A, then there exists a non-zero vector X such that

$$AX = \lambda X \quad ...(1)$$

Now $\quad A(AX) = A(\lambda X) \Rightarrow A^2X = \lambda(AX)$

$$\Rightarrow \quad A^2X = \lambda(\lambda X) \quad \text{[by (1)]}$$

$$\Rightarrow \quad A^2X = \lambda^2 X$$

Similarly $A^3X = \lambda^3X$. In general $A^mX = \lambda^mX$

$\Rightarrow$ λ^m is an eigen value of A^m.

Since λ is an eigen value of A $\Rightarrow$ λ^m is an eigen value of A^m.

$\therefore$ If $\lambda_1, \lambda_2, ..., \lambda_n$ are the eigen values of A, then $\lambda_1^m, \lambda_2^m, ..., \lambda_n^m$ *are the eigen values of* A^m.

Example. *If λ be an eigen value of a non-singular matrix A, show that $\frac{|A|}{\lambda}$ is an eigen value of adj. A.*

Sol. λ is an eigen value of A.

$\Rightarrow$ There exists a non-zero matrix X such that $AX = \lambda X$

$$\Rightarrow \quad (\text{adj } A)(AX) = (\text{adj } A)(\lambda X)$$

$$\Rightarrow \quad ((\text{adj } A)A)X = \lambda(\text{adj } A)X$$

$$\Rightarrow \quad |A| IX = \lambda(\text{adj } A)X \quad [\because (\text{adj } A)A = |A| I]$$

$$\Rightarrow \qquad |A|X = \lambda(\text{adj } A)X \Rightarrow \frac{|A|}{\lambda}X = (\text{adj } A)X$$

$$\Rightarrow \qquad (\text{adj } A)X = \frac{|A|}{\lambda}X \Rightarrow \frac{|A|}{\lambda} \text{ is an eigen value of adj A.}$$

6.17. CAYLEY HAMILTON THEOREM

Every square matrix satisfies its characteristic equation.

i.e., if the characteristic equation of the nth order square matrix A is

$$|A - \lambda I| = (-1)^n\lambda^n + k_1\lambda^{n-1} + k_2\lambda^{n-2} + \cdots + k_n = 0 \qquad ...(1)$$

then
$$(-1)^n A^n + k_1 A^{n-1} + k_2 A^{n-2} + \cdots + k_n I = O$$

Let $\qquad P = \text{adj}(A - \lambda I)$

Since the elements of $A - \lambda I$ are at most of first degree in λ, the elements of P = adj $(A - \lambda I)$ are polynomials in λ of degree $(n - 1)$ or less. We can, therefore, split up P into a number of matrices each containing the same power of λ and write

$$P = P_1\lambda^{n-1} + P_2\lambda^{n-2} + P_{n-1}\lambda + P_n \qquad ...(2)$$

Also, we know that if M is a square matrix, then M (adj M) = | M | × I

$\therefore \qquad (A - \lambda I)P = |A - \lambda I| \times I$

By (1) and (2), we have

$$(A - \lambda I)(P_1\lambda^{n-1} + P_2\lambda^{n-2} + \cdots + P_{n-1}\lambda + P_n) = [(-1)^n\lambda^n + k_1\lambda^{n-1} + k_2\lambda^{n-2} + \cdots + k_{n-1}\lambda + k_n]I$$

Equating co-efficients of like powers of λ on both sides, we have

$$-P_1 = (-1)^n I \qquad [\because \ IP_1 = P_1]$$
$$AP_1 - P_2 = k_1 I$$
$$AP_2 - P_3 = k_2 I$$
....
$$AP_{n-1} - P_n = k_{n-1} I$$
$$AP_n = k_n I$$

Pre-multiplying these equations by $A^n, A^{n-1}, A^{n-2}, \ldots, A, I$ respectively and adding, we get

$$O = (-1)^n A^n + k_1 A^{n-1} + k_2 A^{n-2} + \cdots + k_{n-1} A + k_n I$$

[terms on the LHS cancel in pairs]

or
$$(-1)^n A^n + k_1 A^{n-1} + k_2 A^{n-2} + \cdots + k_{n-1} A + k_n I = O \qquad ...(3)$$

which proves the theorem.

Note 1. Multiplying (3) by A^{-1}, we have $(-1)^n A^{n-1} + k_1 A^{n-2} + \cdots + k_{n-1} I + k_n A^{-1} = O$

$$\Rightarrow \qquad A^{-1} = -\frac{1}{k_n}[(-1)^n A^{n-1} + k_1 A^{n-2} + \cdots + k_{n-1} I]$$

Thus Cayley-Hamilton theorem gives another method for computing the inverse of a matrix. Since this method expresses the inverse of a matrix of order n in terms of $(n - 1)$ powers of A, it is most suitable for computing inverses of large matrices.

Note 2. If m be a positive integer such that $m > n$, then multiplying (3) by A^{m-n}, we get

$$(-1)^n A^m + k_1 A^{m-1} + k_2 A^{m-2} + \cdots + k_{n-1} A^{m-n+1} + k_n A^{m-n} = O$$

showing that any positive integral power A^m $(m > n)$ of A is linearly expressible in terms of those of lower degree.

ILLUSTRATIVE EXAMPLES

Example 1. *Verify Cayley-Hamilton theorem for the matrix* $A = \begin{bmatrix} 2 & -1 & 1 \\ -1 & 2 & -1 \\ 1 & -1 & 2 \end{bmatrix}$.

Hence compute A^{-1}.

Sol. The characteristic equation of A is

$$|A - \lambda I| = 0 \quad i.e., \quad \begin{vmatrix} 2-\lambda & -1 & 1 \\ -1 & 2-\lambda & -1 \\ 1 & -1 & 2-\lambda \end{vmatrix} = 0$$

or $\lambda^3 - 6\lambda^2 + 9\lambda - 4 = 0$ (on simplification)

To verify Cayley-Hamilton theorem, we have to show that

$$A^3 - 6A^2 + 9A - 4I = O \qquad ...(1)$$

Now

$$A^2 = \begin{bmatrix} 2 & -1 & 1 \\ -1 & 2 & -1 \\ 1 & -1 & 2 \end{bmatrix} \times \begin{bmatrix} 2 & -1 & 1 \\ -1 & 2 & -1 \\ 1 & -1 & 2 \end{bmatrix} = \begin{bmatrix} 6 & -5 & 5 \\ -5 & 6 & -5 \\ 5 & -5 & 6 \end{bmatrix}$$

$$A^3 = A^2 \times A = \begin{bmatrix} 6 & -5 & 5 \\ -5 & 6 & -5 \\ 5 & -5 & 6 \end{bmatrix} \begin{bmatrix} 2 & -1 & 1 \\ -1 & 2 & -1 \\ 1 & -1 & 2 \end{bmatrix} = \begin{bmatrix} 22 & -22 & -21 \\ -21 & 22 & -21 \\ 21 & -21 & 22 \end{bmatrix}$$

$$\therefore \quad A^3 - 6A^2 + 9A - 4I = \begin{bmatrix} 22 & -22 & -21 \\ -21 & 22 & -21 \\ 21 & -21 & 22 \end{bmatrix} - 6 \begin{bmatrix} 6 & -5 & 5 \\ -5 & 6 & -5 \\ 5 & -5 & 6 \end{bmatrix}$$

$$+ 9 \begin{bmatrix} 2 & -1 & 1 \\ -1 & 2 & -1 \\ 1 & -1 & 2 \end{bmatrix} - 4 \begin{bmatrix} 1 & 0 & 0 \\ 0 & 1 & 0 \\ 0 & 0 & 1 \end{bmatrix}$$

$$= \begin{bmatrix} 0 & 0 & 0 \\ 0 & 0 & 0 \\ 0 & 0 & 0 \end{bmatrix} = O$$

This verifies Cayley-Hamilton theorem.

Now, multiplying both sides of (1) by A^{-1}, we have

$$A^2 - 6A + 9I - 4A^{-1} = O \quad \Rightarrow \quad 4A^{-1} = A^2 - 6A + 9I$$

$$\Rightarrow \quad 4A^{-1} = \begin{bmatrix} 6 & -5 & 5 \\ -5 & 6 & -5 \\ 5 & -5 & 6 \end{bmatrix} - 6 \begin{bmatrix} 2 & -1 & 1 \\ -1 & 2 & -1 \\ 1 & -1 & 2 \end{bmatrix} + 9 \begin{bmatrix} 1 & 0 & 0 \\ 0 & 1 & 0 \\ 0 & 0 & 1 \end{bmatrix} = \begin{bmatrix} 3 & 1 & -1 \\ 1 & 3 & 1 \\ -1 & 1 & 3 \end{bmatrix}$$

$$\therefore \quad A^{-1} = \frac{1}{4} \begin{bmatrix} 3 & 1 & -1 \\ 1 & 3 & 1 \\ -1 & 1 & 3 \end{bmatrix}.$$

Example 2. *Using Cayley-Hamilton theorem, find A^8 if* $A = \begin{bmatrix} 1 & 2 \\ 2 & -1 \end{bmatrix}$

Sol. The characteristic equation of A is

$$| A - \lambda I | = 0 \quad i.e., \quad \begin{vmatrix} 1-\lambda & 2 \\ 2 & -1-\lambda \end{vmatrix} = 0$$

or $$-(1-\lambda^2) - 4 = 0 \quad \text{or} \quad \lambda^2 = 5 \qquad ...(1)$$

By Cayley-Hamilton theorem. A satisfies its characteristic equation (1)

$$\therefore \quad A^2 = 5I$$

$$\Rightarrow \quad (A^2)^4 = (5I)^4 \Rightarrow A^8 = 625\ I \qquad (\because I^4 = I)$$

Example 3. *Find the characteristic equation of the matrix* $A = \begin{bmatrix} 2 & 1 & 1 \\ 0 & 1 & 0 \\ 1 & 1 & 2 \end{bmatrix}$ *and, hence, find the matrix represented by* $A^8 - 5A^7 + 7A^6 - 3A^5 + A^4 - 5A^3 + 8A^2 - 2A + I$.

Sol. The characteristic equation of A is

$$| A - \lambda I | = \begin{vmatrix} 2-\lambda & 1 & 1 \\ 0 & 1-\lambda & 0 \\ 1 & 1 & 2-\lambda \end{vmatrix} = 0 \quad \text{or} \quad \lambda^3 - 5\lambda^2 + 7\lambda - 3 = 0$$

By Cayley-Hamilton theorem, $A^3 - 5A^2 + 7A - 3I = O$ ...(1)

Now $A^8 - 5A^7 + 7A^6 - 3A^5 + A^4 - 5A^3 + 8A^2 - 2A + I$

$$= A^5(A^3 - 5A^2 + 7A - 3I) + A(A^3 - 5A^2 + 7A - 3I) + (A^2 + A + I)$$

$$= A^2 + A + I \qquad \text{[Using (1)]}$$

$$= \begin{bmatrix} 2 & 1 & 1 \\ 0 & 1 & 0 \\ 1 & 1 & 2 \end{bmatrix}\begin{bmatrix} 2 & 1 & 1 \\ 0 & 1 & 0 \\ 1 & 1 & 2 \end{bmatrix} + \begin{bmatrix} 2 & 1 & 1 \\ 0 & 1 & 0 \\ 1 & 1 & 2 \end{bmatrix} + \begin{bmatrix} 1 & 0 & 0 \\ 0 & 1 & 0 \\ 0 & 0 & 1 \end{bmatrix}$$

$$= \begin{bmatrix} 5 & 4 & 4 \\ 0 & 1 & 0 \\ 4 & 4 & 5 \end{bmatrix} + \begin{bmatrix} 2 & 1 & 1 \\ 0 & 1 & 0 \\ 1 & 1 & 2 \end{bmatrix} + \begin{bmatrix} 1 & 0 & 0 \\ 0 & 1 & 0 \\ 0 & 0 & 1 \end{bmatrix} = \begin{bmatrix} 8 & 5 & 5 \\ 0 & 3 & 0 \\ 5 & 5 & 8 \end{bmatrix}.$$

Example 4. *If* $A = \begin{bmatrix} 0 & 0 & 1 \\ 3 & 1 & 0 \\ -2 & 1 & 4 \end{bmatrix}$, *find A^{-1} by using Cayley-Hamilton theorem.*

Sol. The characteristic equation of A is

$$| A - \lambda I | = 0$$

i.e., $$\begin{vmatrix} 0-\lambda & 0 & 1 \\ 3 & 1-\lambda & 0 \\ -2 & 1 & 4-\lambda \end{vmatrix} = 0$$

or $$\lambda^3 - 5\lambda^2 + 6\lambda - 5 = 0$$

Using Cayley-Hamilton theorem,

$$A^3 - 5A^2 + 6A - 5I = 0$$

On multiplying by A^{-1} both sides

$$A^2 - 5A + 6I - 5A^{-1} = 0$$

This leads to $$A^{-1} = \frac{1}{5}[A^2 - 5A + 6I]$$

Now $$A^2 = A \times A = \begin{bmatrix} 0 & 0 & 1 \\ 3 & 1 & 0 \\ -2 & 1 & 4 \end{bmatrix} \times \begin{bmatrix} 0 & 0 & 1 \\ 3 & 1 & 0 \\ -2 & 1 & 4 \end{bmatrix}$$

$$= \begin{bmatrix} -2 & 1 & 4 \\ 3 & 1 & 3 \\ -5 & 5 & 14 \end{bmatrix}$$

Now $$A^2 - 5A + 6I = \begin{bmatrix} -2 & 1 & 4 \\ 3 & 1 & 3 \\ -5 & 5 & 14 \end{bmatrix} - 5\begin{bmatrix} 0 & 0 & 1 \\ 3 & 1 & 0 \\ -2 & 1 & 4 \end{bmatrix} + \begin{bmatrix} 6 & 0 & 0 \\ 0 & 6 & 0 \\ 0 & 0 & 6 \end{bmatrix} = \begin{bmatrix} 4 & 1 & -1 \\ -12 & 2 & 3 \\ 5 & 0 & 0 \end{bmatrix}$$

$$\therefore \quad A^{-1} = \frac{1}{5}\begin{bmatrix} 4 & 1 & -1 \\ -12 & 2 & 3 \\ 5 & 0 & 0 \end{bmatrix} = \begin{bmatrix} \frac{4}{5} & \frac{1}{5} & \frac{-1}{5} \\ \frac{-12}{5} & \frac{2}{5} & \frac{3}{5} \\ 1 & 0 & 0 \end{bmatrix}.$$

EXERCISE 6.5

1. Find the sum and product of the eigen values of

(i) $\begin{bmatrix} 2 & 3 & -2 \\ -2 & 1 & 1 \\ 1 & 0 & 2 \end{bmatrix}$ (ii) $\begin{bmatrix} 1 & 2 & -2 \\ 1 & 0 & 3 \\ -2 & -1 & -3 \end{bmatrix}$ (iii) $\begin{bmatrix} 1 & 1 & 1 \\ 1 & 2 & 2 \\ 1 & 2 & 3 \end{bmatrix}$.

2. Find the product of the eigen values of $\begin{bmatrix} 7 & 2 & 2 \\ -6 & -1 & 2 \\ 6 & 2 & -1 \end{bmatrix}$.

3. Find the eigen values and eigen vectors of the matrices:

(i) $\begin{bmatrix} 5 & 4 \\ 1 & 2 \end{bmatrix}$ (ii) $\begin{bmatrix} 1 & 2 \\ 2 & 4 \end{bmatrix}$.

4. Find the eigen values and eigen vectors of the matrices:

(i) $\begin{bmatrix} 8 & -6 & 2 \\ -6 & 7 & -4 \\ 2 & -4 & 3 \end{bmatrix}$ (ii) $\begin{bmatrix} 6 & -2 & 2 \\ -2 & 3 & -1 \\ 2 & -1 & 3 \end{bmatrix}$

(iii) $\begin{bmatrix} 2 & 0 & 1 \\ 0 & 2 & 0 \\ 1 & 0 & 2 \end{bmatrix}$ (iv) $\begin{bmatrix} 2 & 1 & 1 \\ 1 & 2 & 1 \\ 0 & 0 & 1 \end{bmatrix}$

(v) $\begin{bmatrix} 3 & 1 & 4 \\ 0 & 2 & 0 \\ 0 & 0 & 5 \end{bmatrix}$

5. Show that the eigen values of a triangular matrix A are equal to the elements of the principal diagonal of A.

6. Show that 0 is a characteristic root of a matrix if and only if the matrix is singular.

7. Show that if λ is a characteristic root of a matrix A, then $\lambda + k$ is a characteristic root of the matrix $A + kI$.

8. If $\lambda_1, \lambda_2, ..., \lambda_n$ are the eigen values of a matrix A, then A^2 has the eigen values $\lambda_1^2, \lambda_2^2, ..., \lambda_n^2$.

9. Prove that the matrices A and $P^{-1}AP$ have the same eigen values, P being an invertible matrix of same order as A.

10. (*a*) Find the characteristic equation of the matrix $A = \begin{bmatrix} 1 & 3 & 7 \\ 4 & 2 & 3 \\ 1 & 2 & 1 \end{bmatrix}$

Show that the equation is satisfied by A and hence obtain the inverse of the given matrix.

(*b*) Find the characteristic equation of the matrix $A = \begin{bmatrix} 1 & 2 & 3 \\ 2 & 1 & 4 \\ 3 & 1 & 1 \end{bmatrix}$

Show that the equation is satisfied by A. Also find A^{-1}.

11. Verify Cayley-Hamilton theorem for the matrix A and find A^{-1} when A is

(*i*) $\begin{bmatrix} 1 & 0 & 2 \\ 0 & 2 & 1 \\ 2 & 0 & 3 \end{bmatrix}$ (*ii*) $\begin{bmatrix} 2 & 6 & 1 \\ 0 & 1 & -6 \\ 3 & 4 & -2 \end{bmatrix}$ (*iii*) $\begin{bmatrix} 7 & 2 & -2 \\ -6 & -1 & 2 \\ 6 & 2 & -1 \end{bmatrix}$.

12. Using Cayley-Hamilton theorem, find the inverse of

(*i*) $\begin{bmatrix} 2 & 3 \\ 3 & 5 \end{bmatrix}$ (*ii*) $\begin{bmatrix} 7 & -1 & 3 \\ 6 & 1 & 4 \\ 2 & 4 & 8 \end{bmatrix}$ (*iii*) $\begin{bmatrix} 1 & 0 & 3 \\ 2 & 1 & -1 \\ 1 & -1 & 1 \end{bmatrix}$.

13. Using Cayley-Hamilton theorem, find A^6 if $A = \begin{bmatrix} 2 & 1 \\ 5 & -2 \end{bmatrix}$.

14. Find the characteristic equation of the matrix $A = \begin{bmatrix} 1 & 4 \\ 2 & 3 \end{bmatrix}$ and use it to find the matrix represented by

$$A^5 + 5A^4 - 6A^3 + 2A^2 - 4A + 7I.$$

15. Verify Cayley-Hemilton theorem for the matrix $A = \begin{bmatrix} 1 & 2 & -2 \\ 1 & 1 & 1 \\ 1 & 3 & 1 \end{bmatrix}$ and hence find A^3.

16. Find A^{-1} using Cayley-Hamilton theorem for the matrix $A = \begin{bmatrix} 1 & 0 & -2 \\ 2 & 2 & 4 \\ 0 & 0 & 2 \end{bmatrix}$ (*K.U. April 2013*)

Answers

1. (*i*) 5, 21 (*ii*) –2, –1 (*iii*) 6, 1 **2.** –21

3. (*i*) 1, 6; (1, –1), (4, 1) (*ii*) 5, 0 ; (1, 2), (2, –1) **4.** (*i*) 0, 3, 15; (1, 2, 2), (2, 1, –2), (2, –2, 1)

(*ii*) 2, 2, 8; (1, 0, –2), (1, 2, 0), (2, –1, 1) (*iii*) 1, 2, 3; (1, 0, 1), (1, 0, –1), (0, 1, 0)

(*iv*) 1, 1, 3; (1, –2, 1), (1, 1, 0) (*v*) 2, 3, 5; (1, –1, 0), (1, 0, 0), (2, 0, 1).

9. [**Hint.** $| P^{-1}AP - \lambda I | = | P^{-1}AP - \lambda P^{-1}P | = | P^{-1}(A - \lambda I)P |$

$= | P^{-1} | \, | A - \lambda I | \, | P | = | A - \lambda I | \, | P^{-1}P | = | A - \lambda I |$]

10. (a) $\lambda^3 - 4\lambda^2 - 20\lambda - 35 = 0, \frac{1}{35}\begin{bmatrix} -4 & 11 & -5 \\ -1 & -6 & 25 \\ -6 & 1 & -10 \end{bmatrix}$ (b) $\lambda^3 - 3\lambda^2 - 14\lambda - 14 = 0, \frac{1}{14}\begin{bmatrix} -3 & 1 & 5 \\ 10 & -8 & 2 \\ -1 & 5 & -3 \end{bmatrix}$

11. (i) $\begin{bmatrix} -3 & 0 & 2 \\ -1 & \frac{1}{2} & \frac{1}{2} \\ 2 & 0 & -1 \end{bmatrix}$ (ii) $\frac{1}{67}\begin{bmatrix} -22 & -16 & 37 \\ 18 & 7 & -12 \\ 3 & -10 & -2 \end{bmatrix}$ (iii) $\frac{1}{3}\begin{bmatrix} -3 & -2 & 2 \\ 6 & 5 & -2 \\ -6 & -2 & 5 \end{bmatrix}$

12. (i) $\begin{bmatrix} 5 & -3 \\ -3 & 2 \end{bmatrix}$ (ii) $\frac{1}{130}\begin{bmatrix} 4 & 20 & -9 \\ -42 & 50 & -3 \\ 20 & -30 & 20 \end{bmatrix}$ (iii) $\frac{1}{9}\begin{bmatrix} 0 & 3 & 3 \\ 3 & 2 & -7 \\ 3 & -1 & -1 \end{bmatrix}$

13. 729I **14.** $\begin{bmatrix} 1861 & 3676 \\ 1838 & 3699 \end{bmatrix}$

15. $A^3 - A^2 - 4A + 4I = 0 ; A^3 = \begin{bmatrix} 1 & 6 & -6 \\ 7 & 6 & 2 \\ 7 & 14 & -6 \end{bmatrix}$ **16.** $A^{-1} = \begin{bmatrix} 1 & 0 & 1 \\ -1 & 0.5 & -2 \\ 0 & 0 & 0.5 \end{bmatrix}$.

6.18. REDUCTION OF A MATRIX TO DIAGONAL FORM

If a square matrix A of order n has n linearly independent eigen vectors, then a matrix B can be found such that $B^{-1}AB$ is a diagonal matrix.

[We prove the result for a square matrix of order 3. The proof can be easily extended to matrices of higher order.]

Let A be a square matrix of order 3. Let $\lambda_1, \lambda_2, \lambda_3$ be its eigen values and

$$X_1 = \begin{bmatrix} x_1 \\ y_1 \\ z_1 \end{bmatrix}, X_2 = \begin{bmatrix} x_2 \\ y_2 \\ z_2 \end{bmatrix}, X_3 = \begin{bmatrix} x_3 \\ y_3 \\ z_3 \end{bmatrix}$$ be the corresponding eigen vectors.

Since eigen vectors are non-trivial solutions of the matrix equation $AX = \lambda X$, we have

$$AX_1 = \lambda_1 X_1, \ AX_2 = \lambda_2 X_2, \ AX_3 = \lambda_3 X_3$$

Let $$B = \begin{bmatrix} x_1 & x_2 & x_3 \\ y_1 & y_2 & y_3 \\ z_1 & z_2 & z_3 \end{bmatrix} = [X_1 X_2 X_3]$$

Consider $AB = A[X_1, X_2, X_3] = [AX_1, AX_2, AX_3] = [\lambda_1 X_1, \lambda_2 X_2, \lambda_3 X_3]$

$$= \begin{bmatrix} \lambda_1 x_1 & \lambda_2 x_2 & \lambda_3 x_3 \\ \lambda_1 y_1 & \lambda_2 y_2 & \lambda_3 y_3 \\ \lambda_1 z_1 & \lambda_2 z_2 & \lambda_3 z_3 \end{bmatrix} = \begin{bmatrix} x_1 & x_2 & x_3 \\ y_1 & y_2 & y_3 \\ z_1 & z_2 & z_3 \end{bmatrix}\begin{bmatrix} \lambda_1 & 0 & 0 \\ 0 & \lambda_2 & 0 \\ 0 & 0 & \lambda_3 \end{bmatrix}$$

= BD, where D is the diagonal matrix $\begin{bmatrix} \lambda_1 & 0 & 0 \\ 0 & \lambda_2 & 0 \\ 0 & 0 & \lambda_3 \end{bmatrix}$

$\Rightarrow$ $AB = BD \Rightarrow B^{-1}AB = B^{-1}BD = ID = D$ which proves the theorem.

Note. The matrix B which diagonalises A is called the **modal matrix** of A and is obtained by grouping the eigen vectors of A into a square matrix. The diagonal matrix D is called the **spectral matrix** of A.

6.18. CALCULATION OF POWERS OF A MATRIX

Given a square matrix A, it is quite tedious to find A^n (n being a positive integer) when n is large. On the other hand, it is quite easy to obtain any positive integral power of a diagonal matrix D, since

$$D = \begin{bmatrix} a & 0 \\ 0 & b \end{bmatrix} \quad \Rightarrow \quad D^n = \begin{bmatrix} a^n & 0 \\ 0 & b^n \end{bmatrix}$$

Therefore, diagonalisation of a matrix is useful for finding positive integral powers of a square matrix.

Let A be a square matrix. Then we know that a non-singular matrix B can be found such that

$$B^{-1}AB = D \qquad ...(1)$$

where D is a diagonal matrix.

From (1), $D^2 = (B^{-1}AB)(B^{-1}AB)$

$= B^{-1}A^2B$ $\qquad [\because \ BB^{-1} = I]$

Similarly, $D^3 = B^{-1}A^3B$

In general, $D^n = B^{-1}A^nB \qquad ...(2)$

To find A^n from (2), pre-multiply (2) by B and post-multiply by B^{-1}. Then

$$BD^nB^{-1} = BB^{-1}A^nBB^{-1}$$

$$= A^n$$

Hence $A^n = BD^nB^{-1}$.

6.19. DIAGONALIZATION OF REAL SYMMETRIC MATRIX

If A is any real symmetric matrix, then we can find an orthogonal matrix B such that B^TAB is a diagonal matrix with diagonal elements as the eigen values of A. The matrix B can be formed in such a way that the columns are the normalised eigen vectors of A.

i.e., $B^TAB = D$, D is a diagonal matrix with diagonal element are the eigen values of A.

Note: To find the orthogonal matrix B, each element of the eigen vector is divided by the norm (length).

ILLUSTRATIVE EXAMPLES

Example 1. *Diagonalise the matrix* $A = \begin{bmatrix} -1 & 2 & -2 \\ 1 & 2 & 1 \\ -1 & -1 & 0 \end{bmatrix}$ *and obtain the modal matrix.*

(K.U. May 2011)

Sol. The characteristic equation of A is

$$|A - \lambda I| = \begin{vmatrix} -1-\lambda & 2 & -2 \\ 1 & 2-\lambda & 1 \\ -1 & -1 & -\lambda \end{vmatrix} = 0 \text{ or } \lambda^3 - \lambda^2 - 5\lambda + 5 = 0 \qquad \text{(on simplification)}$$

Solving it, we have $\lambda = 1, \pm\sqrt{5}$.

When $\lambda = 1$, the corresponding eigen vector is given by

$$-2x_1 + 2y_1 - 2z_1 = 0$$

$$x_1 + y_1 + z_1 = 0$$

$$x_1 - y_1 + z_1 = 0$$

Solving the last two $\frac{x_1}{2} = \frac{y_1}{0} = \frac{z_1}{-2}$ giving the eigen vector (1, 0, –1).

When $\lambda = \sqrt{5}$, the corresponding eigen vector is given by

$$(-1 - \sqrt{5})x_1 + 2y_1 - 2z_1 = 0$$
$$x_1 + (2 - \sqrt{5})y_1 + z_1 = 0$$
$$-x_1 - y_1 - \sqrt{5}z_1 = 0$$

Solving the last two $\frac{x_1}{6 - 2\sqrt{5}} = \frac{y_1}{-1 + \sqrt{5}} = \frac{z_1}{1 - \sqrt{5}}$

or $\frac{x_1}{(\sqrt{5} - 1)^2} = \frac{y_1}{\sqrt{5} - 1} = \frac{z_1}{1 - \sqrt{5}}$ or $\frac{x_1}{\sqrt{5} - 1} = \frac{y_1}{1} = \frac{z_1}{-1}$ giving the eigen vector $(\sqrt{5} - 1, 1, -1)$

Similarly, the eigen vector corresponding to $\lambda = -\sqrt{5}$ is $(\sqrt{5} + 1, -1, 1)$.

$\therefore$ The modal matrix $B = \begin{bmatrix} 1 & \sqrt{5}-1 & \sqrt{5}+1 \\ 0 & 1 & 1 \\ -1 & -1 & 1 \end{bmatrix}$

and $$B^{-1}AB = D = \begin{bmatrix} \lambda_1 & 0 & 0 \\ 0 & \lambda_2 & 0 \\ 0 & 0 & \lambda_3 \end{bmatrix} = \begin{bmatrix} 1 & 0 & 0 \\ 0 & \sqrt{5} & 0 \\ 0 & 0 & -\sqrt{5} \end{bmatrix}$$

is the required diagonal matrix.

Example 2. *Diagonalise the matrix* $A = \begin{bmatrix} 1 & 6 & 1 \\ 1 & 2 & 0 \\ 0 & 0 & 3 \end{bmatrix}$ *and hence find* A^4.

Sol. The characteristic equation of A is

$$|A - \lambda I| = \begin{vmatrix} 1-\lambda & 6 & 1 \\ 1 & 2-\lambda & 0 \\ 0 & 0 & 3-\lambda \end{vmatrix} = 0$$

or $\lambda^3 - 6\lambda^2 + 5\lambda + 12 = 0$ (on simplification)

or $(\lambda + 1)(\lambda - 3)(\lambda - 4) = 0$

$\therefore$ $\lambda = -1, 3, 4$

When $\lambda = -1$, the corresponding eigen vector is given by

$$2x_1 + 6y_1 + z_1 = 0$$
$$x_1 + 3y_1 = 0$$
$$4z_1 = 0$$

Solving the last two, $\frac{x_1}{3} = \frac{y_1}{-1}$, $z_1 = 0$ giving the eigen vector (3, –1, 0).

When $\lambda = 3$, the corresponding eigen vector is given by

$$-2x_1 + 6y_1 + z_1 = 0$$
$$x_1 - y_1 = 0$$

Solving $\frac{x_1}{1} = \frac{y_1}{1} = \frac{z_1}{-4}$ giving the eigen vector (1, 1, –4).

When $\lambda = 4$, the corresponding eigen vector is given by

$$-3x_1 + 6y_1 + z_1 = 0$$
$$x_1 - 2y_1 = 0$$
$$z_1 = 0$$

Solving the last two, $\frac{x_1}{2} = \frac{y_1}{1}$, $z_1 = 0$ giving the eigen vector (2, 1, 0).

$\therefore$ The modal matrix $B = \begin{bmatrix} 3 & 1 & 2 \\ -1 & 1 & 1 \\ 0 & -4 & 0 \end{bmatrix}$

and $$B^{-1}AB = D = \begin{bmatrix} \lambda_1 & 0 & 0 \\ 0 & \lambda_2 & 0 \\ 0 & 0 & \lambda_3 \end{bmatrix} = \begin{bmatrix} -1 & 0 & 0 \\ 0 & 3 & 0 \\ 0 & 0 & 4 \end{bmatrix}$$

is the required diagonal matrix.

To find B^{-1}

$$|B| = \begin{vmatrix} 3 & 1 & 2 \\ -1 & 1 & 1 \\ 0 & -4 & 0 \end{vmatrix} = \begin{vmatrix} a_1 & b_1 & c_1 \\ a_2 & b_2 & c_2 \\ a_3 & b_3 & c_3 \end{vmatrix} \quad \text{(say)}$$

$A_1 = 4, B_1 = 0, C_1 = 4, A_2 = -8, B_2 = 0, C_2 = 12, A_3 = -1, B_3 = -5, C_3 = 4$

$\therefore$ $$|B| = a_1A_1 + b_1B_1 + c_1C_1 = (3)(4) + (1)(0) + 2(4) = 20$$

$$\text{adj. } B = \begin{bmatrix} A_1 & A_2 & A_3 \\ B_1 & B_2 & B_3 \\ C_1 & C_2 & C_3 \end{bmatrix} = \begin{bmatrix} 4 & -8 & -1 \\ 0 & 0 & -5 \\ 4 & 12 & 4 \end{bmatrix}$$

$$B^{-1} = \frac{1}{|B|} \text{adj } B = \frac{1}{20} \begin{bmatrix} 4 & -8 & -1 \\ 0 & 0 & -5 \\ 4 & 12 & 4 \end{bmatrix}$$

Now $$D = \begin{bmatrix} -1 & 0 & 0 \\ 0 & 3 & 0 \\ 0 & 0 & 4 \end{bmatrix} \Rightarrow D^4 = \begin{bmatrix} (-1)^4 & 0 & 0 \\ 0 & 3^4 & 0 \\ 0 & 0 & 4^4 \end{bmatrix} = \begin{bmatrix} 1 & 0 & 0 \\ 0 & 81 & 0 \\ 0 & 0 & 256 \end{bmatrix}$$

$\therefore$ $$A^4 = BD^4B^{-1} = \frac{1}{20} \begin{bmatrix} 3 & 1 & 2 \\ -1 & 1 & 1 \\ 0 & -4 & 0 \end{bmatrix} \begin{bmatrix} 1 & 0 & 0 \\ 0 & 81 & 0 \\ 0 & 0 & 256 \end{bmatrix} \begin{bmatrix} 4 & -8 & -1 \\ 0 & 0 & -5 \\ 4 & 12 & 4 \end{bmatrix}$$

$$= \frac{1}{20} \begin{bmatrix} 3 & 1 & 2 \\ -1 & 1 & 1 \\ 0 & -4 & 0 \end{bmatrix} \begin{bmatrix} 4 & -8 & -1 \\ 0 & 0 & -405 \\ 1024 & 3072 & 1024 \end{bmatrix}$$

$$= \frac{1}{20} \begin{bmatrix} 2060 & 6120 & 1640 \\ 1020 & 3080 & 620 \\ 0 & 0 & 1620 \end{bmatrix} = \begin{bmatrix} 103 & 306 & 82 \\ 51 & 154 & 31 \\ 0 & 0 & 81 \end{bmatrix}$$

Example 3. *Diagonalise* $\begin{bmatrix} 3 & -1 & 1 \\ -1 & 5 & -1 \\ 1 & -1 & 3 \end{bmatrix}$

Sol. Here $\quad A = \begin{bmatrix} 3 & -1 & 1 \\ -1 & 5 & -1 \\ 1 & -1 & 3 \end{bmatrix}$

The characteristic equation of A is $|A - \lambda I| = \begin{bmatrix} 3-\lambda & -1 & 1 \\ -1 & 5-\lambda & -1 \\ 1 & -1 & 3-\lambda \end{bmatrix} = 0$

or $\quad \lambda^3 - 11\lambda^2 + 36\lambda - 36 = 0$ (On simplification)

Solving it, we have $\quad \lambda = 2, 3, 6$

When $\lambda = 2$, the corresponding eigen vector is given by

$$x - y + z = 0$$
$$-x + 3y - z = 0$$
$$x - y + z = 0$$

Solving the first two, $\quad \dfrac{x}{-2} = \dfrac{y}{0} = \dfrac{z}{2}$

giving the eigen vector (–1, 0, 1)

When $\lambda = 3$, the corresponding eigen vector is given by

$$-y + z = 0$$
$$-x + 2y - z = 0$$
$$x - y = 0$$

Solving the first two, $\dfrac{x}{-1} = \dfrac{y}{-1} = \dfrac{z}{-1}$

giving the eigen vector (1, 1, 1) when 6, the corresponding eigen vector is given by

$$-3x - y + z = 0$$
$$-x - y - z = 0$$
$$x - y - 3z = 0$$

Solving the first two, $\dfrac{x}{2} = \dfrac{y}{-4} = \dfrac{z}{2}$

giving the eigen vector (1, – 2, 1)

Since the eigen vectors are linearly independent. They are mutually orthogonal. Corresponding normalized eigen vectors are

$$\begin{bmatrix} \dfrac{-1}{\sqrt{2}} \\ 0 \\ \dfrac{1}{\sqrt{2}} \end{bmatrix}, \begin{bmatrix} \dfrac{1}{\sqrt{3}} \\ \dfrac{1}{\sqrt{3}} \\ \dfrac{1}{\sqrt{3}} \end{bmatrix}, \begin{bmatrix} \dfrac{1}{\sqrt{6}} \\ \dfrac{-2}{\sqrt{6}} \\ \dfrac{1}{\sqrt{6}} \end{bmatrix}$$

$$\therefore \quad \text{The normalized model matrix} = \begin{bmatrix} \frac{-1}{\sqrt{2}} & \frac{1}{\sqrt{3}} & \frac{1}{\sqrt{6}} \\ 0 & \frac{1}{\sqrt{3}} & \frac{-2}{\sqrt{6}} \\ \frac{1}{\sqrt{2}} & \frac{1}{\sqrt{3}} & \frac{1}{\sqrt{6}} \end{bmatrix}$$

and
$$B^{-1} AB = D = \begin{bmatrix} \lambda_1 & 0 & 0 \\ 0 & \lambda_2 & 0 \\ 0 & 0 & \lambda_3 \end{bmatrix} = \begin{bmatrix} 2 & 0 & 0 \\ 0 & 3 & 0 \\ 0 & 0 & 6 \end{bmatrix}$$

is the required diagonal matrix.

Example 4. *Diagonalise* $A = \begin{pmatrix} 6 & -2 & 2 \\ -2 & 3 & -1 \\ 2 & -1 & 3 \end{pmatrix}$

Sol. The characteristic equation of A is

$$|A - \lambda I| = \begin{vmatrix} 6-\lambda & -2 & 2 \\ -2 & 3-\lambda & -1 \\ 2 & -1 & 3-\lambda \end{vmatrix} = 0$$

or $\lambda^3 - 12\lambda^2 + 36\lambda - 32 = 0$

(On simplification)

Solving it, we have $\lambda = 2, 2, 8$.

When $\lambda = 8$, the corresponding eigen vector is given by

$$-2x - 2y + 2z = 0$$
$$-2x - 5y - z = 0$$
$$2x - y - 5z = 0$$

Solving the first two, $\frac{x}{2} = \frac{y}{-1} = \frac{z}{1}$

giving the eigen vector (2, –1, 1)

When $\lambda = 2$, the corresponding eigen vector is given by

$$4x - 2y + 2z = 0$$
$$-2x + y - z = 0$$
$$2x - y + z = 0$$

The only one independent equation is

$$2x - y + z = 0$$

Two variables are to be assigned arbitrary values.

Taking $y = 1$ and $z = 1$, we get $x = 0$.

$\therefore$ The eigen vector corresponding to $\lambda = 2$ is (0, 1, 1)

The third eigen vector is orthogonal to both the eigen vectors corresponding to $\lambda = 8$ and $\lambda = 2$.

$\therefore$ The eigen vector is (1, 1, –1).

Since the eigen vectors are mutually orthogonal. Corresponding normalized eigen vectors are

$$\begin{bmatrix} \frac{2}{\sqrt{6}} \\ \frac{-1}{\sqrt{6}} \\ \frac{1}{\sqrt{6}} \end{bmatrix}, \begin{bmatrix} 0 \\ \frac{1}{\sqrt{2}} \\ \frac{1}{\sqrt{2}} \end{bmatrix}, \begin{bmatrix} \frac{1}{\sqrt{3}} \\ \frac{1}{\sqrt{3}} \\ \frac{-1}{\sqrt{3}} \end{bmatrix}$$

$\therefore$ The normalized model matrix = $\begin{bmatrix} \frac{2}{\sqrt{6}} & 0 & \frac{1}{\sqrt{3}} \\ \frac{-1}{\sqrt{6}} & \frac{1}{\sqrt{2}} & \frac{1}{\sqrt{3}} \\ \frac{1}{\sqrt{6}} & \frac{1}{\sqrt{2}} & \frac{-1}{\sqrt{3}} \end{bmatrix}$

and $$B^{-1}AB = D = \begin{bmatrix} \lambda_1 & 0 & 0 \\ 0 & \lambda_2 & 0 \\ 0 & 0 & \lambda_3 \end{bmatrix} = \begin{bmatrix} 8 & 0 & 0 \\ 0 & 2 & 0 \\ 0 & 0 & 2 \end{bmatrix}$$

EXERCISE 6.6

1. Diagonalise each of the following matrices:

(*i*) $\begin{bmatrix} -1 & 1 & 2 \\ 0 & -2 & 1 \\ 0 & 0 & -3 \end{bmatrix}$ (*ii*) $\begin{bmatrix} 8 & -6 & 2 \\ -6 & 7 & -4 \\ 2 & -4 & 3 \end{bmatrix}$

(*iii*) $\begin{bmatrix} 1 & 0 & -1 \\ 1 & 2 & 1 \\ 2 & 2 & 3 \end{bmatrix}$ (*K.U. May 2010*) (*iv*) $\begin{bmatrix} 7 & -2 & -2 \\ -2 & 1 & 4 \\ -2 & 4 & 1 \end{bmatrix}$

2. Diagonalise A = $\begin{bmatrix} 1 & 2 & -2 \\ -1 & 3 & 0 \\ 1 & -2 & 1 \end{bmatrix}$ and hence find A^4.

3. Diagonalise A = $\begin{bmatrix} 1 & 1 & 1 \\ 0 & 2 & 1 \\ -4 & 4 & 3 \end{bmatrix}$ and hence find A^8.

Answers

1. (*i*) $\begin{bmatrix} -1 & 0 & 0 \\ 0 & -2 & 0 \\ 0 & 0 & -3 \end{bmatrix}$ (*ii*) $\begin{bmatrix} 0 & 0 & 0 \\ 0 & 3 & 0 \\ 0 & 0 & 15 \end{bmatrix}$ (*iii*) $\begin{bmatrix} 1 & 0 & 0 \\ 0 & 2 & 0 \\ 0 & 0 & 3 \end{bmatrix}$ (*iv*) $\begin{bmatrix} -3 & 0 & 0 \\ 0 & 3 & 0 \\ 0 & 0 & 9 \end{bmatrix}$

2. $\begin{bmatrix} -55 & 104 & 24 \\ -20 & -15 & 32 \\ 32 & -40 & -23 \end{bmatrix}$ 3. $\begin{bmatrix} -12099 & 12355 & 6305 \\ -12100 & 12356 & 6305 \\ -13120 & 13120 & 6561 \end{bmatrix}$.

6.21. QUADRATIC FORMS

Definition. A homogeneous polynomial of second degree in any number of variables is called a quadratic form. For example,

(*i*) $ax^2 + 2hxy + by^2$ (*ii*) $ax^2 + by^2 + cz^2 + 2hxy + 2gyz + 2fzx$ and

(*iii*) $ax^2 + by^2 + cz^2 + dw^2 + 2hxy + 2gyz + 2fzx + 2lxw + 2myw + 2nzw$

are quadratic forms in two, three and four variables.

In n-variables $x_1, x_2, ..., x_n$, the general quadratic form is $\sum_{j=1}^{n} \sum_{i=1}^{n} b_{ij}x_ix_j$

In the expansion, the co-efficient of $x_ix_j = (b_{ij} + b_{ji})$.

Suppose $2a_{ij} = b_{ij} + b_{ij}$ where $a_{ij} = a_{ji}$ and $a_{ii} = b_{ii}$

$$\therefore \quad \sum_{j=1}^{n} \sum_{i=1}^{n} b_{ij}x_ix_j = \sum_{j=1}^{n} \sum_{i=1}^{n} a_{ij}x_ix_j \text{ where } a_{ij} = \frac{1}{2}(b_{ij} + b_{ji}).$$

Hence every quadratic form can be written as $\sum_{j=1}^{n} \sum_{i=1}^{n} a_{ij}x_ix_j = X'AX$, so that the matrix A is always symmetric, where $A = [a_{ij}]$ and $X' = [x_1, x_2, ..., x_n]$.

Now writing the above said examples of quadratic forms in matrix form, we get

$$(i)\ ax^2 + 2hxy + by^2 = [x\ y]\begin{bmatrix} a & h \\ h & b \end{bmatrix}\begin{bmatrix} x \\ y \end{bmatrix}$$

$$(ii)\ ax^2 + by^2 + cz^2 + 2hxy + 2gyz + 2fzx = [x\ y\ z]\begin{bmatrix} a & h & f \\ h & b & g \\ f & g & c \end{bmatrix}\begin{bmatrix} x \\ y \\ z \end{bmatrix}$$

and (*iii*) $ax^2 + by^2 + cz^2 + dw^2 + 2hxy + 2gyz + 2fzx + 2lxw + 2myw + 2nzw$

$$= [x\ y\ z\ w]\begin{bmatrix} a & h & f & l \\ h & b & g & m \\ f & g & c & n \\ l & m & n & d \end{bmatrix}\begin{bmatrix} x \\ y \\ z \\ w \end{bmatrix}.$$

6.22. LINEAR TRANSFORMATION OF A QUADRATIC FORM

Let X′AX be a quadratic form in n-variables and let X = PY ...(1)

where P is a non-singular matrix, be the non-singular transformation.

From (1), X′ = (PY)′ = Y′P′ and hence

X′AX = Y′P′APY = Y′(P′AP)Y = Y′BY ...(2)

where B = P′AP. Therefore Y′BY is also a quadratic form in n-variables. Hence it is a linear transformation of the quadratic form X′AX under the linear transformation X = PY and B = P′AP.

Note. (*i*) Here B′ = (P′AP)′ = P′AP = B (*ii*) ρ(B) = ρ(A).

∴ A and B are congruent matrices.

6.23. CANONICAL FORM

If a real quadratic form be expressed as a sum or difference of the squares of new variables by means of any real non-singular linear transformation, then the latter quadratic expression is called a *canonical form* of the given quadratic form.

i.e., if the quadratic form $X'AX = \sum_{j=1}^{n} \sum_{i=1}^{n} a_{ij}x_i x_j$ can be reduced to the quadratic form

$Y'BY = \sum_{i=1}^{n} \lambda_i {y_i}^2$ by a non-singular linear transformation

X = PY then Y′BY is called the canonical form of the given one.

$$\therefore \quad \text{If } B = P'AP = \text{diag}(\lambda_1, \lambda_2, ..., \lambda_n) \text{ then } X'AX = Y'BY = \sum_{j=1}^{n} \lambda_i {y_i}^2.$$

Note 1. Here some of λ_i (eigen values) may be positive or negative or zero.

2. A quadratic form is said to be real if the elements of the symmetric matrix are real.

3. If $\rho(A) = r$, then the quadratic form X′AX will contain only r terms.

6.24. INDEX AND SIGNATURE OF THE QUADRATIC FORM

The number p of positive terms in the canonical form is called the *index* of the quadratic form.

(The number of positive terms)–(The number of negative terms) *i.e.,* $p - (r - p) = 2p - r$ is called *signature* of the quadratic form, where $\rho(A) = r$.

6.25. DEFINITE, SEMI-DEFINITE AND INDEFINITE REAL QUADRATIC FORMS

Let X′AX be a real quadratic form in n-variables $x_1, x_2, ..., x_n$ with rank r and index p. Then we say that the quadratic form is

(*i*) *positive definite* if $r = n, p = r$

(*ii*) *negative definite* if $r = n, p = 0$

(*iii*) *positive semi-definite* if $r < n, p = r$

and (*iv*) *negative semi-definite* if $r < n, p = 0$

If the canonical form has both positive and negative terms, the quadratic form is said to be *indefinite*.

Note. If X′AX is positive definite then $|A| > 0$.

ILLUSTRATIVE EXAMPLES

Example 1. *Reduce $3x^2 + 3z^2 + 4xy + 8xz + 8yz$ into canonical form.*

Sol. The given quadratic form can be written as X′AX where $X' = [x, y, z]$ and the symmetric matrix

$$A = \begin{bmatrix} 3 & 2 & 4 \\ 2 & 0 & 4 \\ 4 & 4 & 3 \end{bmatrix}.$$

Let us reduce A into diagonal matrix. We know that $A = I_3 AI_3$

$$\text{i.e.,} \quad \begin{bmatrix} 3 & 2 & 4 \\ 2 & 0 & 4 \\ 4 & 4 & 3 \end{bmatrix} = \begin{bmatrix} 1 & 0 & 0 \\ 0 & 1 & 0 \\ 0 & 0 & 1 \end{bmatrix} A \begin{bmatrix} 1 & 0 & 0 \\ 0 & 1 & 0 \\ 0 & 0 & 1 \end{bmatrix}$$

Operating $R_2(-2/3)\,R_1$, $R_3(-4/3)R_1$ (for A on LHS and pre-factor on RHS), we get

$$\begin{bmatrix} 3 & 2 & 4 \\ 0 & -\frac{4}{3} & \frac{4}{3} \\ 0 & \frac{4}{3} & -\frac{7}{3} \end{bmatrix} = \begin{bmatrix} 1 & 0 & 0 \\ -\frac{2}{3} & 1 & 0 \\ -\frac{4}{3} & 0 & 1 \end{bmatrix} A \begin{bmatrix} 1 & 0 & 0 \\ 0 & 1 & 0 \\ 0 & 0 & 1 \end{bmatrix}$$

Operating $C_2(-2/3)\,C_1$, $C_3(-4/3)\,C_1$ (for A on LHS and post-factor on RHS), we get

$$\begin{bmatrix} 3 & 0 & 0 \\ 0 & -\frac{4}{3} & \frac{4}{3} \\ 0 & \frac{4}{3} & -\frac{7}{3} \end{bmatrix} = \begin{bmatrix} 1 & 0 & 0 \\ -\frac{2}{3} & 1 & 0 \\ -\frac{4}{3} & 0 & 1 \end{bmatrix} A \begin{bmatrix} 1 & -\frac{2}{3} & -\frac{4}{3} \\ 0 & 1 & 0 \\ 0 & 0 & 1 \end{bmatrix}$$

Operating $R_3 + R_2$, we get

$$\begin{bmatrix} 3 & 0 & 0 \\ 0 & -\frac{4}{3} & \frac{4}{3} \\ 0 & 0 & -1 \end{bmatrix} = \begin{bmatrix} 1 & 0 & 0 \\ -\frac{2}{3} & 1 & 0 \\ -2 & 1 & 1 \end{bmatrix} A \begin{bmatrix} 1 & -\frac{2}{3} & -\frac{4}{3} \\ 0 & 1 & 0 \\ 0 & 0 & 1 \end{bmatrix}$$

Operating $C_3 + C_2$, we get

$$\begin{bmatrix} 3 & 0 & 0 \\ 0 & -\frac{4}{3} & 0 \\ 0 & 0 & -1 \end{bmatrix} = \begin{bmatrix} 1 & 0 & 0 \\ -\frac{2}{3} & 1 & 0 \\ -2 & 1 & 1 \end{bmatrix} A \begin{bmatrix} 1 & -\frac{2}{3} & -2 \\ 0 & 1 & 0 \\ 0 & 0 & 1 \end{bmatrix}$$

or $$\text{Diag}\left(3, -\frac{4}{3}, -1\right) = P'AP$$

∴ The canonical form of the given quadratic form is

$$Y'(P'AP)Y = [y_1, y_2, y_3]\ 0 \begin{bmatrix} 3 & 0 & 0 \\ 0 & -\frac{4}{3} & 0 \\ 0 & 0 & -1 \end{bmatrix} \begin{bmatrix} y_1 \\ y_2 \\ y_3 \end{bmatrix} = 3y_1^2 - \frac{4}{3} y_2^2 - y_3^2.$$

Here $\rho(A) = 3$, index = 1, signature = 1 – 2 = –1.

Note. In this problem the non-singular transformation which reduces the given quadratic form into the canonical form is X = PY *i.e.*, $\begin{bmatrix} x \\ y \\ z \end{bmatrix} = \begin{bmatrix} 1 & -\frac{2}{3} & -2 \\ 0 & 1 & 0 \\ 0 & 0 & 1 \end{bmatrix} \begin{bmatrix} y_1 \\ y_2 \\ y_3 \end{bmatrix}$. *i.e.*, $x = y_1 - \frac{2}{3} y_2 - 2y_3,\ y = y_2 + y_3,\ z = y_3$.

Note. The above example can also be questioned as 'Diagonalise' the quadratic form $3x^2 + 3z^2 + 4xy + 8xz + 8yz$ by linear transformation and write the linear transformation.

Or

Reduce the quadratic form $3x^2 + 3z^2 + 4xy + 8xz + 8yz$ into the sum of squares.

Example 2. *Reduce the quadratic form* $x^2 - 4y^2 + 6z^2 + 2xy - 4xz + 2w^2 - 6zw$ *into sum of squares.*

Sol. The matrix form of the given quadratic form is X′AX, where $X' = [x\ y\ z\ w]$

and
$$A = \begin{bmatrix} 1 & 1 & -2 & 0 \\ 1 & -4 & 0 & 0 \\ -2 & 0 & 6 & -3 \\ 0 & 0 & -3 & 2 \end{bmatrix}$$

Let us reduce A to the diagonal matrix. We know that $A = I_4 A I_4$.

$$\begin{bmatrix} 1 & 1 & -2 & 0 \\ 1 & -4 & 0 & 0 \\ -2 & 0 & 6 & -3 \\ 0 & 0 & -3 & 2 \end{bmatrix} = \begin{bmatrix} 1 & 0 & 0 & 0 \\ 0 & 1 & 0 & 0 \\ 0 & 0 & 1 & 0 \\ 0 & 0 & 0 & 1 \end{bmatrix} A \begin{bmatrix} 1 & 0 & 0 & 0 \\ 0 & 1 & 0 & 0 \\ 0 & 0 & 1 & 0 \\ 0 & 0 & 0 & 1 \end{bmatrix}$$

Operating $R_2 - R_1$, $R_3 + 2R_1$, we get

$$\begin{bmatrix} 1 & 1 & -2 & 0 \\ 0 & -5 & 2 & 0 \\ 0 & 2 & 2 & -3 \\ 0 & 0 & -3 & 2 \end{bmatrix} = \begin{bmatrix} 1 & 0 & 0 & 0 \\ -1 & 1 & 0 & 0 \\ 2 & 0 & 1 & 0 \\ 0 & 0 & 0 & 1 \end{bmatrix} A \begin{bmatrix} 1 & 0 & 0 & 0 \\ 0 & 1 & 0 & 0 \\ 0 & 0 & 1 & 0 \\ 0 & 0 & 0 & 1 \end{bmatrix}$$

Operating $C_2 - C_1$, $C_3 + 2C_1$, we get

$$\begin{bmatrix} 1 & 0 & 0 & 0 \\ 0 & -5 & 2 & 0 \\ 0 & 2 & 2 & -3 \\ 0 & 0 & -3 & 2 \end{bmatrix} = \begin{bmatrix} 1 & 0 & 0 & 0 \\ -1 & 1 & 0 & 0 \\ 2 & 0 & 1 & 0 \\ 0 & 0 & 0 & 1 \end{bmatrix} A \begin{bmatrix} 1 & -1 & 2 & 0 \\ 0 & 1 & 0 & 0 \\ 0 & 0 & 1 & 0 \\ 0 & 0 & 0 & 1 \end{bmatrix}$$

Operating $R_3 + (2/5)R_2$, we get

$$\begin{bmatrix} 1 & 0 & 0 & 0 \\ 0 & -5 & 2 & 0 \\ 0 & 0 & \frac{14}{5} & -3 \\ 0 & 0 & -3 & 2 \end{bmatrix} = \begin{bmatrix} 1 & 0 & 0 & 0 \\ -1 & 1 & 0 & 0 \\ \frac{8}{5} & \frac{2}{5} & 1 & 0 \\ 0 & 0 & 0 & 1 \end{bmatrix} A \begin{bmatrix} 1 & -1 & 2 & 0 \\ 0 & 1 & 0 & 0 \\ 0 & 0 & 1 & 0 \\ 0 & 0 & 0 & 1 \end{bmatrix}$$

Operating $C_3 + (2/5)\ C_2$, we get

$$\begin{bmatrix} 1 & 0 & 0 & 0 \\ 0 & -5 & 0 & 0 \\ 0 & 0 & \frac{14}{5} & -3 \\ 0 & 0 & -3 & 2 \end{bmatrix} = \begin{bmatrix} 1 & 0 & 0 & 0 \\ -1 & 1 & 0 & 0 \\ \frac{8}{5} & \frac{2}{5} & 1 & 0 \\ 0 & 0 & 0 & 1 \end{bmatrix} A \begin{bmatrix} 1 & -1 & \frac{8}{5} & 0 \\ 0 & 1 & \frac{2}{5} & 0 \\ 0 & 0 & 1 & 0 \\ 0 & 0 & 0 & 1 \end{bmatrix}$$

Operating $R_4 + \frac{15}{14} R_3$, we get

$$\begin{bmatrix} 1 & 0 & 0 & 0 \\ 0 & -5 & 0 & 0 \\ 0 & 0 & \frac{14}{5} & -3 \\ 0 & 0 & 0 & -\frac{17}{14} \end{bmatrix} = \begin{bmatrix} 1 & 0 & 0 & 0 \\ -1 & 1 & 0 & 0 \\ \frac{8}{5} & \frac{2}{5} & 1 & 0 \\ \frac{12}{7} & \frac{3}{7} & \frac{15}{14} & 1 \end{bmatrix} A \begin{bmatrix} 1 & -1 & \frac{8}{5} & 0 \\ 0 & 1 & \frac{2}{5} & 0 \\ 0 & 0 & 1 & 0 \\ 0 & 0 & 0 & 1 \end{bmatrix}$$

Operating $C_4 + \frac{15}{14} C_3$, we get

$$\begin{bmatrix} 1 & 0 & 0 & 0 \\ 0 & -5 & 0 & 0 \\ 0 & 0 & \frac{14}{5} & 0 \\ 0 & 0 & 0 & -\frac{17}{14} \end{bmatrix} = \begin{bmatrix} 1 & 0 & 0 & 0 \\ -1 & 1 & 0 & 0 \\ \frac{8}{5} & \frac{2}{5} & 1 & 0 \\ \frac{12}{7} & \frac{3}{7} & \frac{15}{14} & 1 \end{bmatrix} A \begin{bmatrix} 1 & -1 & \frac{8}{5} & \frac{12}{7} \\ 0 & 1 & \frac{2}{5} & \frac{3}{7} \\ 0 & 0 & 1 & \frac{15}{14} \\ 0 & 0 & 0 & 1 \end{bmatrix}$$

i.e., $$\text{diag}\left(1, -5, \frac{14}{5}, -\frac{17}{14}\right) = P'AP$$

$\therefore$ the canonical form of the given quadratic form is

$$Y'(P'AP)Y = Y' \text{ diag}\left(1, -5, \frac{14}{5}, -\frac{17}{14}\right) Y$$

$$= [y_1 y_2 y_3 y_4] \begin{bmatrix} 1 & 0 & 0 & 0 \\ 0 & -5 & 0 & 0 \\ 0 & 0 & \frac{14}{5} & 0 \\ 0 & 0 & 0 & -\frac{17}{14} \end{bmatrix} \begin{bmatrix} y_1 \\ y_2 \\ y_3 \\ y_4 \end{bmatrix}$$

$$= y_1^2 - 5y_2^2 + \frac{14}{5} y_3^2 - \frac{17}{14} y_4^2,$$ which is the sum of the squares.

Note. Here $\rho(A) = 4$, index = 2, signature = 2 – 2 = 0.

6.26. REDUCTION TO CANONICAL FORM BY ORTHOGONAL TRANSFORMATION

Let X′AX be a given quadratic form. The modal matrix B of A is that matrix whose columns are characteristic vectors of A. If B represents the orthogonal matrix of A (the normalised modal matrix of A whose column vectors are pairwise orthogonal) then X = BY will reduce X′AX to Y′DY where D = diag $(\lambda_1, \lambda_2, ..., \lambda_n)$ and $\lambda_1, \lambda_2, ..., \lambda_n$ are characteristic roots of A.

Note. This method works successfully if the characteristic vectors of A are linearly independent which are pairwise orthogonal.

Example 1. *Reduce $8x^2 + 7y^2 + 3z^2 - 12xy + 4xz - 8yz$ into canonical form by orthogonal transformation.*

Sol. The matrix of the quadratic form is

$$A = \begin{bmatrix} 8 & -6 & 2 \\ -6 & 7 & -4 \\ 2 & -4 & 3 \end{bmatrix}$$

The characteristic roots of A are given by $| A - \lambda I | = 0$

i.e., $$\begin{vmatrix} 8-\lambda & -6 & 2 \\ -6 & 7-\lambda & -4 \\ 2 & -4 & 3-\lambda \end{vmatrix} = 0 \quad \text{or} \quad \lambda(\lambda - 3)(\lambda - 15) = 0$$

$\therefore$ $$\lambda = 0, 3, 15$$

Characteristic vector for $\lambda = 0$ is given by $[A - (0)I]\, X = O$

i.e.,
$$8x_1 - 6x_2 + 2x_3 = 0$$
$$-6x_1 + 7x_2 - 4x_3 = 0$$
$$2x_1 - 4x_2 + 3x_3 = 0$$

Solving first two, we get $\dfrac{x_1}{1} = \dfrac{x_2}{2} = \dfrac{x_3}{2}$ giving the eigen vector $X_1 = k_1(1, 2, 2)'$

When $\lambda = 3$, the corresponding characteristic vector is given by $[A - 3I]X = O$

i.e.,
$$5x_1 - 6x_2 + 2x_3 = 0$$
$$-6x_1 + 4x_2 - 4x_3 = 0$$
$$2x_1 - 4x_2 = 0$$

Solving any two equations, we get $X_2 = k_2(2, 1, -2)'$.

Similarly characteristic vector corresponding to

$$\lambda = 15 \text{ is } X_3 = k_3(2, -2, 1)'.$$

Now X_1, X_2, X_3 are pairwise orthogonal

i.e.,
$$X_1 \cdot X_2 = X_2 \cdot X_3 = X_3 \cdot X_1 = 0.$$

$\therefore$ The normalised modal matrix is

$$B = \left[\frac{X_1}{\|X_1\|}, \frac{X_2}{\|X_2\|}, \frac{X_3}{\|X_3\|}\right] = \begin{bmatrix} \frac{1}{3} & \frac{2}{3} & \frac{2}{3} \\ \frac{2}{3} & \frac{1}{3} & -\frac{2}{3} \\ \frac{2}{3} & \frac{-2}{3} & \frac{1}{3} \end{bmatrix}$$

Now B is orthogonal matrix and $|\, B \,| = 1$

i.e.,
$$B^{-1} = B^T \text{ and } B^{-1} AB = D = \text{diag } (0, 3, 15)$$

i.e.,
$$\begin{bmatrix} \frac{1}{3} & \frac{2}{3} & \frac{2}{3} \\ \frac{2}{3} & \frac{1}{3} & -\frac{2}{3} \\ \frac{2}{3} & -\frac{2}{3} & \frac{1}{3} \end{bmatrix} A \begin{bmatrix} \frac{1}{3} & \frac{2}{3} & \frac{2}{3} \\ \frac{2}{3} & \frac{1}{3} & \frac{-2}{3} \\ \frac{2}{3} & -\frac{2}{3} & \frac{1}{3} \end{bmatrix} = \begin{bmatrix} 0 & 0 & 0 \\ 0 & 3 & 0 \\ 0 & 0 & 15 \end{bmatrix}$$

$$X'AX = Y'(B^{-1}AB)\, Y = Y'DY$$

$$= [y_1\, y_2\, y_3] \begin{bmatrix} 0 & 0 & 0 \\ 0 & 3 & 0 \\ 0 & 0 & 15 \end{bmatrix} \begin{bmatrix} y_1 \\ y_2 \\ y_3 \end{bmatrix} = 0y_1^2 + 3y_2^2 + 15y_3^2$$

which is the required canonical form.

Note. Here the orthogonal transformation is $X = BY$, rank of the quadratic form = 2; index = 2, signature = 2. It is positive semi-definite.

Example 2. *Reduce* $6x_1^2 + 3x_2^2 + 3x_3^2 - 4x_1x_2 - 2x_2x_3 + 4x_3x_1$ *into canonical form.*

Sol. The matrix of the quadratic form is $A = \begin{bmatrix} 6 & -2 & 2 \\ -2 & 3 & -1 \\ 2 & -1 & 3 \end{bmatrix}$

The characteristic roots given by $| A - \lambda I | = 0$

i.e.,
$$\begin{vmatrix} 6-\lambda & -2 & 2 \\ -2 & 3-\lambda & -1 \\ 2 & -1 & 3-\lambda \end{vmatrix} = 0$$

or
$$\lambda^3 - 12\lambda^2 + 36\lambda - 32 = 0$$

which on solving gives $\lambda = 8, 2, 2$.

The vector corresponding to $\lambda = 8$ is given by $[A - 8I]X = O$

i.e.,
$$\begin{bmatrix} -2 & -2 & 2 \\ -2 & -5 & -1 \\ 2 & -1 & -5 \end{bmatrix} \begin{bmatrix} x_1 \\ x_2 \\ x_3 \end{bmatrix} = \begin{bmatrix} 0 \\ 0 \\ 0 \end{bmatrix}$$

or
$$-2x_1 - 2x_2 + 2x_3 = 0$$
$$-2x_1 - 5x_2 - x_3 = 0$$
$$2x_1 - x_2 - 5x_3 = 0$$

Solving any two of the equations, we get the vector as $k_1[2, -1, 1]'$.

The characteristic vector for $\lambda = 2$ is given by $[A - 2I]X = O$ which reduces to single equation
$$2x_1 - x_2 + x_3 = 0.$$

Putting $x_2 = 0$, we get $\frac{x_2}{1} = \frac{x_3}{1}$ or the vector is k_2 $[0, 1, 1]'$. Again by putting $x_2 = 0$, we get $\frac{x_1}{1} = \frac{x_3}{-2}$ or the vector is k_3 $[1, 0, -2]'$.

Now $X_1 = k_1 [2, -1, 1]'$; $X_2 = k_2 [0, 1, 1]'$ and $X_3 = k_3 [1, 0, -2]'$.

Here X_1, X_2, X_3 are not pairwise orthogonal

$\because$ $\quad X_1 \cdot X_2 = 0$; $X_2 \cdot X_3 \neq 0$ and $X_3 \cdot X_1 = 0$

To get X_3 orthogonal to X_2 assume a vector $[u, v, w]'$ orthogonal to X_2 also satisfying

$2x_1 - x_2 + x_3 = 0$ *i.e.,* $2u - v + w = 0$ and $0 \cdot u + 1 \cdot v + 1 \cdot w = 0$

Solving $[u, v, w]' = [1, 1, -1]' = X_3$ so that $X_1 \cdot X_2 = X_2 \cdot X_3 = X_3 \cdot X_1 = 0$

$\therefore$ The normalised modal matrix is

$$B = \begin{bmatrix} \frac{2}{\sqrt{6}} & 0 & \frac{1}{\sqrt{3}} \\ -\frac{1}{\sqrt{6}} & \frac{1}{\sqrt{2}} & \frac{1}{\sqrt{3}} \\ \frac{1}{\sqrt{6}} & \frac{1}{\sqrt{2}} & -\frac{1}{\sqrt{3}} \end{bmatrix}$$

Now B is orthogonal matrix and $| B | = 1$

i.e.,
$$B' = B^{-1} \text{ and } B^{-1} AB = D \text{ where } D = \begin{bmatrix} 8 & 0 & 0 \\ 0 & 2 & 0 \\ 0 & 0 & 2 \end{bmatrix}$$

$\therefore$
$$X' AX = Y'(B^{-1} AB)Y = Y'DY$$

$$= [y_1 y_2 y_3] \begin{bmatrix} 8 & 0 & 0 \\ 0 & 2 & 0 \\ 0 & 0 & 2 \end{bmatrix} \begin{bmatrix} y_1 \\ y_2 \\ y_3 \end{bmatrix} = 8y_1^2 + 2y_2^2 + 2y_3^2$$

which is the required canonical form.

Note. Here rank of the quadratic form is 3, index = 3, signature = 3. It is positive definite.

EXERCISE 6.7

1. Write down the matrices of the following quadratic forms

(*i*) $2x^2 + 3y^2 + 6xy$ (*ii*) $2x^2 + 5y^2 - 6z^2 - 2xy - yz + 8zx$

(*iii*) $x_1^2 + 2x_2^2 + 3x_3^2 + 4x_4^2 + 2x_1x_2 + 4x_1x_3 - 6x_1x_4 - 4x_2x_3 - 8x_2x_4 + 12x_3x_4$.

2. Write down the quadratic forms corresponding to the following matrices.

(*i*) $\begin{bmatrix} 2 & 4 & 5 \\ 4 & 3 & 1 \\ 5 & 1 & 1 \end{bmatrix}$ (*ii*) $\begin{bmatrix} 1 & 1 & -2 & 0 \\ 1 & -4 & 0 & 0 \\ -2 & 0 & 6 & -3 \\ 0 & 0 & -3 & 2 \end{bmatrix}$.

3. Reduce the following quadratic forms to canonical forms or to sum of squares by linear transformation. Write also the rank, index and signature.

(*i*) $2x^2 + 2y^2 + 3z^2 + 2xy - 4yz - 4xz$ (*ii*) $12x_1^2 + 4x_2^2 + 5x_3^2 - 4x_2x_3 + 6x_1x_3 - 6x_1x_2$

(*iii*) $2x^2 + 9y^2 + 6z^2 + 8xy + 8yz + 6zx$ (*iv*) $x^2 + 5y^2 + z^2 + 2xy + 2yz + 6zx$.

4. Reduce the following quadratic forms to canonical forms or to sum of squares by orthogonal transformation. Write also rank, index, signature.

(*i*) $3x^2 + 5y^2 + 3z^2 - 2xy - 2yz + 2zx$ (*ii*) $2x_1^2 + 2x_2^2 + 2x_3^2 - 2x_1x_2 + 2x_1x_3 - 2x_2x_3$

(*iii*) $3x^2 - 2y^2 - z^2 - 4xy + 8xz + 12yz$ (*iv*) $x^2 + 3y^2 + 3z^2 - 2yz$.

Answers

1. (*i*) $\begin{bmatrix} 2 & 3 \\ 3 & 3 \end{bmatrix}$ (*ii*) $\begin{bmatrix} 2 & -1 & 4 \\ -1 & 5 & -\frac{1}{2} \\ 4 & -\frac{1}{2} & -6 \end{bmatrix}$ (*iii*) $\begin{bmatrix} 1 & 1 & 2 & -3 \\ 1 & 2 & -2 & -4 \\ 2 & -2 & 3 & 6 \\ -3 & -4 & 6 & 4 \end{bmatrix}$

2. (*i*) $2x^2 + 3y^2 + z^2 + 8xy + 2yz + 10xz$ (*ii*) $x_1^2 - 4x_2^2 + 6x_3^2 + 2x_4^2 + 2x_1x_2 - 4x_1x_3 - 6x_3x_4$

3. (*i*) $2y_1^2 + \frac{3}{2}y_2^2 + \frac{1}{3}y_3^2$; Rank = 3, index = 3, Sig. = 3

(*ii*) $12y_1^2 + \frac{13}{4}y_2^2 + \frac{49}{13}y_3^2$; Rank = 3, index = 3, Sig. = 3

(*iii*) $2y_1^2 + y_2^2 - \frac{5}{2}y_3^2$; Rank = 3, index = 2, Sig. = 1

(*iv*) $-2y_1^2 + 3y_2^2 + 6y_3^2$; Rank = 3, index = 2, Sig. = 1.

4. (*i*) $2y_1^2 + 3y_2^2 + 6y_3^2$; Rank = 3 , index = 3, Sig. = 3

(*ii*) $4y_1^2 + y_2^2 + y_3^2$; Rank = 3, index = 3, Sig. = 3

(*iii*) $3y_1^2 + 6y_2^2 - 9y_3^2$; Rank = 3, index = 2, Sig. = 1

(*iv*) $y_1^2 + 2y_2^2 + 4y_3^2$; Rank = 3, index = 3, Sig. = 3.

APPENDIX—I

Table 1 : NORMAL TABLE

AREAS UNDER THE STANDARD NORMAL

$$\text{CURVE} = \frac{1}{\sqrt{2\pi}} \int_0^z e^{-\frac{z^2}{2}} dz$$

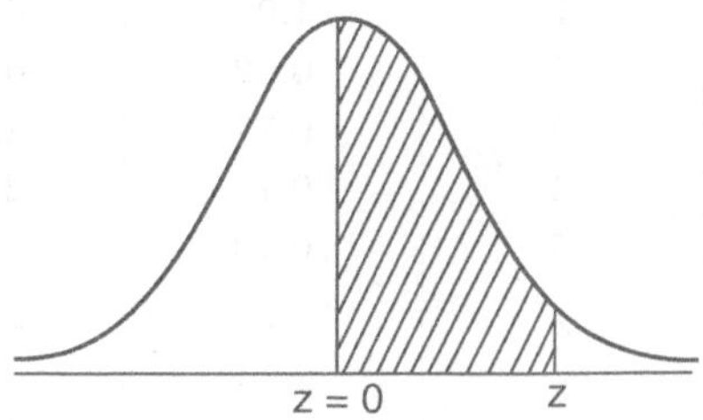

z	*.00*	*.01*	*.02*	*.03*	*.04*	*.05*	*.06*	*.07*	*.08*	*.09*
0.0	.0000	.0040	.0080	.0120	.0160	.0199	.0239	.0279	.0319	.0359
0.1	.0398	.0438	.0478	.0517	.0557	.0596	.0636	.0675	.0714	.0754
0.2	.0793	.0832	.0871	.0910	.0948	.0987	.1026	.1064	.1103	.1141
0.3	.1179	.1217	.1255	.1293	.1331	.1368	.1406	.1443	.1480	.1517
0.4	.1554	.1591	.1628	.1664	.1700	.1736	.1772	.1808	.1844	.1879
0.5	.1915	.1950	.1985	.2019	.2054	.2088	.2123	.2157	.2190	.2224
0.6	.2257	.2291	.2324	.2357	.2389	.2422	.2454	.2485	.2517	.2549
0.7	.2580	.2611	.2642	.2673	.2704	.2734	.2764	.2794	.2823	.2852
0.8	.2881	.2910	.2939	.2967	.2995	.3023	.3051	.3078	.3106	.3133
0.9	.3159	.3186	.3212	.3238	.3264	.3289	.3315	.3340	.3365	.3389
1.0	.3413	.3438	.3461	.3485	.3508	.3531	.3554	.3577	.3599	.3621
1.1	.3643	.3665	.3686	.3708	.8729	.3749	.3770	.3790	.3810	.3830
1.2	.3849	.3869	.3888	.3907	.3925	.3944	.3962	.3980	.3997	.4015
1.3	.4032	.4049	.4066	.4082	.4099	.4115	.4131	.4147	.4162	.4177
1.4	.4192	.4207	.4222	.4236	.4251	.4255	.4279	.4292	.4306	.4319
1.5	.4332	.4345	.4357	.4370	.4382	.4394	.4406	.4418	.4429	.4441
1.6	.4452	.4463	.4474	.4484	.4495	.4505	.4515	.4525	.4535	.4545
1.7	.4554	.4564	.4573	.4582	.4591	.4599	.4608	.4616	.4625	.4633
1.8	.4641	.4649	.4656	.4664	.4671	.4678	.4686	.4693	.4699	.4706
1.9	.4713	.4719	.4726	.4732	.4738	.4744	.4750	.4756	.4761	.4767
2.0	.4772	.4778	.4783	.4788	.4793	.4798	.4803	.4808	.4812	.4817
2.1	.4821	.4826	.4830	.4834	.4838	.4842	.4846	.4850	.4854	.4857
2.2	.4861	.4864	.4868	.4871	.4875	.4878	.4881	.4884	.4887	.4890
2.3	.4893	.4896	.4898	.4901	.4904	.4906	.4909	.4911	.4913	.4916
2.4	.4918	.4920	.4922	.4925	.4927	.4929	.4931	.4932	.4934	.4936
2.5	.4938	.4940	.4941	.4943	.4945	.4946	.4948	.4949	.4951	.4952
2.6	.4953	.4955	.4956	.4957	.4959	.4930	.4961	.4962	.4963	.4964
2.7	.4965	.4966	.4967	.4968	.4969	.4970	.4971	.4972	.4973	.4974
2.8	.4974	.4975	.4976	.4977	.4977	.4978	.4979	.4979	.4980	.4981
2.9	.4981	.4982	.4982	.4983	.4984	.4984	.4985	.4985	.4986	.4986
3.0	.4987	.4987	.4987	.4988	.4988	.4989	.4999	.4989	.4990	.4990
3.1	.4990	.4991	.4991	.4991	.4992	.4992	.4992	.4992	.4993	.4993

Table 2 : SIGNIFICANT VALUES t_v (α) OF t-DISTRIBUTION (TWO TAIL AREAS) $[|\ t\ | > t_v(\alpha)] = \alpha$

d.f.	*Probability (Level of Significance)*					
(ν)	*0.50*	*0.10*	*0.05*	*0.02*	*0.01*	*0.001*
1	1.00	6.31	12.71	31.82	63.66	636.62
2	0.82	0.92	4.30	6.97	6.93	31.60
3	0.77	2.32	3.18	4.54	5.84	12.94
4	0.74	2.13	2.78	3.75	4.60	8.61
5	0.73	2.02	2.57	3.37	4.03	6.86
6	0.72	1.94	2.45	3.14	3.71	5.96
7	0.71	1.90	2.37	3.00	3.50	5.41
8	0.71	1.80	2.31	2.90	3.36	5.04
9	0.70	1.83	2.26	2.82	3.25	4.78
10	0.70	1.81	2.23	2.76	3.17	4.59
11	0.70	1.80	2.20	2.72	3.11	4.44
12	0.70	1.78	2.18	2.68	3.06	4.32
13	0.69	1.77	2.16	2.05	3.01	4.22
14	0.69	1.76	2.15	2.62	2.98	4.14
15	0.69	1.75	2.13	2.60	2.95	4.07
16	0.69	1.75	2.12	2.58	2.92	4.02
17	0.69	1.74	2.11	2.57	2.90	3.97
18	0.69	1.73	2.10	2.55	2.88	3.92
19	0.69	1.73	2.09	2.54	2.86	3.88
20	0.69	1.73	2.09	2.53	2.85	3.85
21	0.69	1.72	2.08	2.52	2.83	3.83
22	0.69	1.72	2.07	2.51	2.42	3.79
23	0.69	1.71	2.07	2.50	2.81	3.77
24	0.69	1.71	2.06	2.49	2.80	3.75
25	0.68	1.71	2.06	2.49	2.79	3.73
26	0.68	1.71	2.06	2.48	2.78	3.71
27	0.68	1.70	2.05	2.47	2.77	3.69
28	0.68	1.70	2.05	2.47	2.76	3.67
29	0.68	1.70	2.05	2.46	2.76	3.66
30	0.68	1.70	2.04	2.46	2.75	3.65
∞	0.67	1.65	1.96	2.33	2.58	3.29

Table 3 : F-Distribution
Values of F for F-Distributions with 0.05 of the Area in the Right Tail

	Degrees of freedom for numerator																		
	1	*2*	*3*	*4*	*5*	*6*	*7*	*8*	*9*	*10*	*12*	*15*	*20*	*24*	*30*	*40*	*60*	*120*	∞
1	161	200	216	225	230	234	237	239	241	242	244	246	248	249	250	251	252	253	254
2	18.5	19.0	19.2	19.2	19.3	19.3	19.4	19.4	19.4	19.4	19.4	19.4	19.4	19.5	19.5	19.5	19.5	19.5	19.5
3	10.1	9.55	9.28	9.12	9.01	9.94	8.89	8.85	8.81	8.79	8.74	8.70	8.66	8.64	8.62	8.59	8.57	8.55	8.53
4	7.71	6.94	6.59	6.39	6.26	6.16	6.09	6.04	6.00	5.96	5.91	5.86	5.80	5.77	5.75	5.72	5.69	5.66	5.63
5	6.61	5.79	5.41	5.19	5.05	4.95	4.88	4.82	4.77	4.74	4.68	4.62	4.56	4.53	4.50	4.46	4.43	4.40	4.37
6	5.99	5.14	4.76	4.53	4.39	4.28	4.21	4.15	4.10	4.06	4.00	3.94	3.87	3.84	3.81	3.77	3.74	3.70	3.67
7	5.59	4.74	4.35	4.12	3.97	3.87	3.79	3.73	3.68	3.64	3.57	3.51	3.44	3.41	3.38	3.34	3.30	3.27	3.23
8	5.32	4.46	4.07	3.84	3.69	3.58	3.50	3.44	3.39	3.35	3.28	3.22	3.15	3.12	3.08	3.04	3.01	2.97	2.93
9	5.12	4.26	3.86	3.63	3.48	3.37	3.29	3.23	3.18	3.14	3.07	3.01	2.94	2.90	2.86	2.83	2.79	2.75	2.71
10	4.96	4.10	3.71	3.48	3.33	3.22	3.14	3.07	3.02	2.98	2.91	2.85	2.77	2.74	2.70	2.66	2.62	2.58	2.54
11	4.84	3.98	3.59	3.36	3.20	3.09	3.01	2.95	2.90	2.85	2.79	2.72	2.65	2.61	2.57	2.53	2.49	2.45	2.40
12	4.75	3.89	3.49	3.26	3.11	3.00	2.91	2.85	2.80	2.75	2.69	2.62	2.54	2.51	2.47	2.43	2.38	2.34	2.30
13	4.67	3.81	3.41	3.18	3.03	2.92	2.83	2.77	2.71	2.67	2.60	2.53	2.46	2.42	2.38	2.34	2.30	2.25	2.21
14	4.60	3.74	3.34	3.11	3.96	2.85	2.76	2.70	2.65	2.60	2.53	2.46	2.39	2.35	2.31	2.27	2.22	2.18	2.13
15	4.54	3.68	3.29	3.06	3.90	2.79	2.71	2.64	2.59	2.54	2.48	2.40	2.33	2.29	2.25	2.20	2.16	2.11	2.07
16	4.49	3.63	3.24	3.01	2.85	2.74	2.66	2.59	2.54	2.49	2.42	2.35	2.28	2.24	2.19	2.15	2.11	2.06	2.01
17	4.45	3.59	3.20	2.96	2.81	2.70	2.61	2.55	2.49	2.45	2.38	2.31	2.23	2.19	2.15	2.10	2.06	2.01	1.96
18	4.41	3.55	3.16	2.93	2.77	2.66	2.58	2.51	2.46	2.41	2.34	2.27	2.19	2.15	2.11	2.06	2.02	1.97	1.92
19	4.38	3.52	3.13	2.90	2.74	2.63	2.54	2.48	2.42	2.38	2.31	2.23	2.16	2.11	2.07	2.03	1.98	1.93	1.88
20	4.35	3.49	3.10	2.87	2.17	2.60	2.51	2.45	2.39	2.35	2.28	2.20	2.12	2.08	2.04	1.99	1.95	1.90	1.84
21	4.32	3.47	3.07	2.84	2.68	2.57	2.49	2.42	2.37	2.32	2.25	2.18	2.10	2.05	2.01	1.96	1.92	1.87	1.81
22	4.30	3.44	3.05	2.82	2.66	2.55	2.46	2.40	2.34	2.30	2.23	2.15	2.07	2.03	1.98	1.94	1.89	1.84	1.78
23	4.28	3.42	3.03	2.80	2.64	2.53	2.44	2.37	2.32	2.27	2.20	2.13	2.05	2.01	1.96	1.91	1.86	1.81	1.76
24	4.26	3.40	3.01	2.78	2.62	2.51	2.42	2.36	2.30	2.25	2.18	2.11	2.03	1.98	1.94	1.98	1.84	1.79	1.73
25	4.24	3.39	2.99	2.76	2.60	2.94	2.40	2.34	2.28	2.24	2.16	2.29	2.01	1.96	1.92	1.87	1.82	1.77	1.71
30	4.17	3.32	2.92	2.69	2.53	2.42	2.33	2.27	2.21	2.16	2.09	2.01	1.93	1.89	1.84	1.79	1.74	1.64	1.62
40	4.08	3.23	2.84	2.61	2.45	2.34	2.25	2.18	2.12	2.08	2.00	1.92	1.84	1.79	1.74	1.69	1.64	1.58	1.51
60	4.00	3.15	2.76	2.53	2.37	2.25	2.17	2.10	2.04	1.99	1.92	1.84	1.75	1.70	1.65	1.59	1.53	1.47	1.39
120	3.92	3.07	2.68	2.45	2.29	2.18	2.09	2.02	1.96	1.91	1.83	1.75	1.66	1.61	1.55	1.50	1.43	1.35	1.25
∞	3.84	3.00	2.60	2.37	2.21	2.10	2.01	1.94	1.88	1.83	1.75	1.67	1.57	1.52	1.46	1.39	1.32	1.22	1.00

Table 4 : CHI-SQUARE

Significant Values χ^2 (α) of Chi-Square Distribution Right Tail Areas for Given Probability α,

$$P = P_r (\chi^2 > \chi^2 (\alpha)) = \alpha$$

And is Degrees of Freedom (d.f.)

Degree of freedom (ν)	*Probability (Level of Significance)*						
	0 = .99	*0.95*	*0.50*	*0.10*	*0.05*	*0.02*	*0.01*
1	.000157	.00393	.455	2.706	3.841	5.214	6.635
2	.0201	.103	1.386	4.605	5.991	7.824	9.210
3	.115	.352	2.366	6.251	7.815	9.837	11.341
4	.297	.711	3.357	7.779	9.488	11.668	13.277
5	.554	1.145	4.351	9.236	11.070	13.388	15.086
6	.872	2.635	5.348	10.645	12.592	15.033	16.812
7	1.239	2.167	6.346	12.017	14.067	16.622	18.475
8	1.646	2.733	7.344	13.362	15.507	18.168	20.090
9	2.088	3.325	8.343	14.684	16.919	19.679	21.669
10	2.558	3.940	9.340	15.987	18.307	21.161	23.209
11	3.053	4.575	10.341	17.275	19.675	22.618	24.725
12	3.571	5.226	11.340	18.549	21.026	24.054	26.217
13	4.107	5.892	12.340	19.812	22.362	25.472	27.688
14	4.660	6.571	13.339	21.064	23.685	26.873	29.141
15	4.229	7.261	14.339	22.307	24.996	28.259	30.578
16	5.812	7.962	15.338	23.542	26.296	29.633	32.000
17	6.408	8.672	15.338	24.769	27.587	30.995	33.409
18	7.015	9.390	17.338	25.989	28.869	32.346	34.805
19	7.633	10.117	18.338	27.204	30.144	33.687	36.191
20	8.260	10.851	19.337	28.412	31.410	35.020	37.566
21	8.897	11.591	20.337	29.615	32.671	36.343	38.932
22	9.542	12.338	21.337	30.813	33.924	37.659	40.289
23	10.196	13.091	22.337	32.007	35.172	38.968	41.638
24	10.856	13.848	23.337	32.196	36.415	40.270	42.980
25	11.524	14.611	24.337	34.382	37.65	41.566	44.314
26	12.198	15.379	25.336	35.363	38.885	41.856	45.642
27	12.879	16.151	26.336	36.741	40.113	41.140	46.963
28	13.565	16.928	27.336	37.916	41.337	45.419	48.278
29	14.256	17.708	28.336	39.087	42.557	46.693	49.588
30	14.933	18.493	29.336	40.256	43.773	47.962	50.892

Note. For degrees of freedom (ν) greater than 30, the quantity $\sqrt{2\chi^2} - \sqrt{2\nu - 1}$ may be used as a normal variate with unit variance.

APPENDIX—II

SOME IMPORTANT CURVES

Cissoid (Imp.)

$y^2(2a - x) = x^3$

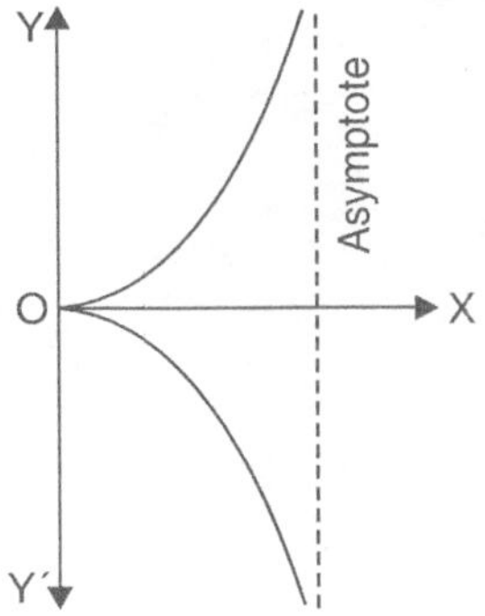

Catenary

$y = a \cosh \frac{x}{a}$

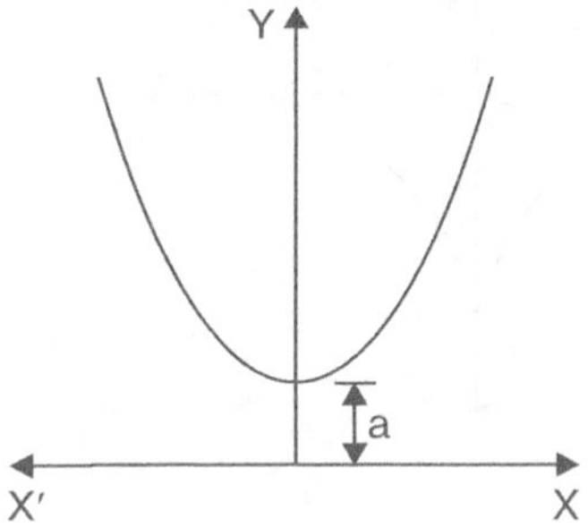

Cubical Parabola

$a^2y = x^3$

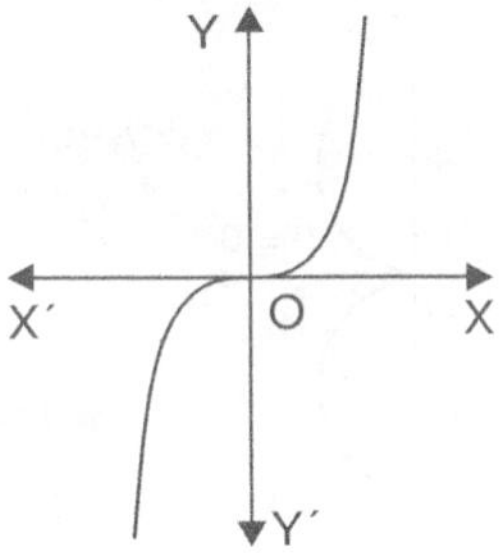

Semicubical Parabola

$ay^2 = x^3$

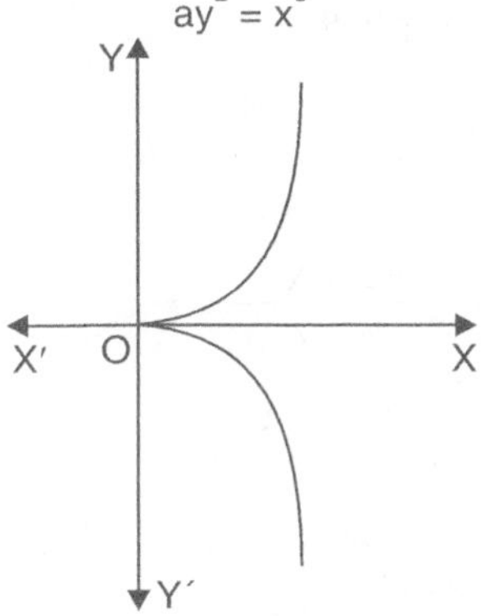

Strophoid

$y^2(a - x) = x^2(a + x)$

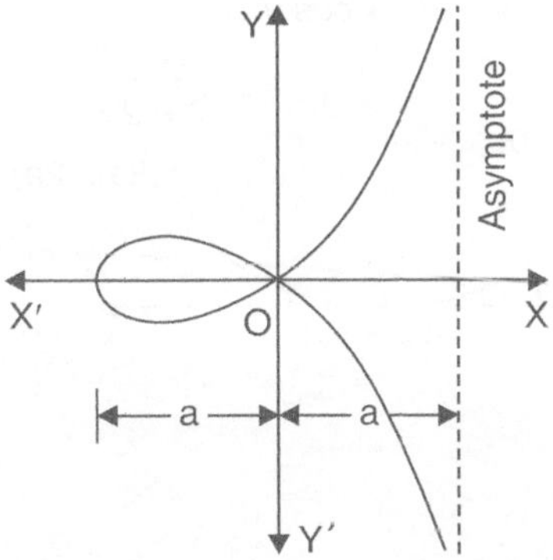

Folium of Descartes

$x^3 + y^3 = 3axy$

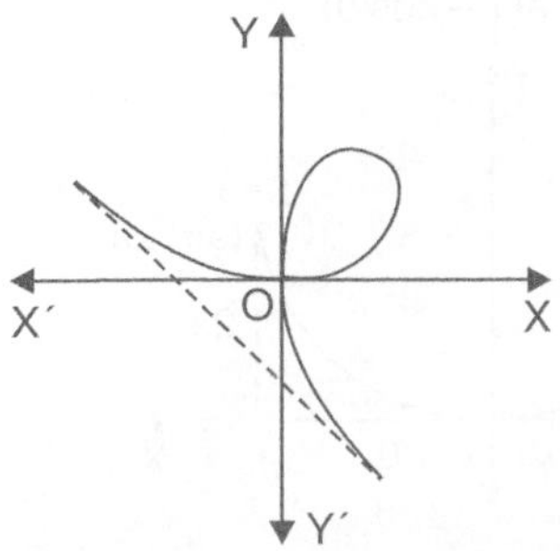

Hypocycloid (Imp.)

$x^{2/3} + y^{2/3} = a^{2/3}$

$x = a\cos^3\theta \quad y = a\sin^3\theta$

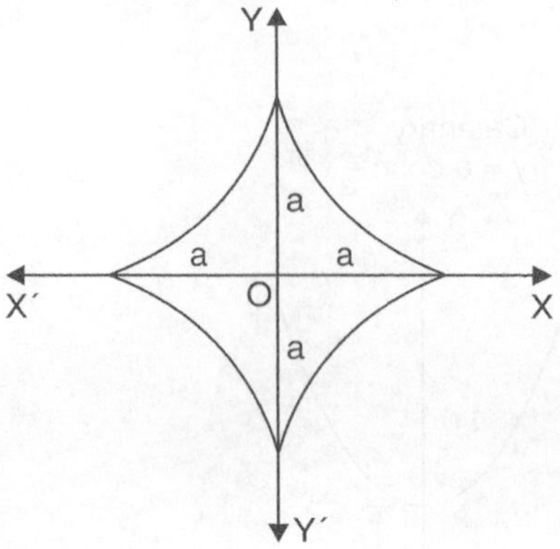

Witon of Agnesi

$xy^2 = 4a^2(2a - x)$

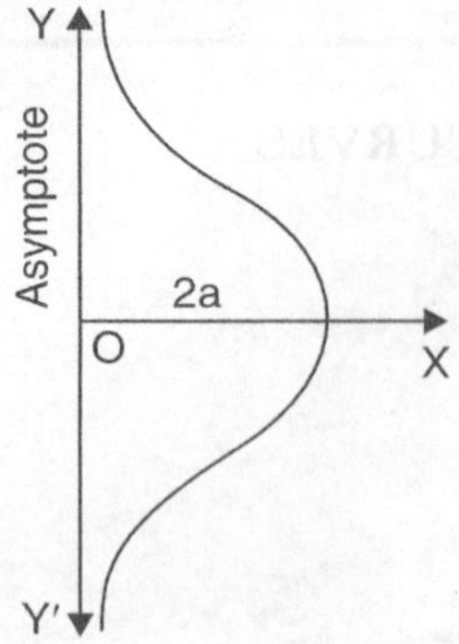

Cardioid (Imp.)

$r = a(1 + \cos\theta)$

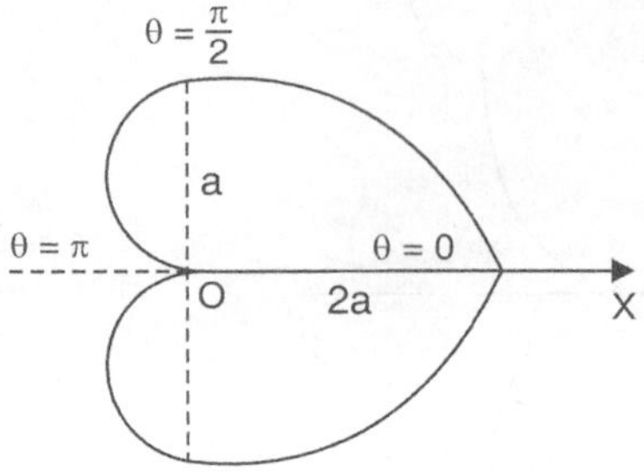

Cardioid (Imp.)

$r = a(1 - \cos\theta)$

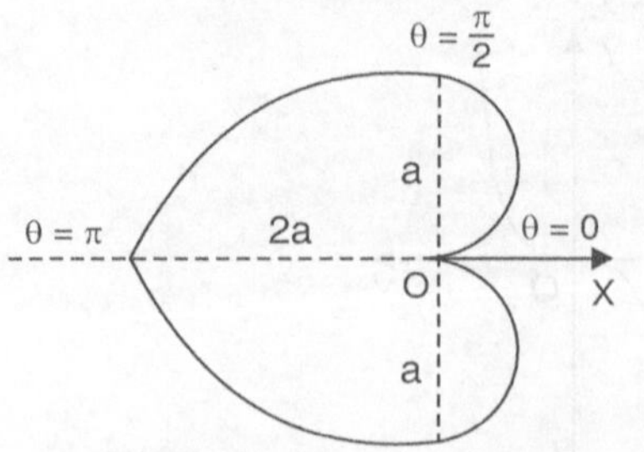

Cycloid (Most Imp.)

$x = a(\theta + \sin\theta)$

$y = a(1 - \cos\theta)$

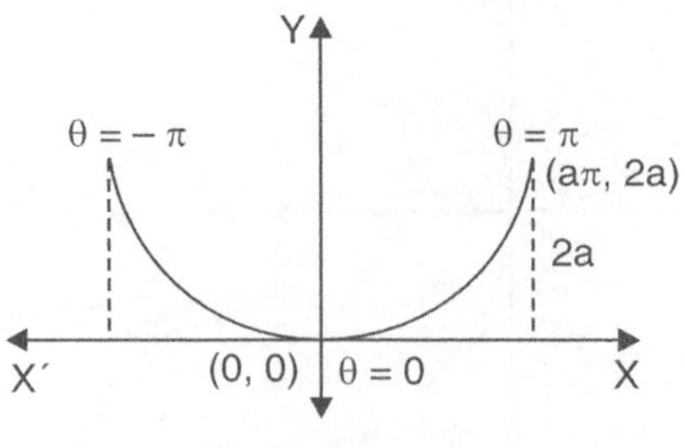

Cycloid

$x = a(\theta - \sin\theta)$

$y = a(1 + \cos\theta)$

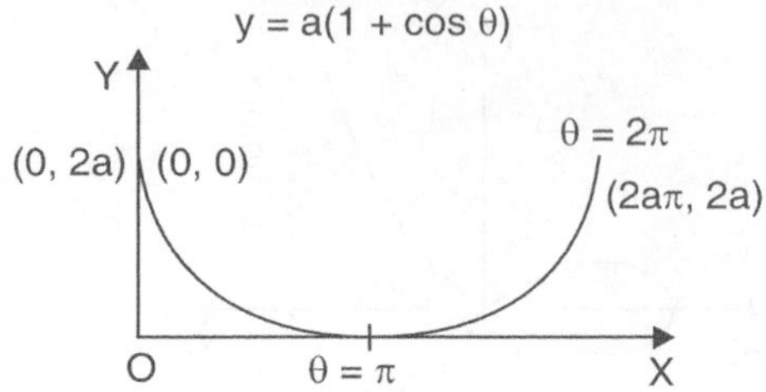

Cycloid (Imp.)

$x = a(\theta - \sin \theta)$

$y = a(1 - \cos \theta)$

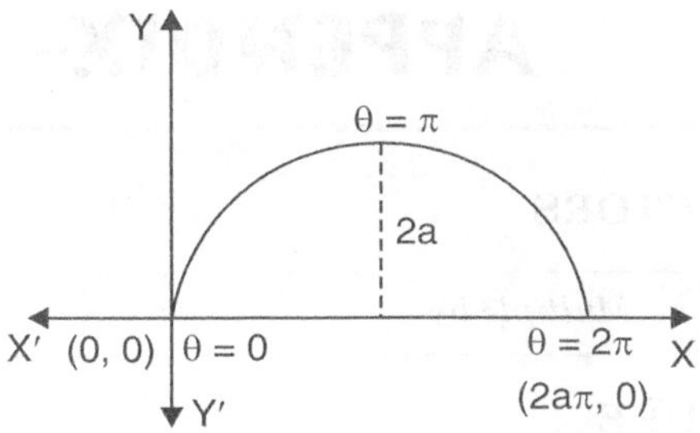

Cycloid

$x = a(\theta + \sin \theta)$

$y = a(1 + \cos \theta)$

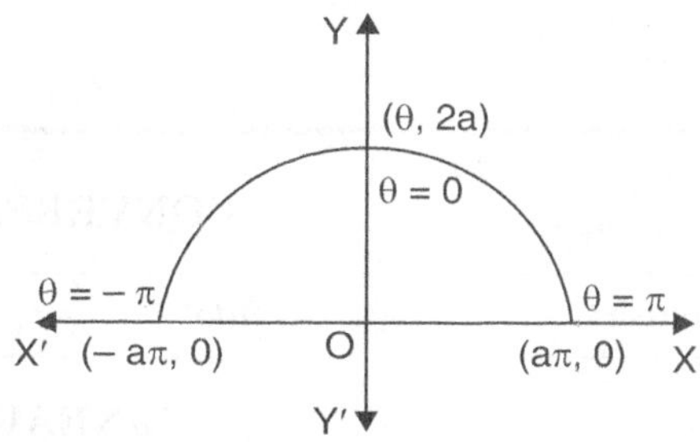

Lemniscate

$r^2 = a^2 \cos 2\theta$

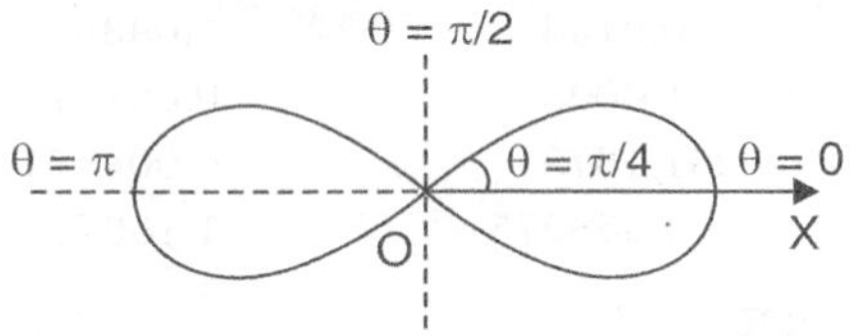

The three leaved rose

$r = a \sin 3\theta$

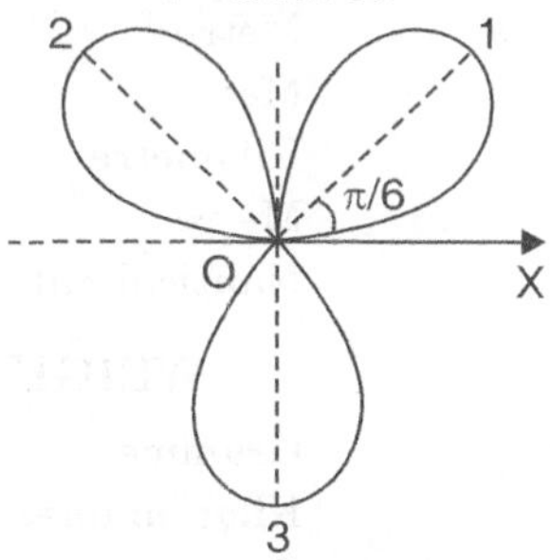

APPENDIX—III

CONVERSION FACTORS

To convert	*into*	*Multiple by*	
	LINEAR MEASURE		
Inch	Millimetre	25.40	0.03937
Inch	Centimetre	2.54	0.3937
Inch	Metre	0.0254	39.37
Feet	Metre	0.3048	3.281
Yard	Metre	0.9144	1.0936
Mile	Kilometre	1.6093	0.62137
Furlong	Metre	201.1678	0.004971
Miles	Nautical miles (int.)	0.858975	1.15078
	WEIGHT MEASURE		
Ounces	Grammes	28.3495	0.035274
Pounds	Kilogrammes	0.4536	2.20462
Cwt.	Quintal (100 kg.)	0.50802	1.9684
Ton	Kilogrammes	1016.05	0.0009842
U.S. Ton	Kilogrammes	907.18	0.0011
Ton (2240 lb)	Metric ton (Tonne)	1.01605	0.9842
Maunds	Ton	0.036735	27.2222
Tola	Ounce	0.411429	2.43056
Tola	Gramme	11.6628	0.087535
	SQUARE MEASURE		
Square inches	Square millimetres	645.16	0.00155
Square inches	Square centimetres	6.4516	0.155
Square feet	Square metre	0.0929	10.764
Square Yard	Square metre	0.836	1.196
Acre	Square metre	4046.9	0.00025
Square mile	Square kilometre	2.59	0.3861
Acre (4840 sq.yd.)	Hectare (10,000 sq. m)	0.40468	2.4711
	CUBIC MEASURE		
Cubic inch	Cubic centimetre	16.387	0.061
Cubic feet	Cubic metre	0.02832	35.315
Cubic Yard	Cubic metre	0.76455	1.308
Cubic foot	Gallon (Imp.)	6.2288	0.16054

Cubic foot	Litres (1000 cc)	28.3161	0.0353156
Gollon (Imp.)	Litres	4.54596	0.219976
Pint (0.124 Imp. Gln)	Litres	0.568	1.76
U.S. Gallon	Litres	3.78533	0.264178
	PRESSURES AND DENSITIES		
Pounds per sq in.	Kg. per sq. cm.	0.070307	14.2233
Pounds per sq. ft.	Kg. per sq. metre	4.88243	0.204816
Tonnes per sq. ft.	Tonnes per sq. metre	10.937	0.091436
Atmospheres	Pounds per sq. in.	14.6959	0.06804
Inch of Hg	Pounds per sq. in.	0.491154	2.036
Inch of Hg	Pounds per sq. in.	0.0345316	28.959
Pounds per sq. ft.	mm. of Hg.	0.035913	2.78450
Pounds per sq. in.	Feet of water	2.30672	0.433516
Feet of water	Kg. per sq. cm	0.0304792	32.81
Pounds per cu. m.	Grammes per cc	27.6799	0.0361273
Pounds per cft.	Kg. per cu. m.	16.02	0.052422
	MISCELLANEOUS		
Pounds per foot	Kg. per metre	1.48816	0.67197
Miles per hour	Feet per metre	1.46667	0.68182
Feet per second	Km per hour	1.09728	0.91134
H.P. (550 ft. lb/sec.)	Metric H.P.	1.01387	0.98632
To obtain	From		Multiple by